IAIN S. MACDONALD

Glencoe and Beyond

The Sheep-Farming Years

1780-1830

JOHN DONALD

First published in Great Britain in 2005 by
John Donald, an imprint of Birlinn Ltd

West Newington House
10 Newington Road, Edinburgh EH9 1QS

www.birlinn.co.uk

ISBN 10: 0 85976 619 5

ISBN 13: 978 0 85976 619 7

British Library Cataloguing-in-Publication Data
A catalogue record for this book is available
on request from the British Library

Typeset by Antony Gray
Printed and bound by Bell & Bain Ltd, Glasgow

Contents

Acknowledgements

Most of the evidence used in this study was found in the National Archives of Scotland (NAS, formerly the Scottish Record Office) and I am grateful for permission to use it and for the unfailing courtesy and help which I received during many hours of searching. The Mackintosh Muniments, the Campbell of Dunstaffnage Papers and the Campbell of Barcaldane Papers are on deposit in NAS and I thank the respective owners or their representatives for permission to make use of them.

Valuable information was obtained from records in the Highland Council Archive, and I am very appreciative of the trouble taken by Mr Robert Steward, until recently Highland Council Archivist, to allow me over several visits to identify and see those records which were relevant.

My thanks are also due to the Trustees of the National Library of Scotland for permission to refer to a number of manuscripts in their possession. The British Library Oriental and India Office Collections has been a source of information about the Glencoe family's connections with India. The letters of John MacDonald of Borrodale in the Glenaladale and Borrodale Papers at the Clan Donald Visitor Centre Library, Isle of Skye, throw light on the affairs of some of the Glencoe people and others on his lands. A letter from Bishop Ranald MacDonald to John MacDonald of Borrodale, held in the Scottish Catholic Archives, has a similar relevance. The Culley Journals are quoted with permission of the Northumberland Record Office.

Mr Colin D. MacDonald in Christchurch, New Zealand, a descendant of the MacDonalds of Drimintorran, has kindly provided me with a copy of the unpublished history of his family which he wrote in 1994, as well as copies of other papers. I am grateful to him for permission to quote from these.

Mrs Jean Murray Cole in Ontario, who has carried out extensive research on the life of her ancestor, Archibald McDonald of the Inverigan family, has allowed me to make use of records of that family. The note about his parents and siblings written by Archibald McDonald at Fort Langley, British Columbia in 1830 has been particularly useful.

Mr G.A. Dixon of Stirling has given me generous assistance in connection with families in Badenoch, and he drew my attention to relevant references in the Gordon Castle Muniments and in published works.

Mr Tearlach MacFarlane of Glenfinnan has kindly provided information about the Biodag McDonalds at Finiskaig in North Morar, apparently

descendants of Alexander Dhu McDonald who has been traced from Tulloch in Brae Lochaber in the early 1780s to Knoydart in 1815.

The late Mrs Ann MacDonell of Spean Bridge who built up over the years an extensive knowledge of families in Lochaber was generous in providing information and encouragement. Turning to her for help was a double pleasure because we found that we shared a common ancestor. Robert MacFarlane, formerly of Spean Bridge and now in South Africa, has shared freely with me his knowledge of people and places in Lochaber and drew my attention to valuable references.

Norman H. MacDonald of the Clan Donald Society of Edinburgh has been very helpful in pointing me in the direction of useful contacts, and I am grateful to him for that.

Professor Alexander Fenton has given encouragement to an amateur and newcomer in this field, for which I am very grateful.

My wife has also been encouraging and has put up with the long periods which I spent searching records and communing with the word processor.

The views which I have expressed are, however, my own, and if any mistakes are found the fault lies with me.

NOTE: All references in the text to money are in sterling unless stated otherwise

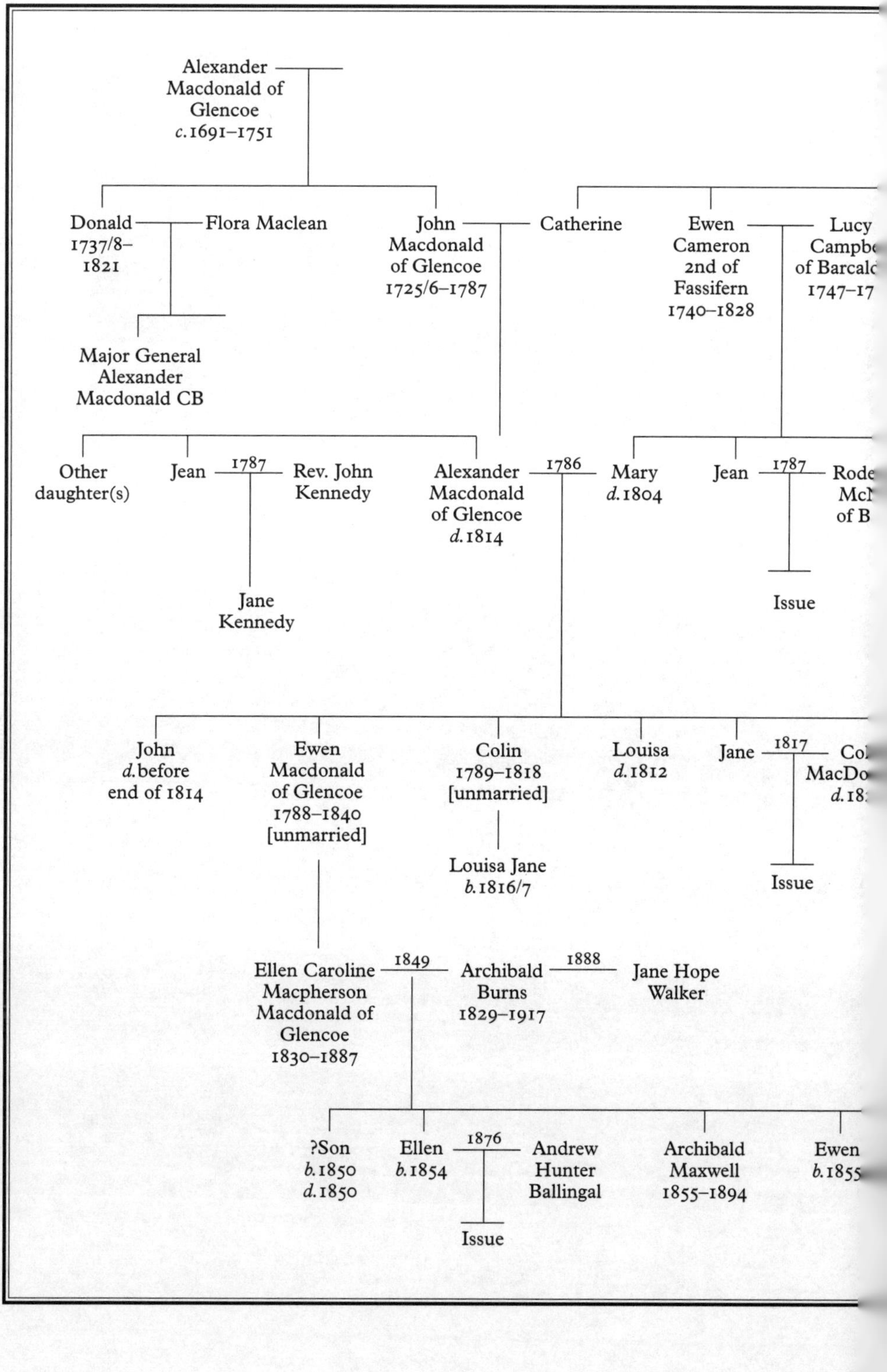

Alexander Macdonald of Glencoe c.1691–1751
Donald 1737/8–1821
Flora Maclean
John Macdonald of Glencoe 1725/6–1787
Catherine
Ewen Cameron 2nd of Fassifern 1740–1828
Lucy
Major General Alexander Macdonald CB
Other daughter(s)
Jean
1787
Rev. John Kennedy
Alexander Macdonald of Glencoe d.1814
1786
Mary d.1804
Jean
1787
Issue
Jane Kennedy
John d.before end of 1814
Ewen Macdonald of Glencoe 1788–1840 [unmarried]
Colin 1789–1818 [unmarried]
Louisa d.1812
Jane
1817
Issue
Louisa Jane b.1816/7
Ellen Caroline Macpherson Macdonald of Glencoe 1830–1887
1849
Archibald Burns 1829–1917
1888
Jane Hope Walker
?Son b.1850 d.1850
Ellen b.1854
1876
Andrew Hunter Ballingal
Issue
Archibald Maxwell 1855–1894
Ewen

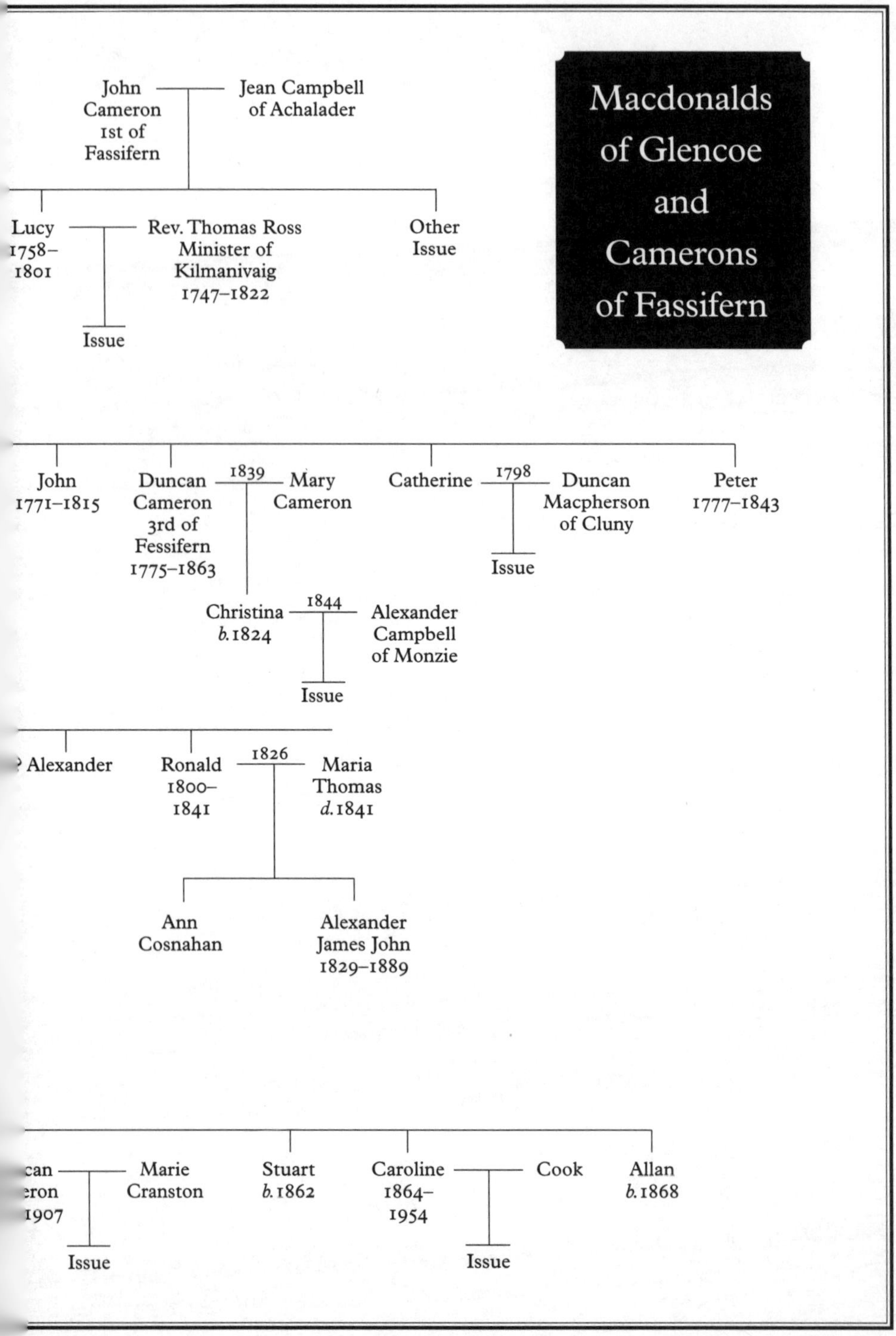
Macdonalds of Glencoe and Camerons of Fassifern
John Cameron 1st of Fassifern
Jean Campbell of Achalader
Lucy 1758–1801
Rev. Thomas Ross Minister of Kilmanivaig 1747–1822
Issue
Other Issue
John 1771–1815
Duncan Cameron 3rd of Fessifern 1775–1863
1839
Mary Cameron
Catherine
1798
Duncan Macpherson of Cluny
Issue
Peter 1777–1843
Christina b. 1824
1844
Alexander Campbell of Monzie
Issue
Alexander
Ronald 1800–1841
1826
Maria Thomas d. 1841
Ann Cosnahan
Alexander James John 1829–1889
Marie Cranston
Issue
Stuart b. 1862
Caroline 1864–1954
Cook
Issue
Allan b. 1868

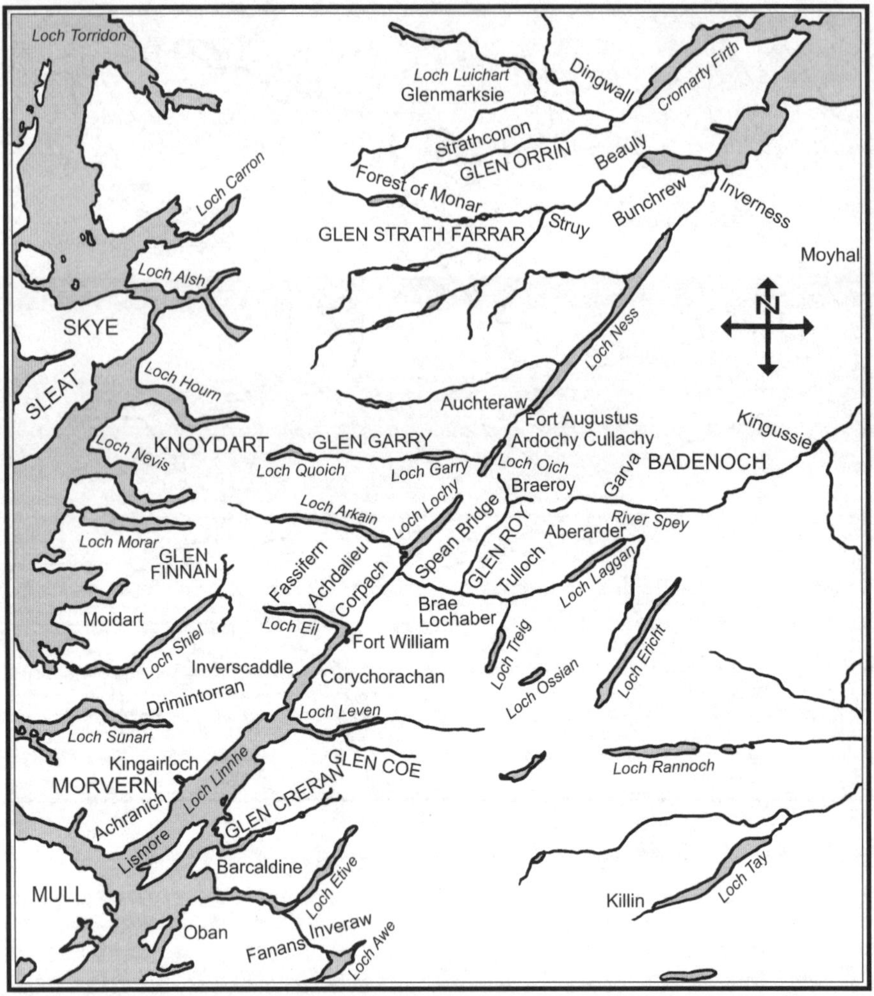

Map 1 An Overview

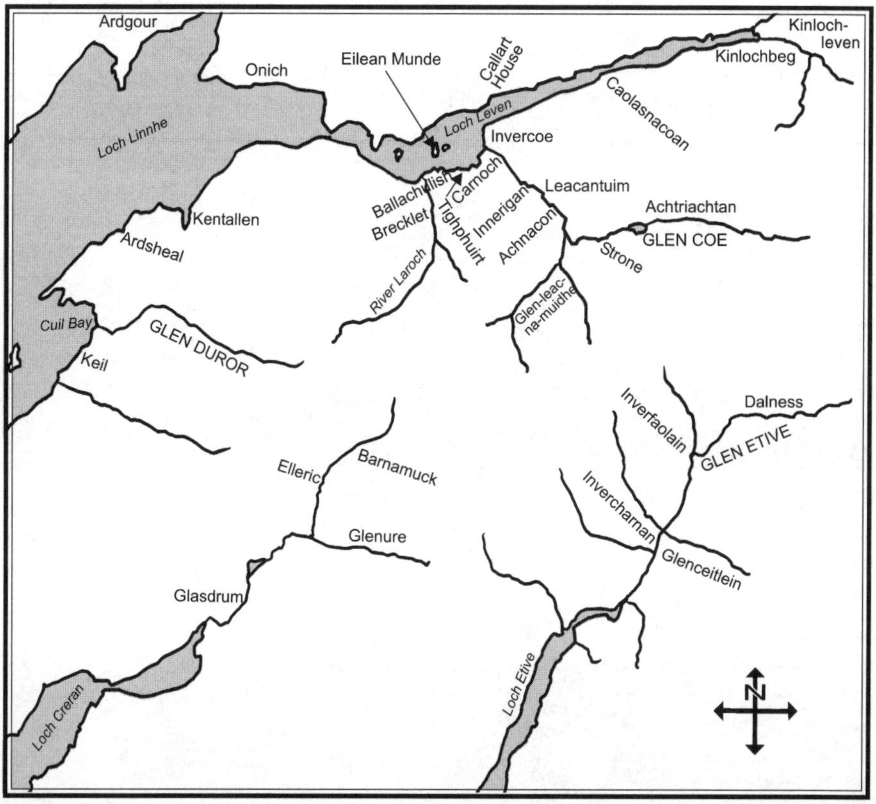

Map 2 Glencoe and Glenetive

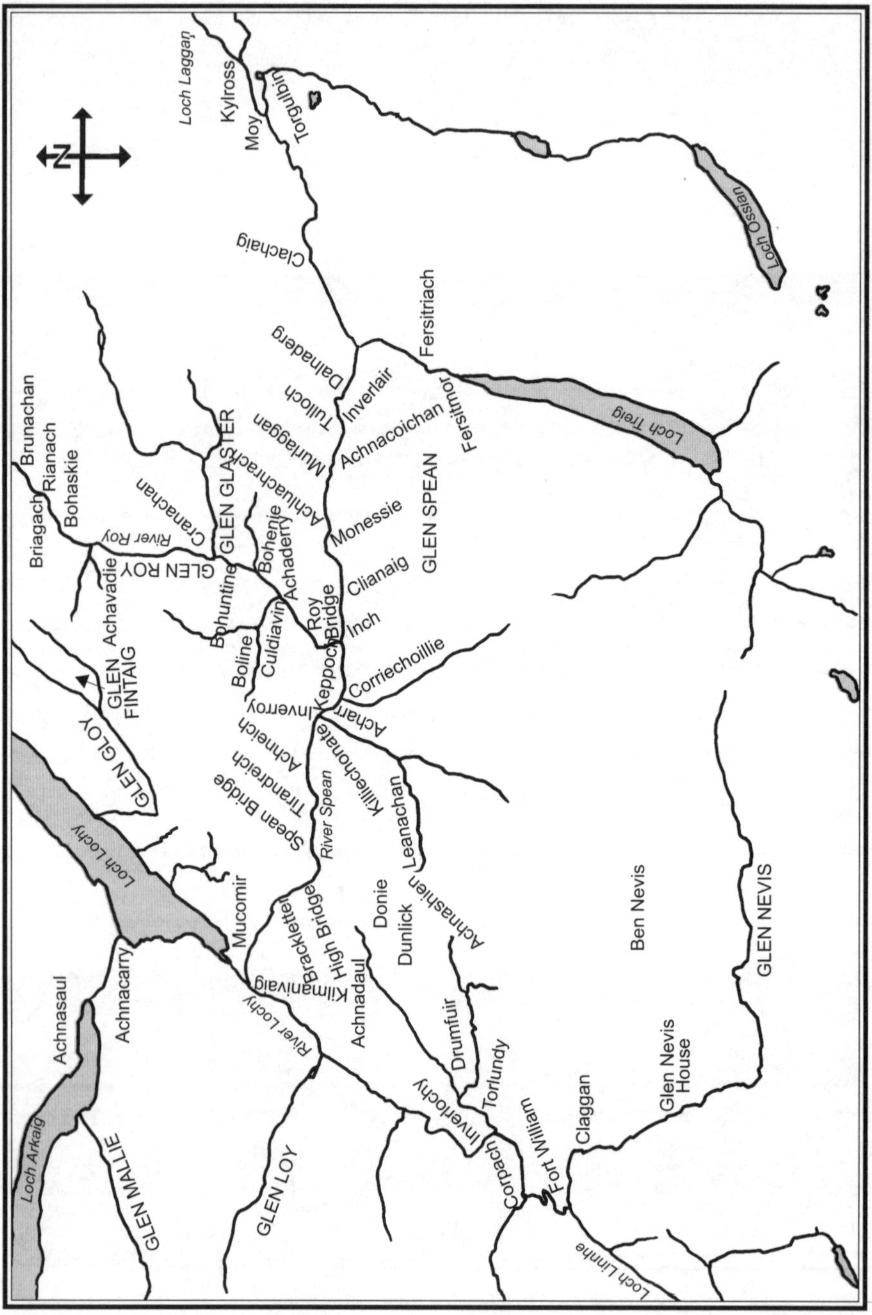

Map 3 Lochaber and Brae Lochaber

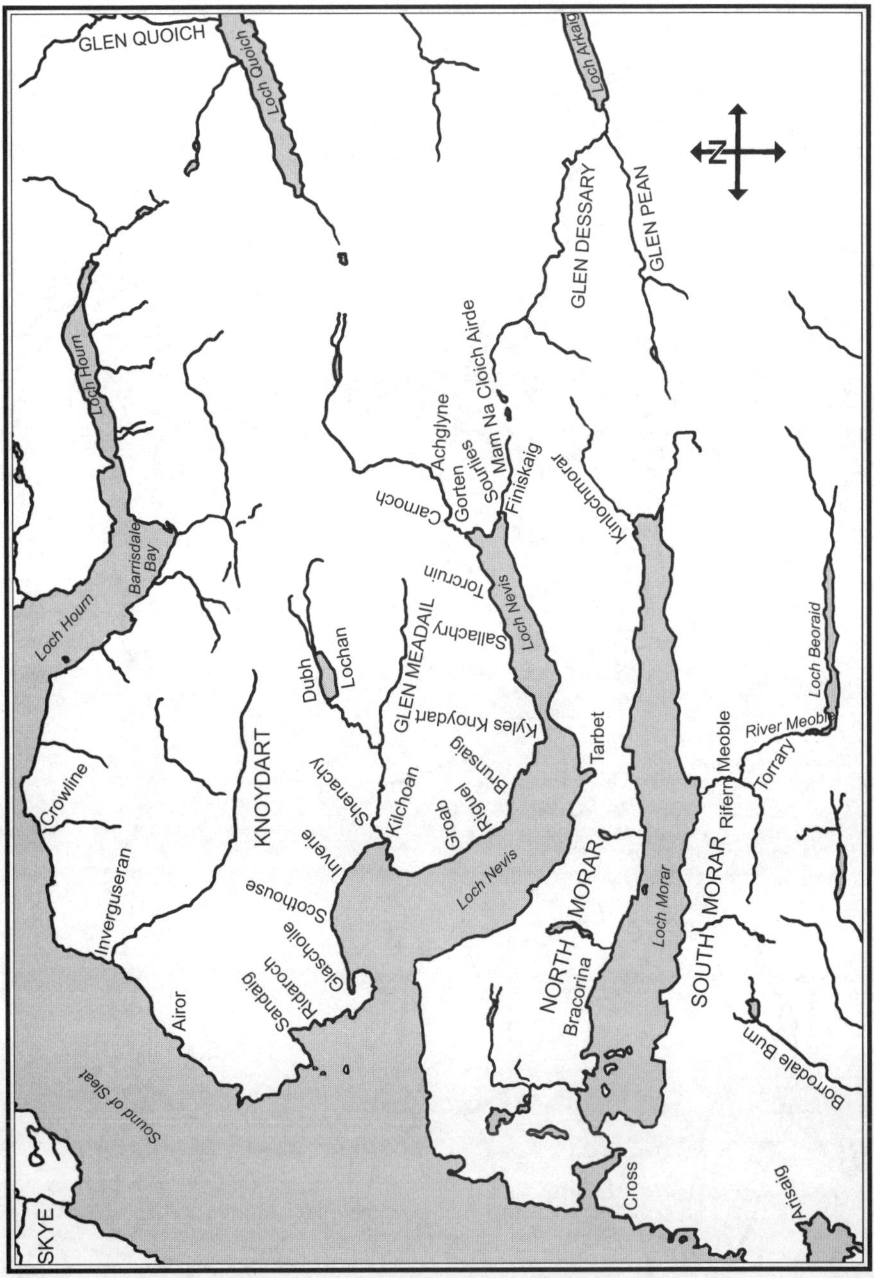

Map 4 Knoydart and Morar

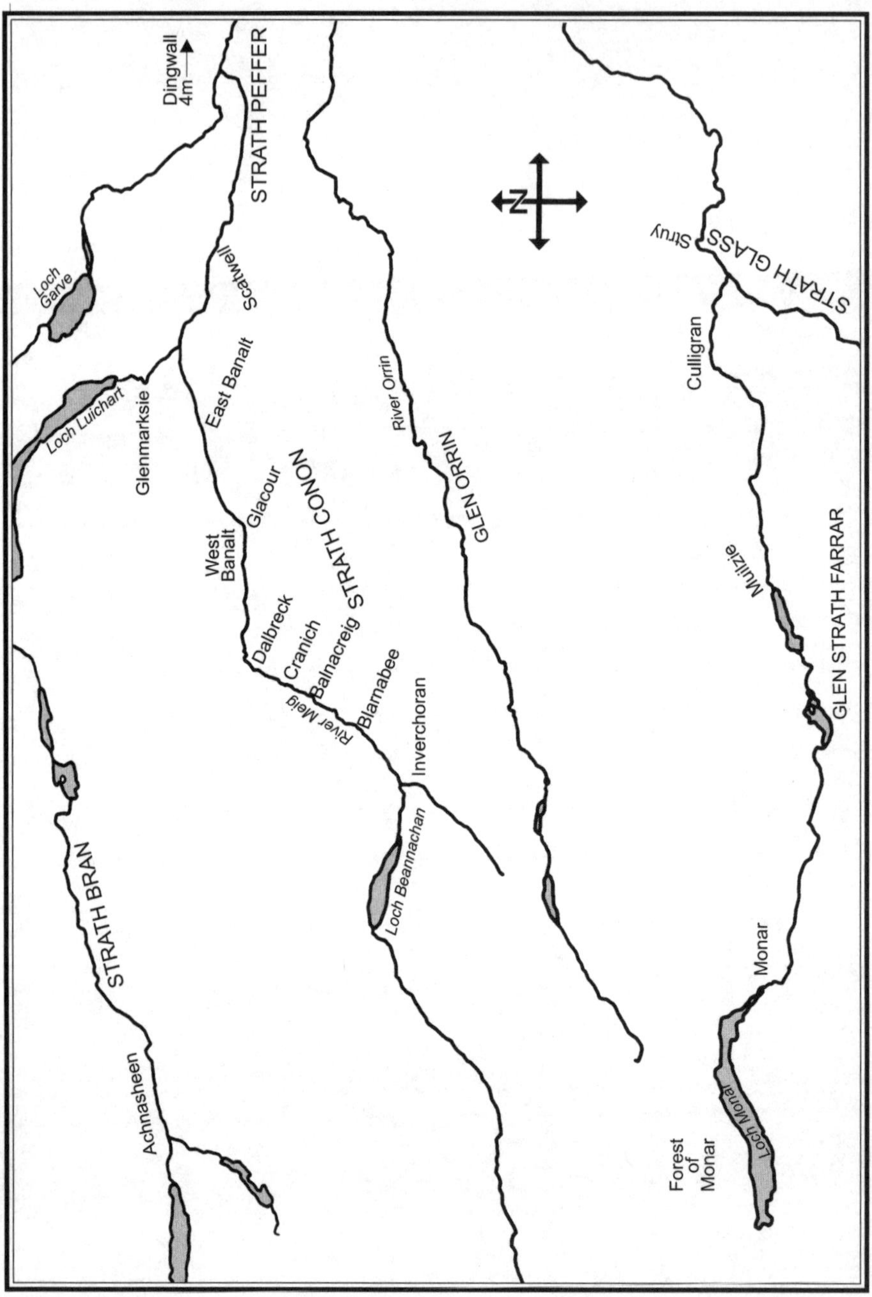

Map 5 Strathconon and Glenstrathfarrar

GLENCOE AND BEYOND

Introduction

> The extent to which beliefs are based on evidence is very much less than believers suppose (Bertrand Russell, 1928).

This book is about Highlanders who were sheep-farmers in the late eighteenth and early nineteenth centuries. It arose from curiosity about two items of information. The first was that my great-great-grandfather, a Macdonald born in Glencoe about 1775, was a farmer in Knoydart in the early 1800s. The second was that around the same time another family of Macdonalds from Glencoe were farming in Morvern. Why should these Glencoe families have been in Knoydart and Morvern in that period?

Alexander Macdonald of Glencoe was known to have leased land in Lochaber and in Brae Lochaber for sheep-farming in the early years of the nineteenth century.[1] That provided a starting point. Could sheep-farming have drawn these families away from Glencoe? It seemed sensible to explore what Alexander Macdonald of Glencoe had done in the hope of finding some clues.

That exploration revealed that Alexander Macdonald of Glencoe had leased numerous extensive tracts of land in Argyllshire, Inverness-shire and Ross-shire from other owners. At the time of his death in December 1814 he was holding Glencreran in North Argyll from Campbell of Barcaldine; in Inverness-shire he had three leases in Brae Lochaber from Mackintosh of Mackintosh, Glendessary to the west of Loch Arkaig from Cameron of Locheil, Kinlochnevis in Knoydart from MacDonell of Glengarry, and the central part of Glenstrathfarrar in the territory of Fraser of Lovat; in Ross-shire he held Glenmarksie near the south of Loch Luichart from Mackenzie of Strathgarve. He was also involved jointly with others in Fersit on the Duke of Gordon's land in Brae Lochaber, and in the Forest of Monar in Ross-shire, which belonged to Mackenzie of Fairburn. At various times he had rented Corpach from Locheil, had been in effect the sub-tenant of Auchteraw near Fort Augustus which belonged to Lovat, and had an interest in a couple of farms in Glengarry.

Sheep-farming probably did explain why the two families mentioned initially had moved from Glencoe to Knoydart and Morvern, but curiosity about this was soon overshadowed by a much wider issue. Exploring the affairs of Alexander Macdonald of Glencoe brought to light many other

Highland sheep-farmers and began to create an impression that in the late eighteenth and early nineteenth centuries most commercial sheep-farming in the Highlands was undertaken by Highlanders.

For example, interest in Knoydart brought Alexander Dhu McDonald to attention. In 1815 he and his sons took a large share in the farms in Knoydart left vacant by the death of Alexander Macdonald of Glencoe. Alexander Dhu was illiterate and unable to make himself understood in English, but he paid almost £1,000 in cash towards the purchase of the sheep on these farms. It was possible to trace him back through South Morar and Achintore to Tulloch in Brae Lochaber, where his father was a sub-tenant of MacDonell of Tulloch in the 1770s. He had a brother, Donald, who was fluent in English and able to read and write. Donald was a drover in the early 1780s and moved confidently among many Highland and Lowland contacts. He then became an important figure in the development of Æneas Mackintosh's Lochaber lands, and by 1795 he was the tenant of Tulloch in place of the old MacDonell family. He had other concerns which will be described in due course, but perhaps the most interesting thing about him is that he was a prolific letter writer and his surviving correspondence in several archival collections provides an intriguing and unfamiliar source of information about the development of sheep-farming.

The careers of these two brothers, and of others who will be identified in succeeding chapters, suggest that some Highlanders had a key role in the rapid economic and social changes associated with the introduction of sheep-farming. These Highland sheep-farmers would move from one farm to another and assume tenancies at appreciable distances from their places of origin. Sometimes they would have more than one farm, or money invested in sheep on the land of other farmers. They could buy and sell flocks of sheep in deals often amounting to several hundred, and occasionally a thousand or more, pounds sterling. They would remove sub-tenants from the land that they were leasing. They would act as cautioners (i.e. guarantors) for one another for the rent of a farm or the price of a flock of sheep. That could on occasion lead to unexpected results. Some were fluent in English and could write well. Others could not write and needed help in transacting business in English. Some could handle numbers competently, but there are occasional references to difficulties over calculations.

The hypothesis that many Highlanders from different backgrounds became sheep-farmers in the late eighteenth and early nineteenth centuries challenges the widely accepted view that sheep-farming in the Highlands was developed and undertaken by southern incomers. Some modern historians have dismissed the possibility that Highlanders could have become sheep-farmers, because they lacked the necessary skill and

capital.[2] [3] [4] Thomas Telford seemed to provide supporting evidence for this view when he reported in 1803 that 'in general, an entirely new People, who have been accustomed to this Mode of Life, are brought from Southern parts of Scotland'.[5] He failed, however, to take account of those Highlanders who were already substantial sheep-farmers in or close to the areas he visited.[6]

A recently expressed view, more in keeping with the findings of the present study, is that Highland landlords encouraged 'indigenous entrepreneurs (often former tacksmen) as well as outsiders' to 'undertake capital-intensive sheepfarming on large tracts of their estates'.[7]

In 1769 Thomas Pennant had noted the beginnings of social and economic changes in the Highlands. He said of Highlanders that

> in many parts of the Highlands their character begins to be more faintly marked; they mix more with the world, and become daily less attached to their chiefs: the clans begin to disperse themselves through different parts of the country, finding that their industry and good conduct afford them better protection (since the due execution of the laws) than any their chieftain can afford; and the chieftain tasting the sweets of advanced rents, and the benefits of industry, dismisses from his table the crowds of retainers, the former instruments of his oppression and freakish tyranny.[8]

These words show that before commercial sheep-farming reached the Highlands there was already some loosening of the old social ties along with the introduction of new commercial influences. Pennant's observations suggest Highlanders would be ready to take up the challenge of sheep-farming.

There are difficulties about Highland sheep-farming in the literature. Nineteenth-century writers such as Stewart of Garth and Charles Fraser-Mackintosh acknowledged reluctantly that there had been active Highland sheep-farmers who had cleared other Highlanders from their lands to make way for sheep, but they treated this like a guilty secret which they would have preferred to ignore. Later writers have argued, as pointed out above, that Highlanders could not have become sheep-farmers at all. Another problem is that many authors on Highland topics have not given precise references, so it is difficult to judge what kind of weight to attach to what they have said. When the same statement appears in the works of different authors, it is not always clear whether this is corroboration or repetition. Many hours can be spent, often fruitlessly, trying to discover the original sources used.

Richards identified an important part of the problem when he said 'much of modern writing on the subject has been a re-cycling of the anthology of material published by Alexander Mackenzie in 1883'.[9] A large part of Mackenzie's material was about events in Sutherland attributable to the

development of sheep-farming there from 1807 onwards. The remainder was about some other parts of the Highlands even later in the nineteenth century. Mackenzie's material is almost entirely about different geographical areas and different periods from the material used in this study.

Richards also drew attention to the wider contexts in which these matters should be seen. The strong economic pressures encouraging the development of sheep-farming were important. From 1780 onwards, wool prices began to increase. In the late 1790s and early 1800s they rose dramatically because of the demand created by the Napoleonic Wars. During the opening two decades of the nineteenth century price levels were between 250 and 400 per cent above the levels recorded in the 1770s and early 1780s.[10] The Highlands were particularly vulnerable to the pressure created by these prices. The land 'had long been understocked, and had been grazed mainly by cattle, so that most of the grazings were in good condition for change'.[11] It is therefore debatable whether Highlanders, either landowners or tenants, were subjected to economic temptation, which they might have resisted, or to economic pressure which was irresistible.

While it is important not to overlook those southerners who did come to the Highlands as sheep-farmers, surprisingly little detail has been recorded about the incomers who have been credited with bringing about so much economic and social change. Some names can certainly be found, but in most instances little has been said about who they were, where they came from or how they were financed. There is probably scope for further work on southern sheep-farmers in the Highlands. That was not the purpose of this study, but sufficient evidence did emerge to raise doubts about how often they actually pioneered sheep-farming. Quite frequently they seized opportunities to take over farms on which sheep-farming had already been established on a substantial scale by Highlanders.

Thomas Gillespie is one of the better known figures. He was the son of a pioneering sheep-farmer, also Thomas, at Caplegill and Carrifearn on the Annandale estate. Initially the stock on those mountainous farms was poor and the wool was coarse. Gillespie senior set about improving the breed, and in due course his stock acquired a good reputation.[12] By 1782 he was trying to find outlets in the Highlands for his sons, and he wrote in March to the Duke of Gordon's factor about the forest of Gaick in Badenoch.[13] However in April he was dealing with Glengarry about Glenquoich, where Thomas junior and a partner, Henry Gibson, leased a large area which they turned into a sheep walk. In 1785, fifty-five tenants, along with their dependants and cottars, were reported to have been cleared to make room for them and their sheep.[14] This earned Gillespie a permanent place in the demonology of Highland sheep-farming.

Gibson soon dropped out, but young Thomas Gillespie persisted.[15] He

was joined by his younger brother, John, and later by James Greig, who was probably the nephew of his wife, Christian Greig.[16] In time Thomas was recognised as an authority on sheep-farming in the Highlands, and his advice was frequently sought. In 1806, when the great investment in sheep-farms in Sutherland was about to begin, it was Thomas Gillespie's advice which was wanted 'to value the sheep-farms and give us proper lines of boundary'.[17]

In contrast to the well known episode at Glenquoich, Thomas Gillespie and his relatives later took over several farms where sheep had already been established by Highlanders. Even on his own farm of Ardochy in Abertarff, with which he was identified for many years, he was not the first to introduce the improved breeds of sheep. Alexander Stewart had moved there from Appin at Whitsunday 1779 with a flock of between eight and nine hundred sheep of the 'south country or Annandale breed'.[18] That was three years before Gillespie senior had even started to try to find a place in the Highlands for his sons. Nevertheless Gillespie has a place in the literature, while Stewart is virtually unknown.

In 1795 Thomas Gillespie took a lease of Scothouse in Knoydart, where sheep-farming had already been established, and placed his brother John there. When Thomas died in 1824 he owned the stock on his own farm of Ardochy and half of the stock at Cullachy, Torlundy, Ballachulish, Tulloch, Brunachan, Annat and Corrychurochan. On most of these it can be shown that sheep-farming had been established by Highlanders. Gillespie also advanced money to a large number of people, including, for example, a landowner such as James Murray Grant of Glenmoriston (£2,300) and other sheep-farmers such as Alexander Cameron of Inverguseran in Knoydart (£1,050). Gillespie's assets were valued at almost £13,000.[19] He was undoubtedly very successful, but was he typical?

In 1794 Lieutenant Evan Macpherson at Cullachy, a well established sheep-farm near Fort Augustus, became bankrupt. That gave Thomas and Robert Oliver and Thomas Stavert from south of Hawick the opportunity to take over that farm. Stavert died soon after. The Oliver brothers continued to hold the lease of Cullachy, but Thomas remained the tenant of a farm in the Borders. Eventually, in 1809, Thomas renounced his interest in Cullachy in favour of his brother Robert. In the same year Robert obtained the farm of Torlundy and the hill grazings of Inverlochy in Lochaber, previously held by a local family of MacBarnetts. Did other sheep-farming families who came to the Highlands from the Borders maintain divided interests in this way?

In 1782 Thomas Stoddart at Killiean at the north-west corner of Loch Lochy appeared as a witness at Inverness Sheriff Court. He was the joint proprietor, along with Henry Butter Esq. of Pitlochry, of a flock of sheep, some of which had allegedly been stolen by Ranald Dow Kennedy. Thomas

was aged thirty, his brother Andrew, also at Killiean, twenty-nine.[20] The surname suggests that they were of southern rather than Highland origin, and the participation of Henry Butter is interesting. He was one of the factors appointed by the Board of Commissioners and Trustees for the Annexed Estates. In that capacity he had lived for a time at Corpach.[21] He had previously been reprimanded for holding more tenancies than the Board thought proper.[22] Perhaps this was a more discreet way of pursuing his private business.

Thomas Stoddart was still at Killiean in 1795, but several people were proceeding against him for debts.[23] [24] By 1797 he had been replaced by Alexander MacTurk, whose name suggests a Galloway origin. MacTurk, the Gillespies, the Olivers and John Hall at Aberchalder seem to have acted together as a group of southerners within the Sheep Farm Association which was set up in the 1790s under the chairmanship of MacLeod of Geanies, Sheriff Depute of the County of Ross. They were a small group, greatly outnumbered by sheep-farmers who were clearly Highlanders.[25]

The greatest impact of southerners during the period of this study may have been that their offers for farms gave Highland landlords a measure of the market value of their land, and also gave prospective Highland tenants a measure of the competition they had to face. There were occasions when Highland landlords were keen to know what southerners would pay, but they preferred Highland tenants if they would match, or come close to matching, the price.

The accounts that follow are based on the localities in which Alexander Macdonald of Glencoe had interests. Each of his scattered activities provided an opportunity to identify other sheep-farmers and examine how sheep-farming had developed there. The accounts generally fall into three parts: the period before his arrival, when sheep-farming may have been developed by others; the period during which he was a tenant and was involved with others; and the period following his death, when his Trustees and others had to take his place. The activities of Alexander Macdonald of Glencoe have therefore determined the geographical and time limits of the study. Within these limits, a surprising amount of information has been uncovered about early developments in commercial sheep-farming. The general rule was not to wander far from the affairs of Alexander Macdonald of Glencoe, but when temptation was strong that rule was sometimes interpreted rather liberally.

There is a noticeable difference in the pace of development of sheep-farming during the period of this study. In the late eighteenth century, the conversion to sheep-farming was encouraged by a relatively moderate and fairly steady growth in the demand for wool, punctuated by an occasional bad year. In the first few years of the nineteenth century the Napoleonic War created a strong surge in demand. Prices of wool and sheep multiplied,

farmers competed for land at high rents and there was a trend towards longer leases. This soon gave way to a long lasting collapse in demand and in prices.[26] It seemed sensible to take account of this when determining the content of chapters; therefore Chapters 1 to 5 are concerned mainly with the late eighteenth century, and Chapters 6 to 12 with the early nineteenth century. Most of the material fits well into this arrangement, but occasionally continuity required material from one century to stray into a chapter concerned mainly with the other century. Chapter 13 explains briefly some of the general problems and tasks which faced the Glencoe Trustees after the death of Alexander Macdonald in December 1814. Finally, a postscript has been included to show how the descendants of Alexander Macdonald of Glencoe coped with the debts that he left them, and how Glencoe came to be sold to Sir Donald Smith (Lord Strathcona) some eighty years after his death.

The early sheep-farming period in the part of the Highlands examined in this study was characterised by a great deal of geographical, social, and financial movement. Two general points can be made. One is that chiefs, tacksmen, clansmen, and even southern sheep-farmers were all individuals reacting to the circumstances in which they found themselves. The other is that these circumstances were characterised by a great deal of economic turbulence.

When the Minister of the Parish of Kilmanivaig wrote about sheep-farming in 1842, he said:

> It is supposed that there are upwards of 100,000 sheep reared in this parish every year . . . Mr. Cameron, Carychvilly [Corriechoille], the most extensive grazier in the north, stated a few years ago, that in the preceding year he had clipped upwards of 37,000 sheep . . . Mr. Greig of Tullach [Tulloch], and the Messrs M'Donell of Keppoch, are supposed to have each near 100 square miles under sheep; the one on the north, and the other on the south banks of the Spean.[27]

Cameron and the MacDonells were Highlanders. Greig was a Lowlander. That must surely tell us something.

1

Glencoe in the Late Eighteenth Century

Ownership of Glencoe

In 1751, before commercial sheep-farming developed in the Highlands, the Valuation Roll for the County of Argyll showed that John Macdonald of Glencoe owned 'Carnuch, Tigh-phuirt, Poluig, and Invercoe Mill', all in the lower part of the glen. Angus Macdonald of Achtriachtan owned Achtriachtan in the upper part of the glen and Kinlochbeg towards the head of Loch Leven. Various Stewarts, including Dugald Stewart of Appin, owned the farms of Strone, Achnacon and Innerigan (or Inverigan, or Inverighan) on the west or south-west side of the River Coe and Leacantuim on the other side of the river.[1]

The subsequent ownership of these Stewart lands was complicated by the bankruptcy and death of Dugald Stewart of Appin, whose property was eventually purchased by the Marquis of Tweeddale.[2] All the Stewart property in Glencoe became identified with the Appin estate. In 1751 Angus Macdonald of Achtriachtan exchanged Kinlochbeg for the Stewart property of Leacantuim in Glencoe.[3] Kinlochbeg was subsequently leased back to him and his son Æneas, along with Strone and Achnacon.[4] The Macdonalds of Glencoe held the lease of the remaining Stewart property in Glencoe, namely Innerigan.

The Macdonalds of Glencoe

John Macdonald of Glencoe was the elder son of Alexander Macdonald of Glencoe who took part in the Jacobite Rising of 1745. Alexander surrendered in May 1746 and was held in captivity until 1749.[5] [6] Glencoe was not forfeited, but it was surveyed in 1748. Young John received the surveyors and explained to them the bounds of Glencoe and the relationships with neighbouring lands.[7] [8] Alexander had referred in 1746 to 'my bad state of health',[9] and he must have died a year or so after his release. By a charter of 1751, the lands of Glencoe were disponed to his son John.[10]

John was twenty-six and not yet married when he gave evidence at the Appin Murder trial in 1752;[11] he must therefore have been born about 1726. His younger brother Donald, who died at Invercoe in 1821 aged eighty-three,[12] must have been born about 1738.

Sometime after 1752 John Macdonald married Catherine Cameron, a daughter of John Cameron of Fassifern.[13] [14] [15] Their only son, Alexander, was probably born about the middle of the 1750s. By 1776 he was old enough to apply for the lease of Fersit[16] and embark on his sheep-farming career. He may have been intended for Lord Macdonald's Regiment which went to New York in March 1779. It was said in 1778 that 'Glenco's son has recovered and now anxiously wishes to be restored to his commission'.[17] However, on 18 December 1780 Alexander signed a bond in Edinburgh in which he was designated 'Alexander McDonald Younger of Glenco' with no indication of military rank.[18]

John and Catherine had several daughters. Mrs Grant of Laggan referred to them in a letter of 17 May 1773 as 'the young ladies of Glencoe, with whom I have a remote connection'.[19] One daughter, Jean, married Rev. John Kennedy of Auchteraw.

The Macdonalds of Achtriachtan

The Macdonalds of Achtriachtan have been called tacksmen of the Macdonalds of Glencoe,[20] [21] but in 1751 they were shown to own their land.[22] Buchan called them 'one of the cadet gentry of Glencoe'.[23]

About 1750 the succession in the Achtriachtan family fell to Angus Macdonald.[24] His farming career of almost fifty years in Glencoe and Lochaber covered the period when large-scale commercial sheep-farming was introduced.

Angus Macdonald of Achtriachtan married Anne Campbell, a daughter of John Campbell of Ballieveolan. A post-nuptial contract was dated 10 December 1753.[25] They had at least six sons and four daughters. Their eldest son, Alexander, joined the Bombay Establishment of the army of the East India Company in 1775. He was promoted to captain in 1784, and died in India in December 1793.[26] The second son, Æneas, was associated with his father at home, and was also factor on the Marquis of Tweeddale's Appin property and an agent purchasing wool for a company in Burnley, Lancashire.[27] [28] He died about the same time as his father. The third son was Adam.[29] He was in the West Indies for a time and succeeded his father due to the deaths of his two elder brothers. After Adam there was John, who by 1794 had 'served his time as a dentist in Edinburgh and London and means to set up in Edinburgh as soon as possible'.[30] Two younger sons, Robert and Hugh, were still alive in 1796. There were two other sons whose places in the family are not known: James died sometime before 1796, and Colin was commissioned as an Ensign in Lord Macdonald's Regiment and sailed for New York on 25 March 1779. The daughters were Isabella, Mary, Janet and Elizabeth. Isabella married John MacLachlan of Aryhoulan. Mary married Allan Stewart who was in Caolasnacoan and later in Shuna in Appin.[31] [32] [33] [34]

Janet (or Jessy) married John Stevenson Esq. of Glenfeochan.[35] Elizabeth married John MacDonald, son of Donald MacDonald (senior) of Drimintorran[36] and will be mentioned again in the account of that family in Chapter 7.

The Beginning of Sheep-Farming in Glencoe

In 1771 Pennant published the first edition of his *Tour in Scotland*, describing a journey made in 1769. He had only a distant view of Glencoe, but in later editions he included an account of it by Mr James Stuart of Killin. He was almost certainly the parish minister there, and his account seems to have been based on a visit for church purposes in 1760. It recorded six farms in Glencoe, growing crops of oats, bear and potatoes. Black cattle were exported, but sheep and goats were kept for the use of private families.[37]

In the years that followed there were strong economic pressures encouraging the development of sheep-farming. The minister of the adjacent parish of Kilmallie, in Lochaber, said that sheep were first introduced in 1764 and that by 1792 they had increased so much that about three-fourths of the country was occupied by them. Most of the wool 'was sent coastwise to Liverpool and other ports in England'. The value of land had tripled and there had been a great augmentation of rents, attributed principally to the stocking of farms with sheep. He referred to 'a strong temptation to proprietors, who value *money* more than *men*, to encourage sheep-farming'.[38]

Other evidence also points to a developing local interest in sheep-farming. Matthew Culley, one of two farming brothers in Northumberland, toured Scotland in 1775. He described how he dined at Fort William 'with some gentlemen who gave me a letter to Mr. Stewart of Balliehuish, a gentleman who was desirous of improving his breed of sheep'. At Ballachulish, 'the seat of John Stewart Esq', he was 'very well received'. Matthew's brother George, a notable Northumbrian sheep breeder, added an explanatory note:

> Mr Stewart and many of his Neighbours have lately adopted the breeding of sheep upon rather a large scale, and the kind of sheep they have been recommended to are the Blackfaced or Short Sheep, a kind perhaps of all others best adapted to live upon a mountainous exposed Country like this, nevertheless, I am persuaded that a cross from the Bakewell breed could be of use, as I apprehend it would both improve the wool, and make them feed more regularly.[39]

One neighbour to the south, Alexander Stewart from Appin, has been mentioned in the Introduction as the owner in 1779 of a large flock of sheep of the 'south country or Annandale breed'. The Macdonalds of Glencoe and of Achtriachtan were close neighbours to the east.

These new commercial trends became apparent in Glencoe. In March 1770 Angus Macdonald of Achtriachtan obtained a bond of credit from the firm of Douglas, Heron and Company in Ayr (commonly known as the Ayr Bank) entitling him to borrow up to £300 at any one time on his account with their Edinburgh branch. His cautioners were his neighbours John Macdonald of Glencoe and John Macdonald of Dalness.[40] This Bank had been formed 1769 with an impressive list of partners including the Dukes of Queensberry and Buccleuch. One of its objects was to support agriculture. It began to trade vigorously but collapsed in 1772.[41] That may have been unfortunate for the enterprising Angus Macdonald, who was probably embarking on sheep-farming in Glencoe and had become tenant of several farms on the Duke of Gordon's land in Lochaber.

In 1790 the minister of the Parish of Lismore and Appin said that very little grain was being raised in the higher parts of Appin (which would include Glencoe), 'since sheep flocks have been found beneficial. Many of the sheep farms are very extensive, and . . . produce excellent grass, and, of course, very good sheep, perhaps the best in these western districts.'[42]

Visitors to Glencoe around the turn of the century commented on the dominance of sheep-farming and noted loss of population from the glen.[43] [44] [45] In 1821 the Commissioners for Highland Roads and Bridges were troubled about the constant blocking of Glencoe by torrents that washed stones down on to the road, saying 'this sort of damage to the Road and to the Valley is said to have been unknown, until the Black Cattle, formerly depastured thereabouts, were supplanted by sheep'.[46]

Commercial sheep-farming had of course been grafted on to an agricultural economy based mainly upon black cattle. Angus Macdonald of Achtriachtan was perhaps ambivalent about the new sheep-farming, although he certainly went along with it. In 1787 Alexander Campbell of Barcaldine asked him to come and take part in a valuation of sheep. Angus said: 'I am obliged to disclose a secret that I intended never to divulge . . . believe me, I am not a judge of sheep . . . tho' I speak learnedly on the subject, I am a pupil of my Tennants in Achtriachtan they are the managers of my Stock and I do nothing in the Sheep way without them.' He proposed to bring a tenant with him to Campbell's estate, saying that two judges would be better than one and that it would be 'my own honour to be oversman, as I'll pretend to be the best judge'. In the event he pleaded the familiar excuse of a pressing engagement elsewhere and sent his son Æneas with one of his tenants.[47]

There was a long gap between Pennant's account, based on observations in 1760 when there were no sheep-farms in Glencoe, and the later descriptions from 1790 onwards, when sheep-farming was well developed and travellers were noting evidence of depopulation in the glen. Had the area as a whole been losing population? In 1790 the parish minister took

a cautious view. 'There is no ascertaining whether population is on the increase or decline here, though, at first view, the immense tracts of sheep-farms might naturally induce us to think it decreasing.' There had been emigration from the Appin part of his parish, i.e. the large mainland part on the east of Loch Linnhe, but he took this calmly. 'The inhabitants are now become so crowded, that some relief of this sort, in one shape or other, seems absolutely necessary.'[48] It seems curious that he did not mention the Ballachulish Slate Quarries in this connection.

Ballachulish Slate Quarries

Travellers who passed through Glencoe were impressed by the activity at the nearby Ballachulish Slate Quarries.[49] [50] Slates had been produced from about 1697 at what was later known as the West Quarry. Another vein of slate from which the larger East Quarry was developed was said to have been discovered in 1780.[51] In 1790, presumably after the East Quarry had been opened, the parish minister observed that there were '74 families in the quarry, containing 322 souls' and that great quantities of slates were sent yearly to the north and east countries, to Leith, Clyde, England, Ireland and even to America.[52]

A demand for large quantities of slates and therefore for labour seems to have coincided fairly closely with the reduction of population on the land to make way for sheep. Unfortunately none of the eighteenth-century observers commented on this juxtaposition, but it does seem reasonable to postulate that many of the inhabitants of Glencoe displaced by sheep-farming would have found employment in the nearby slate quarries at Ballachulish.

Ventures beyond Glencoe

The extensive interests developed by the Glencoe and Achtriachtan families outside Glencoe itself will be discussed in later chapters. Angus of Achtriachtan was the first to venture into more distant parts: by the end of the 1760s he was leasing three farms in Lochaber from the Duke of Gordon. Alexander Macdonald, the son of John of Glencoe, obtained the lease of Fersit in Brae Lochaber in 1778. These developments on Gordon land will be described in Chapter 2.

In October 1786 Alexander Macdonald married Mary Cameron, a daughter of Ewen Cameron of Fassifern,[53] who undertook to pay 'six thousand merks scots money' as tocher.[54] On 6 February 1787 Alexander's father John undertook to provide a tocher of £150 for his daughter Jean who married Rev. John Kennedy, tacksman of Auchteraw near Fort Augustus. This was still unpaid when John MacDonald died later that year. Alexander promised to pay it, but he deferred this commitment and borrowed £450 from his new brother-in-law.[55] He now had some funds at

his disposal and was ready to lease more land. An account of the leases he obtained, and of the Rev. John Kennedy's affairs, is given in Chapter 5.

The Marriage Contracts of Alexander Macdonald of Glencoe and Mary Cameron

There is an indication of Alexander Macdonald's thoughts in the terms of the marriage contract of 1787 and the supplementary contract of 1791.[56 57] Under the marriage contract, he undertook to provide his wife Mary Cameron with an annuity of £50 per annum in the event of his death. He also undertook that 'whatever lands or other heritages may happen to be conquest or acquired by the said Alexander Macdonald during the standing of this marriage he binds and obliges him to provide the fee thereof to the heirs of the marriage and the liferent of the equal half thereof to the said Mrs. Mary Cameron'. (In this context 'conquest' has the same meaning as 'acquired'.) In the supplementary contract of 1791, his wife was to receive an immediate payment of £50 and an increased annuity of £100, and in return she renounced her entitlement to liferent from heritages which might be 'conquest or acquired'. The changes made by the supplementary contract of marriage in 1791 probably indicate that Alexander foresaw even better prospects than in 1787. The level-headed Fassifern family concurred in the terms of the supplementary contract.

The Changing Fortunes of the Achtriachtan Family

As Alexander Macdonald of Glencoe was embarking on his meteoric career as a sheep-farmer, Angus Macdonald of Achtriachtan was beginning to get into difficulties. In June 1778 he had declared his finances 'rather straitened this year'. He attributed this partly to expenditure on his sons, and partly to problems associated with the farm of Tirandreich on the Duke of Gordon's land, which he took over that year.[58] He apparently overcame this difficulty, and it will be shown in Chapter 2 that in the early 1780s he was seeking further commitments in Lochaber and even in Badenoch. However, in the early 1790s he was faced with large debts. The reasons are not clear, but the scale of Achtriachtan's borrowing can be established.

The first sign of heavy borrowing came in 1791 when Angus and his son Æneas borrowed £2,444 10*s.* 0*d.* from the trustees and executors of John Elphinstone, Lieutenant Colonel of the 71st Regiment of Foot. The land at Achtriachtan and Leacantuim was used as security.[59]

Around 1793 there was difficulty in repaying a debt due to Thomas Gillespie. Gillespie was late in paying his rent, presumably to Glengarry. On 23 January, he told Campbell Mackintosh, a writer in Inverness, that: 'I went to Achtriachtan's where I ought to have got a large sum of money due at last Martinmas but could not get a farthing. I got promise of payment to-morrow fortnight and as soon as the money comes to hand will either

see you at Inverness or send down an express. I hope that you will not be impatient.'[60]

It seems unlikely that Gillespie got an early settlement because in March, and again in July, when a diligence was being threatened, his brother John wrote apologetically to Campbell Mackintosh pleading for more time.[61] [62]

Angus Macdonald's problems evidently continued, and in 1794 he had to borrow again. This time he got £1,000 from John Cameron of Camisky, secured again on the lands of Achtriachtan and Leacantuim.[63]

The death of Angus's eldest son, Alexander, in Bombay in December 1793 must have brought some financial relief.[64] He was unmarried, and in 1795 his executors remitted £3,970 18*s.* 11*d.* to the United Kingdom to be divided among his parents, his brothers and sisters, and his uncle, Captain Robert Campbell in Craig of Glenorchy. The largest share went to his father.[65] That should have eased the financial position slightly, but not apparently to the extent of repaying any of the sums already borrowed.

In 1796 Angus made a bond of provision for his wife, Anne Campbell, and his family. The arrangements included a fairly generous annuity of £50 for his wife after his death.[66] About the same time he observed that at Kinlochbeg, Achnacon and Strone, 'I have a suitable stock of horses, Black cattle, and sheep'.[67] This does not create an impression of a man aware that he was facing great financial difficulties, but it was evidently the case.

In September 1796, a few months after making these provisions for his wife and family, Angus borrowed a further £2,000. This time the lender was his son-in-law, John MacLachlan of Aryhoulan, near Inverscaddle on the west side of Loch Linnhe. He was the husband of Angus' daughter Isabella. Again the land at Achtriachtan and Leacantuim was used as security.[68]

In 1798 Thomas Garnett travelled through Glencoe on his way to Fort William. He visited Invercoe House, 'the property of the laird of Glencoe, but occupied by Mr. Macdonald of Achtriachtan, with whom we breakfasted, and from whom we received attention and civility'.[69] That would probably have been Angus Macdonald of Achtriachtan during the last year of his life.

This is an appropriate point at which to leave Glencoe and give attention in Chapters 2 to 5 to developments elsewhere in the late eighteenth century. The account of Glencoe will be resumed in Chapter 6, when the early years of the nineteenth century will be described. The misfortunes of the Achtriachtan family intensified under Angus' third son, Adam, and Alexander Macdonald placed increasing burdens upon his Glencoe estate as security for loans. These caused great problems for his trustees, who became responsible for the management of Glencoe after he died in December 1814.

2

From Glencoe into the Gordon Lands in Lochaber

The Gordon and Mackintosh Lands

The first ventures by the Achtriachtan and Glencoe families beyond Glencoe were in Lochaber, where the Duke of Gordon and Mackintosh of Mackintosh had extensive possessions. The Mackintosh land lay on the north of the River Spean and extended most of the way up Glenroy. The Duke's Lordship of Lochaber lay to the south of the River Spean and to the east, west and north of Mackintosh's land. It included the upper reaches of Glenroy and virtually enclosed the Mackintosh land. The Duke was the feudal superior.

Brae Lochaber, the more easterly part of Lochaber, was dominated by MacDonells. The MacDonells of Keppoch were the senior family, and others such as the MacDonells of Tulloch, of Aberarder, of Fersit and of Cranachan were their cadets. Some families held land from both the Duke and Mackintosh while others held land from one or the other. The removal of these traditional tacksmen by their respective landlords opened the way for others to obtain leases. The Duke made a beginning in 1769 when he took away some, but not all, of his land held by MacDonell of Keppoch.[1]

After Æneas Mackintosh came home from the army in 1783, he began the long process of removing the MacDonell tacksmen from his lands. That eventually provided opportunities for Alexander Macdonald of Glencoe, but not until the next century had begun. What happened on the Mackintosh lands is described in Chapters 3, 4, and 8. This chapter is about the Duke of Gordon's lands in the eighteenth century.

From 1769 Angus Macdonald of Achtriachtan was principal tacksman of three of the Duke's farms: Kilmanivaig, Brackletter and Inverlochy.[2] The MacDonells of Keppoch had held 'a great Part of the Braes of Lochaber', but in 1769 the Duke removed Ronald MacDonell of Keppoch from Kilmanivaig and Brackletter.[3] Angus Macdonald of Achtriachtan then obtained them. The position at Inverlochy is less clear; George Douglass, Master Gunner of Fort William had once held it, but it may have been vacant when Angus Macdonald got it in 1769.[4] [5]

More vacancies on the Duke's farms were due between 1776 and 1778. Fersit was to fall vacant at Whitsunday 1778. On 14 October 1776,

Alexander Macdonald wrote from Glencoe to the Duke's factors, saying that 'having observed in the Edinburgh newspapers the Lordship of Lochaber advertised for set and being desirous of becoming the Duke of Gordon's Tenant in the Farm of Fersit', he was willing to offer a rent of £84 stg. yearly.[6] His bid succeeded, and more will be said about Fersit later.

The Ambitions of Angus Macdonald of Achtriachtan

In late December 1776 Angus Macdonald of Achtriachtan wrote a long letter to James Ross, the Duke's cashier and a senior figure in the estate office at Fochabers. Young Glencoe (Alexander Macdonald) had sought his advice, and Angus Macdonald supported his application for Fersit, commending him as 'an industrious Clever fellow and in good circumstances'.[7] He added, however, that with some encouragement Alexander might be persuaded to consider taking a farm on the Duke's land in Badenoch. Achtriachtan was thinking in expansive terms. He said that an uncle of young Glencoe had 'a Considerable farm from Kingerloch [Kingairloch?]' which he was about to leave. Achtriachtan suggested that Alexander MacDonell (Keppoch's younger brother), Angus MacDonell of Inch (Keppoch's illegitimate brother), 'this Donald McDonald' and young Glencoe might all 'go to Badenoch that is to say if His Grace is still for planting a Colony of strangers there'. Keppoch was said to be advocating this scheme to the Duke and trying to persuade his brothers to take part.[8] There is no indication of what James Ross or any of the proposed 'colonists' thought of this, though, and nothing seems to have come of it.

It is possible that young Glencoe's uncle who was farming at Kingairloch in the 1770s was Donald MacDonald who was later at Drimintorran. That must remain speculative, but it will be considered again in Chapter 7.

Angus Macdonald of Achtriachtan explained his own hopes to James Ross. He wanted to be allowed to continue at Inverlochy; its convenience for the education of his children had brought him there in the first place. He asked if he could also have Corriechoille, which he wanted as a 'Sheep-farm' to be managed in combination with his existing farms of Kilmanivaig and Brackletter.[9] The latter amounted together to only 1,100 acres, but they were low-lying and would serve for the wintering of the sheep. Corriechoille, on the other hand, was over 12,000 acres, mostly hill ground.[10] His partner in this ambitious scheme was to be 'Balnagowan',[11] who has not been clearly identified. Sir John Lockhart-Ross of Balnagowan in Easter Ross pioneered the introduction of Blackfaced sheep into that area in 1774,[12] but no evidence has been found to suggest that he ever had any interest in Lochaber.

Corriechoille and the adjacent Achachar had been held by Donald

MacPherson, the Duke's baron baillie in Lochaber, since the previous occupant, Archibald McDonald of Fersit, had left for America in 1773 (see below). The baillie wanted to keep these farms, and on 4 March 1777 he wrote from Claggan, where he lived, offering £90 rent for a further lease. His longer-term motive was to secure them for one of his sons, Alexander, who was in Edinburgh, where he qualified as a writer and married around this time.[13] [14] In 1779 Alexander was practising his profession in Fort William, but by 1781 he had moved to Inverness.[15] [16]

The Duke's factor must have suggested to Achtriachtan that he should consider Fersit as an alternative to Corriechoille, but Achtriachtan was not willing to compete with the Glencoe family. On 5 March 1777 he said:

> I observe what you say about the farm Glencoe has offered for. Glencoe made me privy to his offers and I gave him all the encouragement possible that he might depend upon being well used by his Grace and you. I cannot see in honour how I can interfere with him in his offers and I [had] much rather be quit of his Grace's lands than be guilty of the least breach of trust, and at the same time Fersit will not answer sheep . . . but since I am to be disappointed at Corrychoily [Corriechoille] I shall give up all thoughts of sheep-farming in Lochaber.[17]

It seems surprising that he dismissed Fersit for sheep-farming, but perhaps he meant that he needed a farm closer to Kilmanivaig and Brackletter for his proposed development with 'Balnagowan'.

Achtriachtan and the baillie may not have known that the factor was trying to find a way of accommodating David Cubison from Ayr, 'who I dare say would be of service in the Country'. He was thinking about letting the baillie continue at Corriechoille if he would give up Claggan to Cubison, or of asking Achtriachtan to cede Inverlochy, 'considering what a number of farms he has'.[18] [19] There is an impression of indecisiveness at this period in the conduct of the Duke's affairs, and it took some time to resolve the competing claims.

In the end the baillie got a further lease of Corriechoille without losing Claggan. Within a couple of years his son Alexander petitioned the Duke for a long lease in his own name to follow the expiry of his father's lease in 1784. He said that there was already a considerable stock of sheep but that it would not be worth investing further without a long lease.[20]

In 1776 Achtriachtan was given a lease of Kilmanivaig and Brackletter for a further year. Highbridge was identified in the lease, and Patrick Johnston was associated with it. Mr Tod, the Duke's factor for the area, was to 'fix the Boundary of Highbridge with a view of it becoming a Settlement for Pat Johnston'.[21] Johnston then became the tenant of Highbridge, where there was an Inn.[22] Achtriachtan continued for some years as tenant of the remainder of Kilmanivaig and Brackletter. The position at Inverlochy is

less clear. In 1776 it was also set to him for one year,[23] and he was still there in 1778.[24] He may, however, have lost it around this time, because his subsequent correspondence came mostly from Kilmanivaig rather than from Inverlochy.

In 1778 Tirandreich also came into the hands of Angus Macdonald. There is no indication that he asked for this farm, and it may not have been a welcome addition. On 15 June 1778, he said to James Ross: 'I am so much drained of money betwixt my sons . . . and this affair . . . that I will be rather straitened this year and my Colleague cannot help me immediately.'[25] 'This affair' was a reference to Donald McDonald of Delfour in Badenoch, who was then in difficulties and was handing over his farm of Tirandreich in Lochaber to Achtriachtan. The mention of 'my Colleague' suggests that Achtriachtan was still thinking of the great sheep-farming venture. Tirandreich, at about 600 acres,[26] would not however serve the purpose for which he had wanted Corriechoille.

The McDonalds of Delfour

In the mid-eighteenth century, Alexander McDonald at Delfour in the parish of Alvie in Badenoch had a son Donald and two daughters, Katherine and Sarah. Around 1764 or 1765 Katherine married Alexander McDonell of the Inverroy family in Lochaber. Her father undertook to pay to his son-in-law a tocher of 500 merks Scots at Martinmas 1767, but he died before it was paid. His son Donald succeeded him in Delfour. Katherine died in January 1768.[27] [28]

Sarah married a shoemaker in Perth. In later years, as the last survivor of the family, she disputed her brother Donald's affairs with her late sister's husband Alexander McDonell. She said that Donald had been 'of an extravagant turn' and was obliged in the early 1770s to sell his Delfour property to the Duke of Gordon for about £600. Sarah also said that Donald was 'frequently in a disordered state of mind' and that his brother-in-law Alexander McDonell in Lochaber 'exercised full influence' over him.[29] In 1769 Donald was tenant of Leanachanbeg on the Duke's land in Lochaber.[30] For a time he was tenant of Achneich,[31] and later of Tirandreich.[32] He seems to have sub-let Leanachanbeg to MacDonell of Keppoch.[33] By 1778 his funds were running out, and in April he was dashing to and fro between Lochaber and Badenoch trying to find out what the factor was going to do. He got a Commission in Lord Macdonald's 76th Regiment, and the factor was hopeful that this would take him abroad immediately, thus avoiding 'the disagreeable necessity of putting him out'.[34] [35] [36] Unfortunately the army failed to oblige in time, and on 15 June 1778 Donald wrote from Tirandreich to the factor saying that 'I have this day given possession of my farm to Mr. McDonald of Achtriachtan'.[37] [38] It was on the same day that Achtriachtan referred to

'Dellyfour' as one of his financial burdens.[39] It would therefore seem that he had stepped in to take Tirandreich off the hands of its embarrassed tenant.

Donald McDonald eventually went with his regiment to America in March 1779, leaving his brother-in-law Alexander MacDonell as his factor in this country, but he died shortly after arrival. His surviving sister Sarah challenged their brother-in-law Alexander over the management of Donald's affairs. Although Alexander was of the Inverroy family, he was then said to have been 'late at Achadourie [Achaderry?], late at Blarour, and latterly at Glenturret'.[40]

Communications between Badenoch and Lochaber

The Delfour family's interests in both districts arouse curiosity about travel between Badenoch and Lochaber in the eighteenth century. The present road by Loch Laggan was not completed until 1818,[41] and the old route via upper Strathspey to Garvamore and thence to Glenroy must have been a well-used if sometimes hazardous way between Badenoch and Lochaber. In 1782 William Tod, who factored the Duke's lands in both districts, used his 'best Endeavours to get a Bridge built over the Burn of Bohuntine in the Braes of Lochaber', because without a bridge 'one cannot depend upon the Communication being open at any time between Badenoch and Lochaber, an Hour's Rain often rendering the Burn impassable'. He had often been detained there while on the Duke's business, and said that 'I once indeed very narrowly escaped with my Life'. As he was trying to engage the support of the Duke, he added wisely that £800 of rent money had also been at risk.[42] His effort was eventually successful, and a bridge was built in the early 1790s. The county contributed £66, and subscribers gave £46, including £20 from the Duke.[43]

The Farm of Fersit

Alexander Macdonald of Glencoe was successful in his bid for Fersit and entered at Whitsunday 1778. He remained tenant for twenty years, and had a complicated involvement with it again in the early nineteenth century.

Fersit was a large farm, much of it hill ground, which straddled the north end of Loch Treig and the short Treig Water which drains that loch into the River Spean. It is often described in two parts: Fersitriach on the east of Loch Treig and the Treig Water, and Fersitmor on the west. Like many farms in Brae Lochaber it was traditionally identified with a particular family of Macdonalds or Macdonells, in this instance the MacDonells of Fersit. (The names Macdonald and Macdonell are the same – see Appendix B.) However, in the late eighteenth century these traditional associations between families and farms were breaking down, and tacks were being offered to the highest bidder. Macdonald is a common name in that part of the country, and it can sometimes be difficult to tell whether the

replacement of one Macdonald by another is maintaining or displacing the original family.

In 1766 the tacksman of Fersit was Archibald McDonald, who was anxious to know if his tack would be renewed. He reminded the Duke of Gordon that he had been recommended to him by the Duke of Atholl, and that by his industry he had made a living out of Fersit, whereas 'the former Possessor altho' he had it at a much cheaper rate went off a Beggar and 1,200 Merks in your Lop^s debt'.[44] That suggests that Archibald was not one of the old Fersit family. His tack was renewed, and by 1771 he also held Corriechoille and Achachar,[45] but unfortunately his success was not sustained. In December 1771 he told William Tod that he was renouncing Fersit, asking him to use his good offices 'to get me clear of it against Whitsunday first and not oblige me to keep by it, which his Grace may, to Whitsunday 1773'.[46] The terms of the Duke's leases gave tenants the right to give up their tacks in 1773, and several took advantage of that.[47] Archibald McDonald went on to say that the hundred-guinea rent for Fersit

> will undoubtedly ruin me . . . considering that I am under a necessity to continue by Corrychoilly for the seven year lease . . . there is little arable ground at Fersit or Corrychoilly . . . the price of cattle is so considerably fallen and [there is] no demand for grass from graziers as formerly . . . and those that have taken grass from me for years past from the South Country have renounced same.[48]

William Tod kept James Ross in Fochabers informed. On 15 April 1772 he sent him letters from 'Fersit and Sandy Blarour' [Archibald McDonald at Fersit and Alexander McDonell at Blarour], who were both declaring an intention to go to America, adding:

> I am certain Fersit is serious . . . he has given orders to a man who . . . is a brother-in-law of his own to make a Purchase for him in America. . . . Neither do I in the least doubt but Blarour will go a second time to the other side of the Atlantic if he does not get his own terms. If these two follow the one already gone, there is not a McDonell behind worth keeping, except Inch who deserves attention on other counts, but not for his being substantial.[49]

Tod saw both of them on 30 April, and reported on 7 May that 'the two MacDonells seem determined on going to America'. He had offered a discount of £35 on the rent of Fersit for the remaining four years of the lease, but Archibald McDonald wanted a discount of £55 backdated for the four preceding years.[50] On 17 June William Tod was lamenting the loss of Archibald McDonald of Fersit, 'who is by odds a man of the most credit in Lochaber'.[51]

Alexander McDonell at Blarour was of course the brother-in-law of Donald McDonald of Delfour mentioned above. If he did go to America he must have returned shortly afterwards, as he was certainly in Lochaber when Donald of Delfour left for America with his regiment in 1779.[52]

MacPherson of Ballachroan as Tenant of Fersit

Archibald McDonald's successor at Fersit was a well known Badenoch man, Captain John MacPherson of Ballachroan near Kingussie. He had been born at Glentruim in Badenoch in 1724, the second son of Alexander MacPherson of Phoness. His mother was one of the MacDonells of Aberarder, and there were members of that family quite close to Fersit. He served in the 82nd Regiment, and when he retired from the army he became tacksman of Ballachroan. He then proved himself to be a competent farmer with interests extending over a wide area.[53] [54] He had financial links with many people in Badenoch and held a bond of credit from the British Linen Company for £500, allowing him to borrow up to that amount.[55]

John MacPherson's farming career has been overshadowed by accounts of his death on a hunting trip to Gaick at the turn of the year 1799/1800, when he and several companions were overwhelmed by a sever snowstorm. After his death, 'distorted and imaginative versions' of his life and military exploits began to appear, among them one by Sir Walter Scott, who was taken to task by Mrs McBarnet, a daughter of Ballachroan. Even more fanciful accounts continued to appear, and Alexander Macpherson of Kingussie responded in 1900 with an account of the Captain's life which he subtitled 'A Counterblast'.[56] Alexander Macpherson was the agent of the British Linen Bank in Kingussie,[57] and his account was published by George A. Crerar, a bookseller there. Highland history would have benefited from more such counterblasts.

MacPherson of Ballachroan got Fersit at a net rent of £60. This was expressed as a rent of £60 and a discount of £45,[58] which equals Archibald McDonald's rent of one hundred guineas. The discount was, however, slightly better than William Tod had offered to Archibald McDonald if he had been willing to stay.

Captain Grant and Captain MacPherson were placed at Fersit by MacPherson of Ballachroan to run it for him. They were sometimes described as the tenants[59] and may have been given a share in the tenancy to reward them for managing the farm and encourage them to maximise the profits. That was evidently a common arrangement. Grant did not remain for long, but MacPherson was Ballachroan's brother and he remained for the duration of the lease.[60]

MacPherson of Ballachroan brought some men of his own to Fersit, again a common practice. Those living at Fersit in May 1778, at the end of

his tenancy, were Donald and Murdoch MacPherson, Alexander MacFarlane and his sons John and Donald, John Cameron, Donald MacGillivantich, John MacDonald and two sons, and two unnamed shepherds.[61] Some of these might have been local men, but the MacFarlanes almost certainly went back to Badenoch after the lease expired. A John MacFarlane was later to be tacksman of Kingussie and was involved with his brother Duncan in various lands there.[62] [63] Duncan was with MacPherson of Ballachroan on the fateful hunting trip to Gaick and was among those who died in the snowstorm.[64]

John MacPherson's tack of Fersit was due to expire at Whitsunday 1778. He apparently wrote in good time saying that he wished to have the tack renewed. He wrote again to James Ross, the Duke's cashier, on 21 March 1777, expressing concern that he had not had a reply. He asked that the tenancy of Fersit be continued for another year beyond 1778 so that he and his tenants might be spared the hardship of having to sell stock at short notice and incur heavy losses. He pleaded that he was entitled to such consideration, 'considering that I took that Farm when no other person would take it either at the price or in such a Neighbourhood'. If the Duke would not let him have Fersit for another year, he was prepared to accept it for a further seven years rather than 'see my Broyr with his . . . Family & the other honest men who came such a Distance set Adrift all at once'.[65]

Perhaps there had been a failure in communication, but when Ballachroan wrote on 21 March 1777 the offer from Alexander Macdonald was about to be accepted. On 11 April 1777 the latter said to James Ross: 'I am informed by Mr. McDonald of Achtriachtan that his Grace the Duke of Gordon and you are pleased to accept of my proposal for the farm of Fersit and that entry is to commence Whitsunday come a year for 80 guineas.'[66] Alexander wrote from Glencoe, but it is interesting that word about his new lease had come to him through Angus Macdonald of Achtriachtan.

Some Problems in Lochaber

Angus Macdonald of Achtriachtan was now well established in Lochaber and played his part in affairs there. For example, on 28 August 1779 he wrote from Kilmanivaig to James Ross recommending 'one Duncan McIntyre, a friend of mine' for the post of Parochial Schoolmaster at Kilmanivaig. He said that he was 'a young man who has finished his philosophy at the College of Aberdeen'.[67] Duncan MacIntyre was the son of John MacIntyre of the Glenoe family and Margaret Macdonald of the Achtriachtan family. He graduated MA at King's College, Aberdeen, in 1779.[68] He did become the schoolmaster in Kilmanivaig,[69] but in later years he was Missioner at Fort William and then Minister of Laggan.[70]

When Angus described him as 'a friend of mine', he was using the phrase in the sense that he was a relative.

There was a relatively lawless situation in Lochaber, and letters from Achtriachtan and others identify some of the troublemakers. The authorities were not helped by suspicions that those who should have maintained good order were in league with some of the main transgressors. Fingers were pointed at Donald MacPherson, the Duke's baron baillie.

Achtriachtan and Baillie MacPherson clashed over Corriechoille, and their letters show that there was an undercurrent of antagonism between them. The baillie referred sarcastically to 'my friend Mr. McDonald', wanting his son out of the area in the hope that he, the baillie, would then follow. Achtriachtan countered that he doubted whether the baillie and his son together could muster what was needed to stock a quarter of Corriechoille.[71]

Towards the end of the 1770s doubts were being expressed about the integrity of Baillie MacPherson. James Ross at Fochabers received a long memorandum about the baillie from the Rev. Thomas Ross, a formidable character who had been ordained to the parish of Kilmanivaig in May 1776. In 1780 he married Lucy Cameron of the Fassifern family,[72] an aunt of both Alexander Macdonald of Glencoe and his future wife Mary Cameron. The content of Mr Ross's memorandum suggests that it was written in 1778, and it alleged that the baillie was colluding with the wood-keeper in selling the Duke's wood 'for their mutual interest', conniving with the inhabitants of the riverbanks to kill fish – which they sold in barrels – and taking bribes to conceal acts of theft. Mr Ross had doubts about the baillie's relationships with one troublesome family, the MacMartins (Camerons) in Braeroy.[73] Suspicion also arose that he was over-friendly with another, the Kennedys in Glengarry.

The Camerons of Braeroy

The Camerons in the upper reaches of Glenroy were MacMartins, presumably one of the offshoots in Lochaber of the MacMartins of Letterfinlay, who later took the name Cameron.[74] On 22 February 1778, William Tod wrote cheerfully that one of the MacMartins in Braeroy had got a Commission in Colonel Gordon's Regiment, 'by which means we shall get rid of *him* – and half a dozen of the greatest Blackguards in the Highlands'.[75] He was referring to Donald Cameron, a nephew of Donald more og Cameron, tacksman of Annat in Glenroy. His joy was premature. Six days later he reported that a gang from Braeroy, led by Ensign Cameron, the newly commissioned MacMartin, had taken possession of the Inn at Highbridge, seized one of the servants of Patrick Johnston and carried away by force six fisherman from Nairn on their way south to work for Campbell of Shawfield. The Ensign wanted to

force them to enlist in the army. William Tod told Baillie MacPherson to raise a criminal prosecution, but the men were released and nothing came of that. Mr Tod also told the baillie to execute a removal against 'the whole tribe' as security for their future behaviour.[76] However, the Rev. Mr Ross thought that the baillie had had some personal involvement in the episode.[77]

Ensign Cameron continued his enthusiastic recruiting. On 13 April he carried off by force an unfortunate Allan Cameron, who had previously been recruited and 'dispensed with' because he was unfit.[78] In 1781 he was again recruiting in Lochaber, but he was now Lieutenant Cameron of the 95th Regiment. One of his recruits was Ewen MacMillan, late tenant in Kinlocharkaig, who was being held in the tolbooth in Inverness, accused of theft by John MacDonald of Meoble. MacMillan protested his innocence but could not afford the expense of defending himself. To avoid this he enlisted as a private with Lieutenant Cameron, and together they petitioned the sheriff to release him on condition that he joined the 95th Regiment serving in Jersey. Meantime, Lieutenant Cameron would give bail for him. Alexander MacPherson, the baillie's son, who was by then a writer in Inverness, was the procurator for both petitioners.[79]

While recruiting in Lochaber, Lieutenant Cameron found time for local squabbles. The tacksman of Blarachaorin had refused to sign a certificate of good character for a former Cameron sub-tenant and his son because they were 'persons of bad fame' who had been guilty of several acts of theft. The Lieutenant, along with his uncle, Donald more og Cameron, and James Cameron, uncle of Angus Cameron of Kinlochleven, descended on the unfortunate tacksman and forced him to do so by threats of violence.[80] All of the parties involved seem to have been MacMartin Camerons.

Lieutenant Cameron's regiment was disbanded in 1783, and for the next couple of years he led a gang of thieves who stole livestock from the Highlands and disposed of it in the Lowlands.[81] A few years later it emerged in proceedings involving his cousin Donald McDonald, drover at Tulloch, that Alexander MacPherson, the baillie's son, had been 'in familiar habits' with Lieutenant Cameron and had advanced money to him.[82] [83] Mr Ross's suspicions seem to have been well-founded.

The Kennedys from Glengarry

The Kennedys from Glengarry were a notorious family who 'lived by tribute or blackmail over a wide range of country'.[84] References have been found to several of them, but it would be impossible to establish all of the relationships between them with confidence. The authorities found it difficult to get any of them into custody, keep them there and secure convictions against them. One of them escaped from the tolbooth in Inverness with remarkable ease in 1779, apparently after some message

had been passed to him by a servant of George Bean, writer in Inverness.[85]

In 1780 another of them had been arrested, and on 24 April Alexander Cameron of Glenturret wrote to William Tod 'respecting Kennedy who is to be tried at the approaching circuit'. Cameron had lost cattle and suffered at the hands of the Kennedys. He was afraid, with good reason, that the Kennedy who had been arrested might not be brought to trial, and he hoped that the Duke might be persuaded to intervene. He knew that the Kennedys had held a meeting at Laggan-ach-drum in Glengarry and had contributed money to defend their colleague who was in custody. There was even talk of having bribed lawyers in Inverness. Donald MacPherson, the baron baillie, had collected evidence in preparation for the trial and had shown Cameron a list of possible witnesses. Cameron was alarmed that very few of the potential witnesses were to be called. He said 'there is reason to think the cause was mismanaged and that the precognition was not in whole sent to the Crown lawyer'.[86]

On 12 May 1780, Angus Macdonald of Achtriachtan wrote from Kilmanivaig to William Tod. He supported Cameron of Glenturret and added that there had been reports in the locality of some 'friendly settlement' between Baillie MacPherson and the Kennedys 'since the precognition taken by his son at Fort William took place'.[87]

The Kennedy in custody in 1780 probably did escape justice, but there seem to have been repercussions. Cameron of Glenturret said in his letter of 24 April 1780 that Campbell of Delnies, who was then sheriff in Inverness, had resigned and would be replaced by Simon Fraser of Farraline.[88] That did indeed happen, although Sheriff Fraser was not appointed until May 1781.[89] He added that Baillie MacPherson had also resigned, and this can be corroborated. On 10 February 1781 John Macdonald of Glencoe wrote to James Ross: 'I am informed that Mr. MacPherson, the Duke's Baron Baillie in Lochaber is quit of his charge.' He commended Donald Cameron in Dunlick in Lochaber for consideration as a possible successor.[90] There is a hint of some link between Glencoe and Dunlick. A list of arrears in 1785 shows that Glencoe owed £59 4*s.* 9*d.* to the Duke for 'Balance of Rents for Torlundy and Donleck'.[91]

Sheriff Fraser of Farraline was made of stern stuff, and he was ably assisted by the redoubtable John MacKay, constable and messenger in Fort Augustus. The name Kennedy began to appear frequently in the records of the Circuit Court and of the Sheriff Court, although not always successfully. Several Kennedys were absent from their intended trial at the Circuit Court in Inverness on 16 September 1782. Angus Bain Kennedy, John Bain Kennedy, John Roy Kennedy and Archibald Bayne Kennedy, all residing in Shian in the Parish of Kilmanivaig, were declared outlaws and fugitives.[92] On 20 November 1783, Ronald Dow Kennedy, formerly in Shian, was tried in the Sheriff Court on a charge of theft and reset, acting

with Angus Bain Kennedy, John Roy Kennedy, John Buie Kennedy and Archibald Buie Kennedy. All were formerly in Shian, and all but the last were prisoners in Inverness. Sheep and a bull had been stolen from various farms, but a verdict of 'not proven' was returned.[93]

On the same day, the court did rather better with two of Ronald's sisters-in-law, Katherine Kennedy, wife of his brother Angus Bain, and Katherine MacDonald, wife of his brother John Bain. They had set about Dugald Macdonald, a sheriff officer trying to arrest one of Ronald's servants, who had had to be rescued by his colleagues John MacKay and Hugh Chisholm. The Kennedy wives were found guilty of the crime of deforcement. Alexander MacPherson represented Katherine Kennedy, who appealed but found Lord Braxfield unsympathetic.[94]

Alexander Breck Kennedy was the head of a family of Kennedys at Cullachy near Fort Augustus. He had at least two sons, Angus Bain at Shian and Donald. At Whitsunday 1779 Alexander Stewart brought a flock of between eight and nine hundred sheep of the 'south country or Annandale breed' from Appin to Ardochy on the other side of the River Tarff from Cullachy. Alexander Breck Kennedy was alleged to have stolen some of these sheep. A precognition in 1781 seems to have come to nothing, but evidence given by Duncan Mackenzie, one of Stewart's shepherds, was interesting. He said that sixteen feet had been found a little above the Bridge of Doe. He was 'positive they are of the south country or Annandale breed of sheep and that [his] master and Mr Fraser of Gortulag are the only persons that have that breed of sheep in this part of the country'.[95] [96]

Archibald Bain Kennedy, lately residing at Cullachy, who may have been another of this family, was examined by the sheriff depute on 14 April 1784 and put on trial in 1785 for thefts occurring before 1778. He had been grasskeeper at Ardochy when Lieutenant Fraser of Errogie had it, but Alexander Stewart dismissed him when he took over in 1779. Archibald was found guilty, sentenced to be taken from the tolbooth, given twelve strokes, and then drummed through the streets of Inverness and given a further nine strokes in each street. After that he was to be taken to Fort Augustus, then to Fort William, and the performance was to be repeated in each place. He was then to be banished.[97] Several others were said to have been banished, and one hanged.[98]

The environment for the peaceful development of sheep-farming was no doubt improved by these measures. However, Baillie MacPherson's dubious relationships with the Camerons of Braeroy and the Kennedys of Glengarry were to have repercussions. Donald McDonald, drover at Tulloch, was well aware of the relationships and knew that the baillie's son Alexander had been drawn into them. So when Alexander was establishing himself as a writer in Inverness in the 1780s, he was apprehensive that Donald might cause him embarrassment and made strenuous efforts to

ensure that he could not use the knowledge against him. The bitter tussle between them coincided with the early stages of Donald's remarkable involvement in the development of Æneas Mackintosh's land, a fuller description of which will be found in Chapter 3.

Achtriachtan's Later Years in Lochaber

By the early 1780s, Angus Macdonald of Achtriachtan seemed to have put his money troubles of 1778 behind him and was once again in an expansive frame of mind. On 26 September 1781, he wrote to James Ross offering himself as a candidate for the vacant position of baron baillie.[99] Nothing came of that, but on 9 May 1782 he wrote expressing interest in the Forest of Gaick in Badenoch. He had left it rather late and said that he had never seen the place, but if James Ross could 'with propriety stop the sett' he would go and look at it and make an offer.[100] He may have been too late, or James Ross may have decided that he could not delay the matter, but it shows that Achtriachtan was confident about his position then.

Achtriachtan's interests in Lochaber were not confined entirely to the Duke's land. A record in 1785 shows that he was tacksman of Mucomir and Torness on the land of George MacMartin of Letterfinlay at a rent of £50 per annum.[101] Geographically, that was close to the farms of Kilmanivaig and Brackletter which he held from the Duke, but it is not known when he obtained the tack from MacMartin or how long he held it.

Relatively little information has been found about Achtriachtan's final years in Lochaber. In the early 1790s he and a Mr Duncan Stewart carried out a souming on the Duke's farms in Lochaber, for which they were paid a fee and expenses.[102] He was also involved in the appointment of one of the foresters on the Duke's land in 1792.[103] However, an examination of his financial affairs (see Chapter 1) show that by this time he was accumulating large debts and getting into serious difficulties. It seems probable that by the middle of the decade he had withdrawn or been removed from his tacks in Lochaber. When Captain Alexander MacDonell of Moy, one of the Aberarder family, offered £140 rent for Kilmanivaig, Brackletter and Highbridge at the Lochaber Sett of 1804, they were said to be 'presently occupied by him'. Similarly, the Rev. Thomas Ross made an offer for another year of 'his farm' of Tirandreich and other lands.[104] These wordings suggest that Achtriachtan had vacated these farms some years earlier, and there is no indication that any of them were in his name when he died at the end of 1798.

Alexander Macdonald of Glencoe Quits Fersit

Alexander Macdonald's first lease of Fersit was for seven years from 1778 and must have been renewed several times, because his final lease did not expire until 1797. By then the rent had increased from the original £84

to £160, and the Duke had decided that at the next lease it would be increased to £300.[105]

Alexander Macdonald was due to vacate Fersit at Whitsunday 1797, and he acknowledged to William Tod on 20 April that 'he was obliged to Remove at that term'. He added that he 'would be sorry to keep possession any longer', a comment perhaps related to the intended increase in rent. The Duke then 'sett the said lands to other tenants', who were promised entry at Whitsunday 1797. In spite of what he had said, Alexander Macdonald and his sub-tenants continued to keep possession and refused to remove themselves. Accordingly, in 1798 the Duke resorted to legal action and obtained a decreet of removal in the Court of Session.[106]

That process reveals names of Alexander Macdonald's sub-tenants at the end of his lease. The court was asked to rule that 'Alexander MacDonald and his sub-tenants and particularly the after-named, viz., Angus MacDonald, John MacDonald, John Bain MacDonald and Donald MacDonald, sub-tenants in Fersitriach, and Donald Rankine, Donald Rankine MacNeill, John Rankine, John Rankine alias Stalker, Allan Robertson alias Stalker, and Alexander Cameron, sub-tenants in Fersitmor, and all other sub-tenants of the said Alexander MacDonald' should remove themselves.[107]

All of the named sub-tenants in Fersitriach (i.e., east of the Treig Water) were Macdonalds. It is quite probable that some or all of them had come from Glencoe at the behest of Alexander Macdonald. There is little doubt about the first named, Angus MacDonald. It will be shown in Chapter 5 that in 1798 Auchteraw was sub-let by the Rev. John Kennedy to Angus MacDonald residing at Fersit. Alexander Macdonald of Glencoe was his cautioner.[108] Kennedy described Angus MacDonald as a 'ffolanen of Macdonald of Glencoe'.[109] Whatever the precise meaning of that curious word may be, in this context it probably means that Angus was from Glencoe.

Quite different names occur in Fersitmor (west of the Treig Water). A clue to their origins may be found by comparing these names with those of the men listed twenty years earlier, in May 1778, as residing on the land belonging to Angus Cameron of Kinlochleven, i.e. James Cameron, Donald Stalker, Allan Stalker, Allan Stalker, Allan Stalker, Donald Rankine, Donald Henderson, Donald Henderson, Archibald Henderson, Donald MacInnes, Evan MacInnes, Duncan MacInnes, Donald MacDonald, Angus MacDonald, Archibald MacDonald and Malcolm Stalker.[110] The name Stalker is fairly unusual in Lochaber. Its appearance at Fersit, along with some Rankins, suggests that Angus Cameron of Kinlochleven may have had an interest there. He is often mentioned in the affairs of Alexander Macdonald of Glencoe and was involved jointly with him in borrowing certain sums of money.[111] There is no clear identification of

their joint ventures, but Fersit could be part of the explanation, even though the lease was in the name of Alexander Macdonald of Glencoe alone. Someone else could have been involved in the funding of such a venture without his name appearing in any formal lease.

The tenants who succeeded Alexander Macdonald of Glencoe at Fersit were Captain Alexander MacDonell of Moy, Alexander McVean and James McCaul, who undertook to pay the increased rent of £300 per annum for a seven-year lease.[112] This proved to be too much for them, and after a few years Alexander Macdonald of Glencoe tried to obtain the tenancy once again. His offers were rejected, but he seized an opportunity to involve himself as a cautioner for others. That part of his concern with Fersit will be explained, along with other nineteenth-century affairs, in Chapter 12.

By the close of the eighteenth century, Angus Macdonald of Achtriachtan was dead and Alexander Macdonald of Glencoe had ceased to have any direct connection with the Duke of Gordon's land in Lochaber.

3

The Mackintosh Lands in Lochaber before 1786

Alexander Macdonald of Glencoe became a substantial tenant on Mackintosh land in Lochaber in the early years of the nineteenth century. Long before that happened, Æneas Mackintosh of Mackintosh began the transformation of his Lochaber lands to meet the new economic conditions. This chapter discusses the early stages of that lengthy process.

Background to Change

When Æneas Mackintosh succeeded his uncle in 1770, he was a captain in the army and served in the American War of Independence.[1] While he was away, the management of his estates was undertaken by his brother-in-law, Captain Lachlan Mackintosh of Balnespick.[2] Æneas's regiment, the 71st (Frazer's Highlanders), was disbanded in 1783[3] and he then took control of his own estates.

The family seat of the Mackintoshes was at Moyhall, near Inverness, but their extensive possessions in Inverness-shire included an area of land in Lochaber. As explained in Chapter 2, this lay on the north of the River Spean and extended most of the way up Glenroy. Few of Æneas Mackintosh's own clan lived in Lochaber. In the fourteenth century, a predecessor had been granted lands in Badenoch, and most of his clansmen migrated there. The Lochaber lands were left to be occupied mainly by MacDonells. In 1688, under the leadership of MacDonell of Keppoch, they inflicted heavy losses on the Mackintoshes in the fiercely fought Battle of Mulroy.[4]

In the early 1780s the MacDonell families were a cohesive group claiming a common ancestry and linked in more recent times by several marriages. John MacDonell, the eldest son of Ronald MacDonell of Aberarder, was married to Catherine, a daughter of Alexander MacDonell of Keppoch who had died at Culloden.[5] Margaret, a daughter of Ronald MacDonell of Aberarder, was married to Angus, the eldest son of Alexander MacDonell of Tulloch, and a daughter of the Tulloch family was married to Donald, son of John MacDonell of Cranachan.[6] Following these families can be confusing because their territorial designations could remain with them long after they had ceased to hold the corresponding lands.

The finances of the MacDonell tacksmen were mostly in a poor state.

The Keppoch family were in debt, and after the death of Ronald MacDonell of Keppoch in 1785 there was constant bickering about the management of the affairs of his successor, who was a minor (see Chapter 4). John MacDonell of Cranachan had tried his hand as a merchant in Maryburgh (one of the names by which Fort William was known at this time), but had failed. By 1788 he was incarcerated in the tolbooth of Inverness at the instance of one of his creditors, Colin MacLeran, merchant in Stirling,[7] while Ranald MacDonell of Aberarder was bankrupt as early as 1769.[8] The position at Tulloch was less clear. Angus MacDonell of Tulloch claimed in 1802 that his family had never been 'a sixpence in arrears'.[9] Nevertheless, Æneas Mackintosh said, 'the truth is I don't wish his being one of my Tenants, his wife's connections being disagreeable to me and had nearly ruined him'.[10] His wife was an Aberarder MacDonell, and Æneas was probably referring to events in the early 1780s.

Mackintosh's Lochaber tacks were due to expire in 1786.[11] He could then begin to dislodge the old tacksmen, but he needed others to put in their places. Two men emerged for this purpose. One was Alexander Mackintosh, a merchant in Maryburgh. He replaced the MacDonells at Keppoch in 1795 and consolidated a large tack on the east of the River Roy in 1804, in both instances paving the way for Alexander Macdonald of Glencoe. The other was Donald McDonald, a drover at Tulloch who replaced the MacDonells of Tulloch in two stages and then became a substantial sheep-farmer. He also became a kind of sub-factor over the whole of Mackintosh's Lochaber estate, in which capacity he developed intriguing relationships with Æneas Mackintosh and Campbell Mackintosh, his lawyer in Inverness. Donald, however, had to shake off some problems before he could make progress.

As mentioned in Chapter 2, Donald McDonald had incurred the enmity of Alexander MacPherson, writer in Inverness. There was also a bitter dispute in progress between Donald's family and the MacDonells of Aberarder and their relatives. In quieter times, when these were well in the past, Æneas Mackintosh was to say of Donald that 'he has been a useful man to me in breaking the McDonald [MacDonell] combination and so raising my rents in his quarter, which otherwise would never have been effected to the present extent – I do not admire his Character, but from these Circumstances find it proper to countenance him'.[12] Who were Alexander Mackintosh and Donald McDonald?

Alexander Mackintosh, Merchant in Maryburgh

Alexander Mackintosh was born about 1756. In 1834, when he was seventy-eight, he wrote a long memorial to Æneas Mackintosh's successor pleading for financial assistance in his old age. This sheds useful light on some obscure events, but it has to be interpreted with caution. It was based on

his recollection of events that happened up to fifty years earlier, and it was presented with a deliberate intention to impress. When comparison can be made with what he wrote at the time, his Memorial is much more exciting and colourful.

He was related to Sir James Mackintosh (1765–1832) of the Kyllachy family, a doctor, writer, lawyer, Recorder of Bombay, and Member of Parliament for Nairn. Alexander was well established as a merchant in Maryburgh before he became a tenant on Æneas Mackintosh's land. He had a vessel of his own trading at sea and said that he was 'accumulating money fast'. His interests were varied, which was typical of merchants at that time. He made much of his loyalty to Æneas Mackintosh as his chief, and liked it be known that he had become involved in his chief's land to help and oblige him. He claimed that he lost money in this service;[13] perhaps he did, but in later life he may have exaggerated the amount.

Donald McDonald, Drover in Tulloch

Donald McDonald, sometimes called Donald Dhu (or du or Dow), was one of the sons of Archibald MacDonell alias McAlister.[14] The family were at Tulloch in Brae Lochaber in 1773 when Donald gave evidence identifying an allegedly stolen horse. He was unmarried, and his age was given as '25 or thereby'.[15] On another occasion, in 1787, his age was given as 'about 30'.[16] The 1773 record seems more credible.

A precognition following a disturbance at Tulloch in 1783 shows that Archibald was then aged sixty and was a sub-tenant of Alexander MacDonell of the old Tulloch family, who still held the whole of Tulloch from Æneas Mackintosh. Archibald's wife was Sarah McGlaserich alias Campbell, also aged sixty. Two of Archibald's sons, John and Alexander, figure in the precognition. John, aged about forty, was also a sub-tenant in Tulloch, with a house of his own there. He had been a soldier, and by 1783 was an invalid pensioner. His wife was Mary McLean, aged thirty-nine. Alexander, often called Alexander Dhu, was apparently living in his father's house at Tulloch.[17] Donald was not mentioned, but by 1783 he was active as a drover, collecting animals from the Highlands to take to the great fairs in Doune, Stirling, Falkirk and elsewhere. From time to time he would send for his brother Alexander to help him when a drove had to be moved.[18] Sarah McGlaserich said that John and Alexander were her stepsons,[19] but there is no indication of whether she was Donald's mother. McGlaserich was a well-known name in Brae Lochaber, and there were several of that name in and around Tulloch.[20] [21] It is therefore probable that Sarah McGlaserich was a local woman, while Archibald may have come from elsewhere. The affairs of his son Alexander Dhu, to be described in later chapters, suggest a link with Glencoe, although the nature of such a link is not clear.

Quarrels with the MacDonells of Aberarder

In the early 1780s there was a bitter feud in progress between this McDonald family at Tulloch and the MacDonells of Aberarder. In the precognition of 1783 there are allusions to Alexander Dhu having given 'informations to the Sheriff Depute against Ronald McDonald of Aberarder and his Sons'.[22] His brother Donald said on another occasion that he had taken an active part in a precognition against Ronald MacDonald of Aberarder for the crime of sheep-stealing and theft.[23]

Ronald was the head of the MacDonells of Aberarder, a veteran of the 1745 rising who had been present at Culloden.[24] The family were removed from Aberarder, on the west shore of Loch Laggan, in 1770, but their tentacles stretched in various directions and they appear in many situations in Lochaber and Badenoch. In 1769 they had been given Torgulbin by the Duke of Gordon,[25] who tended to be more tolerant of them than his neighbour Æneas Mackintosh. Around that time they also held Kylarchill, Moy, Kylross and the Forest of Ben Alder.[26] Their association with Garvamore in upper Strathspey dated from a Wadset of 1725,[27] and Ronald's brother Alexander was the tenant there. He also kept the inn at Garvamore[28] and was an active drover.[29] On Mackintosh land, Ronald held the tack of Blarnahanin in Glenroy, which was due to expire in 1786.[30]

Ronald had three sons, John, Alexander and Archibald. Alexander got a commission in the army in 1778[31] and was away from Lochaber during the next few years when his father and two brothers had several brushes with the law. In 1781, 'an Information of Sundry Articles of Charge of a Criminal Nature' was laid against them and a precognition was taken.[32] It seems probable that it was on this charge that Donald McDonald at Tulloch and his brother Alexander Dhu provided information to the authorities. Donald's comments indicated that the crime charged was sheep-stealing and theft.[33] Archibald McKillop in Cranachan and a Donald McDonald in Laggan also gave information against the Aberarders. Ronald's son-in-law, Angus McDonald in Monessie, and other supporters subsequently beat them up on the road between Rannoch and Lochaber.[34]

Legal representation for the Aberarders was arranged on behalf of the Duke of Gordon. In August 1781 George Bean, writer in Inverness, was instructed by William Tod, the Duke's factor for the area, to attend to the affairs of Ronald MacDonell of Aberarder. Within a couple of days he had written a petition and memorial for Ronald and his sons John and Archibald to be laid before the sheriff depute. Ronald and his sons claimed that they were 'no ways guilty but altogether innocent'. They were concerned that in the precognition already taken the witnesses examined were those giving information against them. It is not clear if a prosecution was intended. Nevertheless, George Bean submitted a list of sixty-one witnesses and

petitioned for an exculpatory proof. He meant by this that these witnesses for the MacDonells should also be examined 'in the Way of Precognition'. The sheriff depute was reluctant, but a copy of the petition was sent to the crown lawyers in Edinburgh and an instruction apparently given that the Aberarders were to get their way. Mr Bean's fees were paid by William Tod on behalf of the Duke.[35] [36] It is not clear why the Duke went to such lengths.

The Aberarders were in trouble again in 1783. In March, Ronald, John and a servant were charged in the Sheriff Court with an assault on John Stewart, tenant of Shirramore in the Parish of Laggan. The verdict was not proven,[37] but worse was to come. On 12 May 1783, Ronald and his sons John and Archibald appeared before the Circuit Court in Inverness 'for the Crime of Deforcement of a Messenger employed in the execution of his office'. Ronald was found guilty and sentenced to be confined in the tolbooth of Inverness until 13 June and then to find bail of 500 merks Scots. The crown witnesses included John MacKay and Hugh Chisholm, messengers at Fort Augustus, but there is no indication of what they were about when the Aberarder people interfered with them. Ronald's daughters Margaret and Grizel had failed to appear, presumably as witnesses, and sentence of fugitation was passed against them. As Margaret was the wife of Ensign Angus MacDonell, eldest son of the tacksman of Tulloch, he was dragged into the proceedings and held responsible for her costs.[38]

Ronald died around 1785, and after that his sons seem to have kept on the right side of the law. John was at Torgulbin and then at Killiechonate. Alexander returned from the army and was for long identified with Moy, although he also had other farms. His interest in Fersit is described in Chapter 12. Archibald farmed at Lassintullich to the east of Loch Rannoch.[39] Some of the animosities of the early 1780s were to persist, however.

It is curious that in spite of all the abrasive interaction between Archibald's family and the Tulloch and Aberarder MacDonells, there is no evidence of any threat by the Tulloch family to remove Archibald and his sons. Instead the Tulloch MacDonells and their Aberarder relatives tried to isolate and ostracise them. Ensign Angus MacDonell of Tulloch warned another of the tenants there, Angus McDonald alias Conadach, that he would have him removed because he was on friendly terms with Archibald and his sons. His father-in-law, Ronald of Aberarder, made it abundantly clear to Allan MacDonald, tenant in Brunachan in Glenroy, that he disapproved of him having Archibald and his sons so frequently in his house. Ronald MacDonell's daughter Margaret, wife of the ensign at Tulloch, told Margaret MacDonald, the wife of Donald MacDonald in Murlagan, that she did not like her or her father 'because they are in such good terms with Archibald McDonald, tenant in Tulloch who was an enemy to her father Aberarder'.[40]

Archibald and his family nevertheless survived unharmed. Perhaps this was an indication that the power of the old tacksmen families was slipping away, but there may have been another explanation. Æneas Mackintosh was soon to be remarkably supportive towards Donald while he fended off repeated threats from Alexander MacPherson, writer in Inverness. Later, when Alexander Dhu was put out of Tulloch in 1796, Alexander Macdonald of Glencoe interceded on his behalf with Locheil and obtained a place for him at Achintore. Support seemed to be available when it was needed.

Fraser-Mackintosh has given a colourful account of misdeeds attributed to the family of Archibald MacDonell alias McAlister at Tulloch in 1779 and 1780. He said that this family was probably cleared from Tulloch,[41] but this was not the case. It was the old Tulloch family who were cleared, and Donald McDonald became the tenant. Donald was then able, in a complete reversal of roles, to allow Angus MacDonell of Tulloch to come back as his sub-tenant from 1796 to 1799 (see Chapter 4).

There is ample evidence that Donald was a capable man. One of his opponents conceded in court that he was 'considerably above the common line of business, and capable of managing, with accuracy and attention, these affairs in which he was engaged'. Another was more guarded but said that he was 'descended of indigent parents without education or any visible kind of subsistence . . . he . . . obtained some credite and commenced a Drover, which business he carried on for several years'.[42] Alexander Mackintosh, the merchant in Maryburgh, said that Donald was 'an experienced active man who was familiar with the Mackintosh estate'. He added that Donald 'had a little of the Mackintosh blood in his veins'.[43] A drop or two of Mackintosh blood probably accounts for anything good that he felt able to say of Donald.

Donald McDonald was fluent and literate in English and wrote many letters. In contrast, his father was barely able to scratch his signature, and neither of his brothers could write.[44] [45] Most of Donald's letters were to lawyers in Inverness and to factors, and many have survived in various collections in the National Archives of Scotland, with a few in the Highland Council Archive in Inverness. He was punctilious in dealing with the business of his landlords and his own affairs, but he would often comment on such matters as the weather, the price of stock, troubles with his neighbours and so on. His style was respectful and he never missed an opportunity to show what a diligent fellow he was and how well he was looking after the interests of his betters. He was not, however, so deferential that he was unable to make a blunt comment, or an occasional malicious one. It is unusual to find such a large body of written material from someone who was not of the landowning or professional classes, and if allowance is made for his foibles his letters are informative and useful.

Legal documents and processes in the courts are another productive

source of information about Donald's life and his relationships with his contemporaries. He acquired enough experience of the courts to become something of an amateur lawyer, and he would not hesitate to explain in great detail to his legally qualified correspondents the procedures that they were to follow. When instructing a lawyer he was doing just that. Tact was probably not one of Donald's strengths.

Donald McDonald's Affairs as a Drover

Donald McDonald's first known dealings with the droving fraternity were in 1779. In June of that year, John McIntyre, a Perthshire drover then living at Ballibeg of Ruskie, to the north of Flanders Moss, was in Lochaber selling young sheep.[46] At that time young sheep were being driven from Perthshire to the Highlands and sold there as stock. In 1791/92, the minister of Balquhidder, in west Perthshire, lamented the passing of this trade which had benefited his area. He explained that the demand for sheep-hogs and lambs from the west and north Highlands was decreasing, 'as these lands are now mostly supplied from adjacent farms that are already stocked'.[47] McIntyre had probably brought the sheep from Perthshire to sell in Lochaber as breeding stock. He was about to return to Perthshire and did not want to have to take 212 unsold sheep back with him, so he offered them to Donald McDonald at Tulloch and John McDonald at Acharr (or Achachar) at seven shillings each. They looked at the sheep but claimed that they did not like what they saw and declined to buy them. It was agreed, however, that each would keep half and graze them for John McIntyre at his expense until he came back to Lochaber later in the summer. McIntyre convinced himself that they would buy the sheep then, but they were certain they had not committed themselves.[48]

John McDonald at Acharr does not seem to have been related to Donald McDonald at Tulloch. He had been a lieutenant in the 105th Regiment, which was raised in 1761 and disbanded in 1763.[49] [50] He then became a farmer and a dealer in cattle[51] and was at Acharr by August 1764.[52] He married a daughter of James Small, one of the factors on the annexed estates, and his eldest son, Angus, was born at Acharr in 1766 or 1767.[53]

John McIntyre did not return to Lochaber as intended, but he wrote to Donald and John McDonald on 12 April 1780. He told them that Peter McIntyre of Gartnafuaran in Balquhidder was coming to Lochaber and that he had authority to deal with the matter. He assumed that Donald and John McDonald would buy the sheep for a total of £67. Peter McIntyre was then to give this money to Donald Kennedy, a drover in Maryburgh to whom John McIntyre was deeply indebted.[54]

However, Donald Kennedy was one step ahead of John McIntyre. He had already gone to the two McDonalds and demanded that they give him the sheep or the value of them. As they had no intention of buying the

sheep they agreed to let him have them. On 3 June 1780, Kennedy sent two men, John Kennedy jnr. from Fort William and William McIntyre from Drumfuir, to collect the sheep that Donald McDonald had at Tulloch. Donald was away from home, but one of his brothers insisted on a bill for £26 16*s.* 0*d.*, payable to Donald. This was said to be to give 'proper security against John McIntyre and his creditors'. Presumably the sheep that John McDonald was keeping at Acharr were also collected from him.

John McIntyre was declared bankrupt, and John Graham, tenant of Cambushinie, a short distance north-west of Dunblane, was appointed trustee for the creditors. He believed that Kennedy was a wealthy man and was concerned that he may have secured an improper share of McIntyre's assets by getting possession of the sheep shortly before an application for the sequestration of these assets. If so, other creditors, perhaps poorer and more deserving, would suffer.[55] He was probably correct about Kennedy's wealth. A few years later, in 1785, Donald Kennedy, drover, was living in the house with the second highest rental in Maryburgh, far above the average in the village.[56]

In the early summer of 1781, Donald McDonald came south and lodged for some weeks with Charles McDonald, at Torbrex, near Stirling.[57] Charles McDonald was probably a butcher.[58] This gave John Graham a chance to have Donald called before the Sheriff Court in Stirling to explain what had happened. Initially he was quite cooperative. He explained that he was going back to Lochaber on 9 June 1781, and asked if he could make his statement before he left. A special arrangement was made for him to be examined on oath on the following Monday. However, Donald Kennedy turned up mysteriously in Stirling on the preceding Saturday and spoke to Donald McDonald, who then failed to appear for examination. He had made himself scarce by going back to Tulloch and then disappearing even further north. John Graham was convinced that Donald Kennedy and Donald McDonald had acted in collusion to thwart his attempt to challenge Kennedy's possession of McIntyre's sheep.[59]

Donald McDonald had another trick to play. In 1783 he went to the Sheriff Court in Inverness and obtained a decreet against the men who had collected the sheep for Kennedy for payment of a debt of £26 16*s.* 0*d.* arising from an unpaid bill dated 3 June 1780.[60] The amount and date correspond exactly with the bill which the Sheriff Court in Stirling had been told was taken instead of a receipt and simply to provide 'proper security'. Donald appears to have secured an order for payment for sheep which were not his to sell in the first place, and it is difficult to believe that Kennedy ever intended that to happen.

Donald McDonald was very active as a drover in the early 1780s and spent a good deal of time in the Lowlands, where he made extensive

contacts. It was alleged that for a time he 'found ways and means to reside between Edinburgh and Stirling in a pretty comfortable stile'.[61] However, he was poised precariously between people who were helpful to him and others who were to bring him trouble and come close to undoing him completely.

One of the mysteries about Donald was how he managed to find the substantial amounts of money needed to purchase livestock. The explanation probably lies in his relationship with David Mitchell, a merchant in Doune, Perthshire, who had 'large transactions' with him. Because of his ability as a drover, Donald

> was taken into a sort of co-partnery for a time by David Mitchell in Down [Doune], in the view of introducing his son as a drover. Accordingly, previous to 1783 he was possessed of very considerable credit, and engaged in extensive business as a drover. This is deponed to by many . . . witnesses.[62]

David Mitchell's business as a merchant was mostly along the usual lines,[63] but Doune was then prominent as a centre for the droving trade.[64] He probably saw an opportunity to expand into another kind of business and wanted his son David to become familiar with it.

David Mitchell advanced money to John McDonald, the drover at Acharr who had been associated with Donald McDonald in the matter of John McIntyre's sheep. In 1781, John McDonald's father-in-law, James Small, secured for him the lease of half of the farm of Gart, just outside Callander on the annexed estate of the Perth family. According to his son Angus, John McDonald was 'in the habit of borrowing from Mitchell' and owed him £140 when he failed at Gart in October 1783.[65] That lends support to the supposition that David Mitchell was a source of credit for the purchase of animals in the Highlands.

David Mitchell jnr. has left an account of some of his dealings in the Highlands with Donald McDonald in 1784. He said that when they went to Lochaber they 'were conjunct', but Donald had £50 of his own with which he bought some cattle for himself. Donald's brother, Alexander, served as topsman, i.e. he would take charge of a drove when it had to be moved. Donald would send for him when he was needed and remunerate him accordingly. Once he gave him £5 for taking a drove south. David Mitchell jnr. understood that Alexander had formerly been in service in the low country; his usefulness was limited because he spoke only broken English and had difficulty in making himself understood in the south. He arrived once in Doune with a drove of cattle and went to the house of David Mitchell for instructions. Janet McLeish, the wife of William Miller, a shoemaker in Doune, had to be fetched to interpret for him. David Mitchell jnr. gives brief glimpses of Donald McDonald travelling widely,

for example disposing of cattle at Glenfalloch and then turning up at a market in Sheriffmuir. He saw Donald in Edinburgh in July or August 1784 when he told him that he had just sold a few cattle at the Falkirk Tryst and made a profit of £20. He was then anxious to go back to the Highlands to buy some more.[66]

David Mitchell's books show that on 5 November 1784 he advanced £291 4*s.* 0*d.* to his son David 'to take to the Highlands'.[67] If he advanced money on anything like that scale to Donald McDonald and allowed him a share in the profits, Donald should indeed have prospered. As for the Mitchells, David snr. appears to have died about the end of the 1780s[68] and David jnr. must have abandoned his father's interests, because in later life he was said to have become a writer in Stirling,[69] although this has not been corroborated from any other source.

One of Donald McDonald's contemporaries in the droving business was Alexander McDonald, tacksman of Garvamore in upper Strathspey. He was a brother of Ronald MacDonell of Aberarder, the family with whom Donald and his family were in constant dispute. Alexander followed the fairly common practice of combining droving with farming, and one of his experiences shows that people in the Lowlands were wary of parties of Highland drovers in their midst. In November 1783 he was in the south with Allan McPherson of Shanvall, which lies close to the River Spey just before it is joined by the River Truim. On 4 November they arrived, with several servants, at the inn kept by Colin Lennox at Bridgend, Dunblane, intending to lodge there for the night. One of the servants, John McDonald, asked for his supper and a pint of ale. This was alleged to have been enough to cause Robert Lennox, son of the proprietor, to pick up a chair and assault him. The party of Highlanders made much of John's injuries and the extent of his blood loss, going immediately to Patrick Stirling, the procurator fiscal for the western district of Perthshire, and persuading him to have Robert Lennox lodged in the tolbooth in Dunblane and charged on the same evening. Robert of course gave a different account of events, claiming that he had to take strenuous action because the Highlanders were causing riot and disturbance.[70]

Alexander MacPherson, Writer in Inverness

Donald McDonald was well aware that Donald MacPherson, the Duke of Gordon's baron baillie in Lochaber, was alleged to have improper relationships with the Camerons in Braeroy and the Kennedys from Glengarry (see Chapter 2). The baillie's son, Alexander, was also implicated, and when he became a writer in Inverness he was concerned that Donald's knowledge could be a source of embarrassment. His anxiety was probably enhanced when one of the Camerons of Braeroy reappeared and started keeping company with Donald.

In 1783, the 95th Regiment of Foot was disbanded and Donald McDonald's cousin german, Lieutenant Donald Cameron, came back to Scotland. Both were nephews of Donald Cameron of Annat in Braeroy. Donald McDonald associated with Lieutenant Cameron and lent him money. The amount varies in different reports, from £60[71] up to £137 10*s.* 7*d.*[72] Lieutenant Cameron claimed that his credit with the agents of the 95th Regiment was good, but that was probably not true. A bill drawn by him on 31 December 1782 for £25 on Messrs. Cox, Mair and Cox of Craig's Court, London was dishonoured.[73]

Lieutenant Cameron was reported to have become 'the head of a gang of thieves who for two or three years before the end of the year 1785, infested the whole country, by carrying off numbers of horses, black cattle and sheep from the highlands of Inverness-shire, Perth and Argyle, which they sold in the low countries'.[74] He lived briefly at Old Callendar House, a short distance north of Crieff, and then at Muthill.[75] His visits to Annat seem to have been infrequent, but he was there in early February 1784,[76] and his wife was living there at the end of 1784.[77] At one stage he lived 'in the Heights of Stirlingshire'.[78] That probably refers to Kippen, where he lodged for a time in 1785 with a Mrs Martin alias Cameron.[79] Eventually a Justiciary Warrant was issued for his arrest, but he made his escape and was reported to have gone to the East Indies. Donald McDonald accompanied him from Stirlingshire to Edinburgh, and John MacKay, the messenger and constable at Fort Augustus, suspected that it was Donald McDonald who 'gave him the hint to keep out of the way'. Before Cameron left, Donald bought his horse, which turned out to have been stolen.[80]

In the early part of 1784, Alexander MacPherson took part in the first of several efforts to discredit Donald McDonald and presumably ensure his silence. In February 1784 some of Donald's creditors met and decided that formal meetings of all his creditors should be called. Alexander MacPherson, writer in Inverness, drew up the note of that meeting, which suggests that he was behind this attempt to embarrass Donald McDonald over his financial affairs. As an active drover, Donald was bound to be owing money at any given time, but there is reason to doubt whether he was really in difficulties. Two meetings of creditors were held: one in Stirling at the house of James Stewart, Vintner, on Tuesday 16 March 1784; the other in Maryburgh, at the house of Allan McLean, Vintner, on Thursday 1 May 1784.[81] The need for two meetings shows that Donald's affairs were extensive.

Donald said later that he made an offer which his creditors were willing to accept. That was not the only surprising feature of the business. On 17 May 1784, sixteen days after the second meeting of his creditors, Donald was able to pay £57 10*s.* 0*d.* to Henry Laing, an innkeeper at Torwood, near Larbert.[82] And, as shown above, David Mitchell was content to

continue to operate with him throughout 1784 and into 1785. If this was indeed an attempt by MacPherson to entrap Donald, it failed. He had to wait a couple of years for another opportunity.

Æneas Mackintosh Introduces the New Tenants

In the early months of 1786, Æneas Mackintosh gave out new tacks. That allowed him to introduce Alexander Mackintosh, the merchant in Maryburgh, and Donald McDonald into his plans for developing his estate in Lochaber.

Alexander Mackintosh and a partner, John Cumming, were given a nine-year lease of Briagach and parts of two other farms in Glenroy.[83] Alexander explained in typical fashion that: 'I was honoured with my Chief's countenance when he had his first set in Lochaber at which time he was pleased to express his wish that I should become one of his tenants . . . I took the farm of Briagach.'[84]

Donald McDonald's troubles were far from over, and Æneas Mackintosh did not give him a tack in his own name. Instead he gave a tack to Donald's father, Archibald MacDonell alias McAlister, and his older brother, John McDonald. The tack consisted of one third of the lands of Tulloch 'as lately possest by Alexander MacDonell of Tullich . . . also all and whole the lands of Blarnahanin . . . as lately possest by Ronald MacDonald of Aberarder . . . and that for the space of nine full and compleat years from and after the Term of Whitsunday 1786'. However, Æneas Mackintosh accepted Donald McDonald, who had allegedly been bankrupt two years before, as cautioner for his father and brother. He also allowed Donald 'the sum of £10 in consideration of his taking care of the wood, planting and copses in Lochaber and otherways attending to his interest and obeying his orders and Directives'.[85] That was quite a generous allowance. The Duke of Gordon paid £12 a year to his baron baillie in Lochaber.[86] It was an important provision, because Donald applied himself to this task with great zeal and developed into a kind of sub-factor on Mackintosh's Lochaber estate.

Alexander MacDonell of Tulloch had lost his position as tacksman there. Archibald McDonell (or McDonald) and his son John, who had been his sub-tenants, now became the tenants of Æneas Mackintosh. They were able have sub-tenants themselves, and were alleged to have drawn a surplus yearly rent of £12 from them.[87] Ronald MacDonell of Aberarder, who had held Blarnahanin, probably died about this time, but his family were denied the succession there. In this way Æneas Mackintosh reduced the holdings of two long-established families of tacksmen, the MacDonells of Tulloch and the MacDonells of Aberarder. The tacks of 1786 will be considered in more detail in the next chapter.

Alexander MacPherson Attacks Again

While the arrangements for the new tacks were being made, three alleged members of Lieutenant Donald Cameron's 'gang of thieves' were apprehended. John MacPherson, Duncan Cameron and Angus Cameron were tried at the Circuit Court in Inverness in May 1786. John MacPherson pleaded guilty to a charge of sheep-stealing, which he said was the first crime he had ever committed. He was sentenced to be hanged. The two Camerons pleaded not guilty and were acquitted. MacPherson was represented by Alexander MacPherson,[88] who was present at John MacPherson's execution on 14 July 1786. Alexander MacPherson said that immediately before he was hanged John MacPherson revealed to him the names of his accomplices, who included Donald McDonald. Mr Fraser, minister of Inverness, was also present but was not within earshot and so could not corroborate this. It seems surprising that Alexander MacPherson had not taken care to ensure that Mr Fraser heard this important statement. He passed the information immediately to the sheriff, Simon Fraser of Farraline,[89] but nothing happened. He had failed again.

Alexander MacPherson's next move depended upon an extraordinarily complicated manoeuvre in which he must have had the cooperation, willingly or otherwise, of Lieutenant Cameron. The effect was to make Donald McDonald liable to Alexander MacPherson for a bill for £49 19*s.* 0*d.* This bill had been drawn on Donald by John Rankine & Son, merchants in Falkirk, on 6 June 1783, payable three months after that date. The Rankines said that they had drawn the bill on Donald McDonald at his request in order to let him have cash, which he needed. They said that he was then 'reputed in pretty good credit', which suggests they knew him fairly well. The bill was subsequently indorsed, genuinely, to Henry Laing, innkeeper at Torwood. Then, according to the Rankines, it somehow got into the hands of Lieutenant Cameron, who forged an indorsement to 'Donald Cameron of Annat'. A scheme was then hatched 'to keep the bill in negotiation in the name of a Trustee'. Alexander MacPherson, writer in Inverness, turned out to be the trustee. According to the Rankines, whose intelligence must have been good, he and Lieutenant Cameron 'were in familiar habits', and MacPherson had advanced £50 to him to get a step up in his regiment.[90] Cameron's debt to MacPherson was, however, greater than that. It consisted of a bill for £50 dated 4 April 1783, and another for £60 dated 26 August 1784.[91] Lieutenant Cameron was probably well and truly in his power.

Alexander MacPherson was now able to move against Donald McDonald, on the contrived grounds that Donald now owed to him the value of the Rankines' bill. Donald refused to pay, claiming that he was

unable to do so. At the beginning of September 1786, Alexander MacPherson obtained letters of caption against him and had him incarcerated as a debtor in the tolbooth of Inverness. The messenger who was sent to Tulloch to apprehend Donald and bring him to Inverness observed that there was a very large stock of cattle on the farm, 'directed and managed by the pursuer [Donald McDonald] as much at least as by any of the other nominal tacksmen [his father and brother John]'. Donald was busy 'superintending the building of a larger and better farm-house than is usual in that part of the country, or would probably have been thought of by the other pretended tenants'. Alexander MacPherson was in no doubt that 'the tack granted by the Laird of McIntosh to the pursuer's father and brother, and in which he is cautioner, is a tack for the behoof of the pursuer [Donald McDonald]'. He thought it 'a very idle piece of form in the pursuer to bind himself as cautioner in this lease, and of the Laird of McIntosh to require his security'.[92] All the evidence suggests that Alexander MacPherson's suspicions were understandable. Æneas Mackintosh had knowingly been a party, no doubt for his own good reasons, to a very odd arrangement.

When the messenger appeared, there was a commotion. Donald had 'got about him a number of people armed with durks and batoons who threatened violence against the messenger'. However, he seems to have calmed down, and

> having equipped himself in his boots, spurs, and a handsome suit of clothes, with silver buttons upon them, he surrendered himself to the messenger, and agreed to go to prison, upon condition that he was furnished with a saddle-horse to carry him to Inverness. Accordingly, in that form, he was conducted to prison, where he has lived as if he was in no want of money, if he thought proper to disclose it; and, in fact, he has occasionally, since his confinement, produced bank-notes to several persons he had occasion to converse with.[93]

This episode is quite revealing about Donald. First, there was the tendency to react violently, which he probably learned to control to some extent as he got older. Second, there was the streak of personal vanity, which may have been exaggerated in this report but was probably real enough.

Having failed to pay MacPherson, Donald maintained the fiction that he had no money. On 7 September 1786, he submitted 'Unto the Honourable the Baillies of Inverness' a petition for aliment, claiming that he was unable to pay the debt for which he was incarcerated, or other debts due by him, and was unable to support himself while a prisoner. He therefore sought the benefit of an act of parliament of 1696, and his request for aliment was immediately granted by Baillie John Mackintosh. Donald was then entitled to an allowance of eight shillings Scots per day. This had to be paid by those

who incarcerated him, and they were required to deposit sufficient security for this purpose in the hands of the town clerk.[94] That must surely have given Donald some pleasure.

On 2 October 1786 Donald raised an action of cessio bonorum in the Court of Session to try to obtain his liberty. The Edinburgh lawyer whom he engaged for this purpose was William Blair WS.[95] Mr Blair's services were also used by David Mitchell of Doune in an action which he raised in 1787 against Angus McDonald, the son of John McDonald of Gart (formerly at Acharr).[96]

Meantime, Donald was trying from within the tolbooth in Inverness to recover money owed to him by people in Badenoch and Lochaber. On 2 November 1786 he wrote from the tolbooth to Alexander Macdonell, writer in Inverness, asking him to handle these matters for him. Donald was annoyed because the day of compearance, when they would come up in court, had been fixed for 15 November, which did not give him enough time for preparation. There were documents to be brought from Edinburgh, and he 'expected my nephew to come here with other papers I had in the South Country'. He had written to John MacKay, the messenger in Fort Augustus who had some seniority in the County, and asked him the tell Alexander Marshall, the messenger at Maryburgh, 'how much he misbehaved in making the day of compearance so soon'.[97] Donald was not one to mince his words when he was displeased.

Æneas Mackintosh and his lawyer in Inverness, Campbell Mackintosh, were both well aware of Donald's situation. Some of the formalities of his incarceration passed through Campbell Mackintosh's hands in his capacity at that time as town clerk depute. On 29 November 1786 Æneas asked: 'Pray is McDonald's Business going on favourably – and do you think he will get out?'[98] Æneas Mackintosh was also in touch with his Edinburgh lawyer, Charles Mackintosh WS, about Donald's position. Charles Mackintosh told him on 12 February 1787 that he had paid, as instructed, £15 to Mr Blair 'for McDonald's Cessio. . . . The Court have allowed Donald McDonald a prooff in his process of Cessio – I find there is a great prejudice against him on account of his supposed connection with Lieutenant Cameron and the gang of Thieves.'[99] So Æneas Mackintosh was well aware of that unfortunate association.

The £15 paid to Mr Blair on Æneas Mackintosh's instructions was no doubt passed to him discreetly to avoid interfering with the pretence that Donald had no money at his disposal. On 16 February 1787 Donald petitioned for the benefit of the poors roll,[100] claiming that 'the petitioner is utterly unable by reason of poverty' to proceed. He got the benefit of the poors roll.[101]

Early in 1787 the Court of Session required Donald's presence in Edinburgh. In order to be released from custody in Inverness, he had to

find two persons willing to give a bond of caution that he would return to custody. Æneas Mackintosh agonized over this. On 11 March 1787 he said cautiously to Campbell Mackintosh:

> As to Donald McDonald . . . [letter torn] . . . not chuse to go Cautioner for him on so precarious a footing or desire another to do it for me – without knowing what the penalty may be – or what benefit he might derive from getting out of Prison – of these particulars you can get information and acquaint me without saying to Donald what I wrote on the subject.[102]

He still held back from becoming directly involved, and a bond of caution was granted by Alexander Mackintosh of Ballispardan, who was his factor, and John McDonald at Tulloch, Donald's brother who was an old soldier on a pension and unlikely to possess much money. Donald was released temporarily from the tolbooth in Inverness in April 1787. On 12 June 1787 he returned there and became a prisoner again.[103] [104] [105]

Alexander MacPherson of course opposed Donald's attempt to get the benefit of cessio bonorum. He did this not only in his own name but in the names of several others to whom Donald owed money. He made the mistake of not obtaining authority from all of them, and several submitted petitions to the court dissociating themselves from the action. This may have helped Donald's cause, but Alexander Fraser of Struie, to whom Donald owed £49,[106] and James MacBarnet of Torlundy rallied behind Alexander MacPherson.

Donald McDonald had to persuade the court that his debts had arisen from misfortune and not from fraud or extravagant living. He explained that he had commenced early in his life in the business of a drover. For several years he had had good fortune and regularly kept good credit. That was confirmed by witnesses. Then he had the misfortune to become a creditor of several other drovers who became bankrupt about 1783 while owing him substantial sums of money. John MacDonald, a drover in Sutherland, Angus MacMillan, a drover in Maryburgh, John Cameron, a drover in Inchery, and another John McDonald were all mentioned. The first two were alleged to owe Donald £400. Another debtor, Allan Cameron, conceded that he owed Donald £214 and confirmed that the drovers mentioned above had all stopped payment at the time alleged by Donald McDonald.[107]

Donald claimed that his debts amounted to about £300, and he produced bills showing that his creditors owed him £568 2*s.* 5*d.* His adversaries immediately countered that the bills were fictitious, arguing 'it is somewhat remarkable that all the persons with whom he has been connected . . . became Bankrupt and fled the country'.[108] That may not, however, have been particularly unusual for a drover. In 1814, John

MacDonell, son of a ground officer to MacDonell of Glengarry, who became a drover, said, 'it is a remarkable thing for a Drover, the most accute and who has the advantage of Education, not to fail at some period of his Life and there is no instance of it in the Highlands of Scotland for 50 years back except two – viz. Mr. MacKay of Erriboll and Mr. MacPherson of Ralia'.[109] As Donald McDonald put it himself, droving was a business 'most subject to sudden reverses of fortune . . . from the nature of the subject and the need to give considerable credits'.[110]

Donald McDonald was required to 'surrender his whole means and estate to his creditors'. MacPherson and the others complained bitterly that 'instead of giving any proper account of his stocking of cattle, money or Bills . . . [his] whole effects consisted of a Sword, a Durk, and two ffowling pieces'. They were of course convinced that there were valuable assets at Tulloch alleged to be the property of his father and brother which really belonged to Donald. They were, however, quite unable to make anything of this in court. In spite of their best endeavours, and no doubt to their great chagrin, Donald's action was successful, and those who had opposed his application were required to reimburse his legal expenses amounting to £40.[111]

That was the end of Alexander MacPherson's attempts to ruin Donald McDonald. Perhaps the most intriguing feature of the whole affair was Æneas Mackintosh's sustained loyalty to him. He must have been tempted in the winter of 1786/87 to wash his hands of Donald and leave him to his fate. In later years, he gave some hints about why he did not do this.

In 1799, when Campbell Mackintosh was finding Donald a trifle over-zealous, Æneas said, 'Donald I confess was of infinite service to me in 1785 and I have not been unfriendly to him, however he will be better of a hint not to forget himself'.[112] Alexander Mackintosh, the merchant in Maryburgh, gave a clue in his memorial of 1834 about what this great service might have been. He explained in colourful terms how in 1785 the Keppoch MacDonells were refusing to accept any new tack on Mackintosh's terms and were parading the streets of Fort William with their followers, threatening and terrorising others into following their example. He and Donald McDonald then devised a plan to get 'responsible experienced Shepherds from the Southern Borders' to come and have a look at the lands. 'The southern gentlemen came forward, perambulated, and examined the lands which pleased them well.' That softened the attitude of the Keppochs, and the new tacks were accepted.[113] Donald's extensive contacts in the south had probably been essential, and perhaps he had been the prime mover.

Æneas Mackintosh's remark in 1802 about Donald having been useful to him in breaking the McDonald combination and so raising his rents has been mentioned above.[114] Even so, the extent of his support for a man who

was so close to serious trouble remains surprising. Donald must indeed have been valuable to him.

Now that his process of cessio had been successful, Donald McDonald had one or two scores to settle. He caused Mr Blair to raise a summons of defamation against Alexander MacPherson, writer in Inverness (now serving as procurator fiscal), and against John MacDonald, tacksman of Torgulbin, as well as his brother-in-law, Lieutenant Alexander MacDonald, tacksman of Blarour. John MacDonald was the eldest son of Ronald MacDonell of Aberarder. Alexander MacDonald was a younger brother of Major Ronald MacDonell of Keppoch, who died in 1785.[115] [116] His sister Catherine was the wife of John MacDonald at Torgulbin.[117]

Donald McDonald showed in great detail how they had wronged him. He alleged that Alexander MacPherson had said in the presence of John MacDonald, watchmaker in Inverness, Andrew Munro, clerk to William Sharp, bookmaker in Inverness, and William Ross Munro Esq. of Newmore that Donald was 'the greatest villain and scoundrell in the County of Inverness, and that he was art and part concerned with Lieutenant Donald Cameron and the gang of thieves stealing horses [etc] . . . in the highlands and [was] . . . seen selling these horses in Greenock, Glasgow [etc]'. MacPherson said similar things 'in the houses of Allan McLean and Donald Cameron, Vintners in Maryburgh, in the presence of many respectable persons, also in the house of George Beverley, Vintner in Inverness'. He also said that 'he would have Donald McDonald hanged or whipped'.[118]

Lieutenant Alexander MacDonald and John MacDonald were accused of making similar statements in the house of Charles MacPherson, tacksman of Shirrabeg in upper Strathspey, on 4 October 1786. They had also 'on many different occasions applied to sundry persons, particularly to Andrew Mitchell, Tacksman of Aberarder [who had replaced the MacDonells there] and William Mitchell, his son at Tullochcrom forbidding them to have any dealings or transactions with the pursuer [Donald McDonald]'. Donald claimed that 'the defenders did entertain an inveterate grudge against the pursuer' and attributed this to his active involvement in a precognition against the Aberarders 'for the crime of sheep stealing and theft' some years earlier.[119] Donald's brother Alexander had also given 'informations to the sheriff depute against Ronald McDonald of Aberarder and his Sons'.[120] The quarrel between Donald's family and the Aberarder MacDonells was not yet over.

Once these preliminaries were out of the way, Donald was at last able to devote his attention to Æneas Mackintosh's affairs. It will be shown in Chapter 4 that the process of squeezing the MacDonell tacksmen off the Mackintosh lands continued step by step.

This chapter concludes with a glimpse of Donald in another context. In

December 1787, while his case against Alexander MacPherson and the two MacDonells was still in progress, he was at the Sheriff Court in Inverness as a witness for Angus McKillop in Achluachrach. McKillop was accused of stealing a butter churn at a market at Knappach in Badenoch. Donald said that he had known him for twenty years and had a good opinion of his character.[121] That did not help the unfortunate McKillop, but it shows that in spite of all his troubles Donald remained confident and resilient.

4

The Mackintosh Lands in Lochaber, 1786–1804

This chapter takes the development of the Mackintosh lands into the early years of the nineteenth century. It covers the period of two setts, one from 1786 to 1795, and the other from 1795 to 1804. When Æneas Mackintosh introduced Alexander Mackintosh and Donald McDonald to his land in 1786 he began clearing out the MacDonell tacksmen. The process was virtually completed at the 1795 sett. In 1802, Alexander Macdonald of Glencoe took over Keppoch and the Inverroys and gained his first foothold on Mackintosh land.

Tulloch (1786–1795)

Alexander M'Donald or MacDonell of Tulloch was the last of the old Tulloch family to hold his traditional lands from Mackintosh of Mackintosh. Until 1786 he held Tulloch and Dalnaderg to the east; from 1786 Tulloch was let to others in three parts. As explained in Chapter 3, Donald McDonald's father and his brother John became tenants of one third of Tulloch, with Donald as their cautioner.[1] The Rev. Thomas Ross, minister of the Parish of Kilmanivaig, got a quarter of Tulloch. Samuel, John, Donald and Dougald MacGlashan (sometimes called McGlaserich and sometimes Campbell) got five twelfths.[2] These changes brought in an increased rent. Before 1786, Tulloch and Dalnaderg together were leased for £18 per annum. After 1786 Tulloch alone produced a total of £48: £16 from the McDonalds, £12 from Rev. Thomas Ross and £20 from the MacGlashans.

Ensign Angus MacDonell was now head of the Tulloch family. In 1786 he and his brother-in-law, Lieutenant Alexander MacDonell of Moy (a son of Ronald of Aberarder), were given the tack of Dalnaderg and some other lands on the east of Tulloch.[3] [4] [5] Both men were often called 'captain'. This seems to have been a courtesy title for gentlemen of a certain standing who had held commissioned rank in the army, even if that rank had been at a lower level. Donald McDonald usually gave Angus MacDonell of Tulloch his courtesy title of captain, but when he was feeling annoyed with him he would revert to ensign.[6]

Cranachan and Blarnahanin (1786–1795)

Cranachan and Blarnahanin lie together on the east side of Glenroy. Before 1786 they were let separately. Cranachan was held by John MacDonell of

Cranachan.[7] As shown in Chapter 3, he had set up as a merchant in Maryburgh [Fort William] and had failed.[8] He was nevertheless allowed to continue for a time at Cranachan.[9]

Before 1786 Blarnahanin had been leased to Ranald MacDonell of Aberarder at a rent of £8 11*s.* 0*d.*[10] When Donald McDonald's family got Blarnahanin in 1786, the rent was increased to £20. However, £10 of that was remitted to Donald as an allowance for looking after the landlord's interests in Lochaber.[11]

For a few years after 1786 Donald McDonald lived at Blarnahanin.[12] Alexander Mackintosh, the merchant from Maryburgh, was a fairly close neighbour at Briagach, on the west side of Glenroy. In later years, when he was feeling bitter about his removal from the Mackintosh lands, he explained how he had helped Donald McDonald in his early days.

> I took the farm of Briagach and so tenacious was I of the Interest of any person who I could imagine supported the Interest of Mackintosh that when Donald McDonald was so obnoxious to all his friends, mistrusted by all hands, and a professed Bankrupt, I on that account, I mean on account that he apparently forwarded Mackintosh's Interest to his own prejudice, stocked the farm of Blarnahinven for him and I supported him with my credit and cash which at that time was pretty extensive.[13]

As usual, he makes much of his commitment to his landlord, but he also gives a glimpse of Donald's unpopularity.

Keppoch and the Inverroys (1787–1795)

Traditionally the MacDonells of Keppoch held the tack of Keppoch and the Inverroys and lived at Keppoch House. All was not well with their finances. In 1775, Major Ronald of Keppoch owed £414 7*s.* 6*d.* to Mackintosh.[14] He died in 1785,[15] just before his tack was due to be renewed. This caused delay, but in 1787 Æneas Mackintosh gave a tack for eight years to Ronald's son and heir Alexander, who was then a minor.[16] The Duke of Gordon also allowed him to succeed to his lands which Ronald had held. There was uncertainty about where responsibility rested for the management of young Alexander's affairs. Three curators had been appointed, Ewen Cameron of Fassifern, Colin Campbell, Collector of Customs at Fort William, and Lieutenant Alexander MacDonell, his father's younger brother, who was then at Blarour.[17] [18] In spite of that, young Alexander's mother, formerly Sarah Cargill, who was living in Edinburgh, continued to involve herself in her son's business. In 1789, Colin Campbell described her as 'the Principal Manager for her son'.[19] Nevertheless, in 1794 it was the curators who borrowed £173 4*s.* 0*d.* from James McCaul, a steelbow tenant of theirs on the farm of Loch Treig, in order to

buy a Commission in the army for young Keppoch.[20] The confusion continued for some time.

Donald McDonald and the Management of the Lochaber Estate (1786–1795)

Donald McDonald took up his task of looking after Mackintosh's interests in Lochaber with great enthusiasm. In July 1786 he went to see Cluny Macpherson, armed with a letter from Æneas Mackintosh, to resolve how to preserve the woods on Loch Laggan side while still supplying timber to Mackintosh's tenants. Cluny said, 'I have seen McDonald and explained to him the Plan to be followed by him and my Forrester'.[21]

On 29 July 1786 Donald asked Campbell Mackintosh, writer in Inverness, for an extract of the tack given to his father and brother, which would 'let me see what Every Tack Given by McIntosh reins upon so as to enable me to give proper instruction to McIntosh's Tacksmen agreeable to their said tacks'.[22]

On 1 August 1786 he took three witnesses with him to inspect damage to the woods, especially at Cranachan, caused by tenants cutting great quantities of timber. On 28 September 1786, when he must have been in the tolbooth in Inverness at the instigation of Alexander MacPherson (see Chapter 3), he sent a memorandum to Æneas Mackintosh telling him what he had found.[23]

None of the letters which Donald presumably wrote in the years immediately following his release from the tolbooth has been found. However, on 26 February 1791 he wrote a long letter to Campbell Mackintosh in Inverness, while Æneas Mackintosh and his wife were visiting London.

> Dear Sir, I send the bearer hereof to you with two boxes of Muir Fowls [grouse] that must be sent to London by one of the Smacks from your place in Consequence of an order I had lately from Lady McIntosh to that purpose. One of the boxes is for Lady Grant Dalvey Containing 16 Muir Fowls and the other for Mr. James MacPherson and Containing 18 Muir Fowls and one Black Cock . . . I would have sent some Tarmichan – But the Bad weather prevented my going to the high mountains and of course I got none sent.

Donald raised several other matters, as he usually did, in the same letter. He enclosed a bill against Donald Shaw at the Haugh and George McLeod at Inverness, wool merchants, on which he was due an unpaid balance. He wanted them prosecuted unless they paid the outstanding sum immediately. Finally he said, 'I ordered the Bearer to Bring me a load of whisky on the horse I will be obliged if you Direct him where he will be properly served'.[24]

On 18 April 1794 Donald wrote to Campbell Mackintosh with his usual

mixture of personal and estate business. He had been in Inverness and had left two bills with him. One was for a payment due to him from William MacGillivray and John Grant of Gartmore. That lies south of Loch Garth, now part of Loch Mhor, created by the dam built to provide water power for the former Foyers aluminium works. The other bill was for a payment due from William McIntyre in Baldow in Badenoch. Donald wanted decreets against all of them. He also had some outstanding business in Ross-shire, and said, 'I also hope that you ordered a proper Lawier at Dingwall to prosecute Dallas at my instance according to instructions'. Then came estate business. On his way home from Inverness he had met a farmer called McCulloch in Gask, in lower Strathnairn, who was willing to sell up to sixty bolls of meal. He asked Campbell Mackintosh to negotiate a price with this man, so that he could arrange to have the meal transported to Lochaber. The great attraction was that the meal could be delivered at Bona at the north end of Loch Ness and taken to Fort Augustus by boat. Donald could have it collected there, saving the trouble of 'sending horses to the low country'.[25] A month later he said that he would not pursue the deal;[26] McCulloch's price must have been too high.

In June 1795, more debtors had to be pursued. Campbell Mackintosh was asked to 'write Mr. Wm. McGregor in Sutherland who is due me as ballance on his bill £12 & that I will expect payment against the Marymass market & that he cannot expect any longer delay'. Donald also wanted a warrant of arrestment against a William McIntyre, perhaps the man in Baldow, 'as I know where he has some subject & unless you give me by the bearer a warrant I never expect a penny from him'.[27] In December 1795 Campbell Mackintosh was being asked to pursue yet another debtor, 'John Grant Avymore for £31 12*s.* 0*d.* . . . please . . . execute a diligence against him . . . he has behaved very unhandsome to me therefore I shall do to him as law directs unless I am paid'.[28]

At the Maryburgh Market in November 1794, Donald tried to prevent the son of one of Mackintosh's tenants from becoming the victim of a recruiting sergeant.

> John McGilivantich a native of the Laird of McIntosh's people happened to be in company with a recruiting Serjeant belonging to the Duke of Gordon's Fencibles & he got himself useless Drunk & it appeared that he got two shillings from the party and Upon his friends seeing him kept with [the] party [a] Great many interfered to take him from them.

Donald tried several times to get him released. Later correspondence shows that John McGilivantich's father was given some land as a result of his son's enlistment. He did not share Donald's enthusiasm for getting the lad out; his concern was that he was being charged too high a rent for the land given to him.[29] [30] [31]

Preparations for New Tacks from 1795

By the later years of the eighteenth century, landlords like Mackintosh had a sett every few years when all the tacks were renewed for the same number of years. A sett of the Mackintosh farms was due in 1795. On 4 December 1792, Donald McDonald put to Æneas Mackintosh a plan for a new division of his Lochaber estate. In his covering letter he said:

> Annexed to this is a state of the different lots that your Estate in Lochaber should be divided into. By which measure Every lot has proper Share of the hill Ground & the marches more Commodious than formerly – Which in the Grassing ways will Enable Industrious Tenants in good Circumstances to pay the rent annexed to Every Lot provided the prices of Cattle &c shall continue as ffavourable as it is at present. I assure you that it is not Difficult for you to get a sett of very agreeable Tenants to take that Lands at the said rent and in my oppinion it is as much as ought to be Expected - and I beg leave to add that there is very few of your present Tenants that could be able to stand their grounds at such high rent Because they are not good managers particularly your Small Tenants. Everything in my power to promote your Interest shall not be wanted – & I have the honour to be.

Donald suggested eight lots, the three smaller ones to have rents of £70 or £75, and the five larger ones to have rents of £140, £145 or £150.[32] The total yield would have been £935. This scheme influenced the subsequent division of the estate, but inflation ensured that his proposed rents, which were well above the previous ones, would soon be overtaken.

Two of Donald's cautionary comments are worth noting. The first is that tenants could be found to pay the rents proposed, providing the favourable prices for livestock continued. That shows that he realised that prices might fall and that the rents could prove to be on the high side. Perhaps he had been told to aim for a high yield. The second point is his cavalier dismissal of any prospect that Mackintosh's present tenants, particularly the small ones, could manage these large farms. Underlying this is a theme which emerges from time to time: some Highlanders could cope with the new market-driven world, while others could not.

Some tenants had their own ideas about preparations for the new sett. Donald reported that they were removing the old thatch and using it to make dung 'seeing it now nears the end of their lease'. They replaced the thatch with soft divots 'that will scarcely hold the wind out & by drops the whole timber suffers and goes to almost nothing leaving their bigging in this situation they think it of Service in order to discourage strangers to offer for their farms'.[33]

To clear the way for the new tacks, Æneas Mackintosh obtained a

decreet of removing at the Sheriff Court in Inverness dated 14 April 1795 against all of his tenants in Lochaber. The decreet listed not only those who held tacks from him but many who were sub-tenants or cottars.[34] This cumbersome procedure followed current legal practice and opinion that a tenant was not bound to remove at the end of a fixed period unless he had received warning from the landlord by raising an action of removing at least forty days before Whitsunday of the year concerned. If the landlord failed to give such warning, the tenant could remain for another year on his current terms.[35] The formal removal procedure would ensure that each individual named in the citation received a copy delivered by a messenger of the court. The delivery would be witnessed and a record kept.

The names of sub-tenants and cottars were probably included because experience had shown the need to err on the side of caution. Lovat was given legal advice in 1803 about the difficulties in removing tacksmen. Problems could arise when tacksmen had powers to sub-let, when heirs might be entitled to succeed, and when the original tacksman might have disappeared or died. Lawyers were concerned not to overlook anyone who might be able to claim rights which had not been extinguished. The advice to Lovat concluded with an ominous warning that the smallest inaccuracy could be fatal.[36] It was probably better to include a name which might not matter than to risk omitting one which did.

Bringing a sett to an end with a sweeping decreet of removal did not mean that a landlord intended to clear his land of its tenants. It gave him a free hand, unencumbered by any existing commitments to tenants. He could then, for example, rearrange the tacks, as in this instance into fewer and larger ones. He could alter the conditions attaching to them, perhaps imposing more onerous requirements about the provision of buildings and dykes. He would then invite bids for these tacks. His former tenants could, and usually did, make offers, often hoping to get whichever new tack was most like their old one. There might not be room for all of them, and others could make offers and add to the competition. It was an anxious time for the tenantry.

Tulloch (1795–1804)

In 1795, a decreet of removing was served on all of the tenants, sub-tenants and cottars in Tulloch. Archibald MacDonell alias McAlister is not mentioned, but he may have died by then. His sons Donald and John were 'removed'. Those who had shared Tulloch in the 1786 tacks were also 'removed' and apparently got other leases elsewhere on Mackintosh land. Donald then secured in his own name the whole of Tulloch and Dalnaderg for nine years at a rent of £165.[37] [38] [39]

Donald McDonald was now on the threshold of becoming a fully fledged farmer with a large sheep-farm. In March 1796 he obtained from the Bank

of Scotland a bond of credit for £300. This was like an overdraft facility, of indefinite duration, allowing him to borrow up to £300 at the usual rate of interest from the Inverness Office of the Bank. His two cautioners were Rev. Thomas Ross, minister of Kilmanivaig, and Rev. John Macdonell, minister of Forres.[40] Thomas Ross was the local parish minister, but it is not clear how Donald came to be acquainted with Mr Macdonell in Forres. He had been born in Glenmoriston in 1756, was educated at Marischal College, Aberdeen, and was a Missionary at Fort Augustus from 1785 to 1791.[41]

Donald also became a member of the Sheep Farm Association, based in Inverness and presided over by Sheriff MacLeod of Geanies.[42] There he would mix with well established sheep-farmers, mostly Highlanders but including some incomers from the south.

However, he still had some troubles to resolve, with his brothers, with Captain (or Ensign) Angus MacDonell of Tulloch, who now had no land at all,[43] and with Alexander McIver, John McArthur and several McKillops at Murlagan and Achluachrach, who persisted in allowing their animals to graze on Tulloch land.

Donald was anxious that his disputes with Captain Angus at Tulloch and the tenants of Murlagan and Achluachrach might upset Æneas Mackintosh and jeopardise his future. His worries on that score were unnecessary. Mackintosh had a poor opinion of both the Captain and the tenants of Murlagan and Achluachrach, and he was usually content to back Donald. His only reservation was that Donald could be over-enthusiastic and occasionally had to be restrained.

Donald McDonald Removes his Brothers from Tulloch

Soon after his new tack commenced in 1795, Donald set about putting his two brothers, John and Alexander, out of Tulloch. Rev. Thomas Ross said of Donald in 1800 that 'he could never agree with his own brothers – this . . . part of his conduct is notorious'.[44] His brother John had been a sub-tenant when the MacDonells of Tulloch still held the tack, and his name had been used in the first tack given to the family by Æneas Mackintosh in 1786.[45] [46] Donald appeared to be putting him out when he no longer needed him.

Initially it had seemed that John was going to remain at Tulloch, and his son John came to the farm as a shepherd. Donald said that he was willing to give them a quarter of the farm if they could advance the money to stock it and lodge rent for a year, as he had had to do himself. However, they were not able to do that.[47]

Alexander had assisted Donald from time to time by acting as a topsman and moving droves for him. However, he was probably more impetuous than Donald and less able to control his temper. Any opportunity to go to

the courts was a temptation which Alexander seemed unable to resist. He would embark on litigation against his neighbours over the most trivial matters, and on at least one occasion Donald felt obliged to intervene in 'Sandy's unfair business'.[48]

It is impossible to find a favourable comment on Alexander. In 1783 he had been said to be 'a man disliked in the Country'.[49] In 1803 Angus McDonald of Kinchreggan referred to 'his litigious and troublesome disposition'.[50] When he tried to get a tack on Mackintosh's land at the sett of 1804, Campbell Mackintosh had no hesitation in saying to Æneas Mackintosh, 'I do not suppose you will be inclined to introduce him again upon the Estate'.[51] In 1810 John MacDonald of Borrodale remarked that 'Sandy was never yet blessed with the most flattering reputation'.[52] Perhaps Donald feared that Alexander would embarrass him in his new role as one of his landlord's major tenants.

In the spring of 1796 Donald had a summons of removing issued against his brothers. At about the same time he arranged to sub-let Tulloch, or part of it, to Captain Angus MacDonell of Tulloch, who had lost all of his land at the sett of 1795. His brothers reacted furiously and prepared to plough all of the low ground on the farm 'to hurt the pasture of sheep'. Some of that ground had not been ploughed for sixty years.[53]

Alexander could hardly be expected not to rise to the challenge of a summons of removing, and he set about having defences prepared. However, at the beginning of May 1796 he relented and told Æneas Mackintosh's factor that he and John withdrew their objections. It was agreed that they would submit a claim for compensation to be settled by Alexander McDonald at Garva and William Mitchell at Achnadaul.[54] As a result, Donald had to pay Alexander £50 13*s.* 6*d.*,[55] but it is not known what compensation John received.

Alexander then got 'a small piece of the Farm of Achintore', on Locheil's land just to the south of Fort William. He 'obtained this small farm by the intercession of Mr. MacDonald of Glencoe'.[56] Alexander Macdonald of Glencoe was not yet a tenant on Mackintosh's land, so he was not simply acting as a helpful neighbour. Alexander Dhu must have been known to him in some other capacity, and he must have had sufficient reason for using his influence with Locheil on behalf of a man for whom no one was able to find a good word to say.

Donald and Alexander were on reasonably good terms within a few years. Donald then helped Alexander either by writing letters for him or by rallying support when he got himself into tight corners, as he did from time to time.

Nothing more is known about John until he and Alexander tried to claim the right to succeed to Tulloch after Donald's death in 1808. Perhaps he

was of a quieter disposition than his two brothers. Details of Alexander's later life will be found in Chapters 9 and 10.

Dealing with Captain Angus MacDonell of Tulloch

At the 1795 sett, Captain Angus MacDonell of Tulloch and his brother-in-law Captain Alexander MacDonell of Moy lost Dalnaderg, which was given to Donald McDonald along with Tulloch. Captain Angus thus held no land, but he was not going to be shifted easily. Æneas Mackintosh had to obtain another decreet of removing in April 1796 to get the two MacDonells out of Dalnaderg, Tollie, etc.[57] Alexander MacDonell at Moy is not mentioned again in this connection.

Captain Angus' departure from Dalnaderg coincided with Donald McDonald's move to get his brothers out of Tulloch. Donald then allowed Captain Angus to have land at Tulloch as a sub-tenant. This may have been a tactical move on Donald's part to help him to get his brothers out. Alternatively, it may have suited Donald to use Captain Angus to fill a gap in his plans created by the loss of his brother John's expected contribution to the rent.

Donald apparently lived himself at Dalnaderg while Captain Angus was at Tulloch. After getting rid of his brothers he was patient for a year or two, but it was not long before he was manipulating the situation to get the Captain out of Tulloch. Matters came to a head towards the end of 1798. He wrote to Æneas Mackintosh on 30 December 1798.

> Then Capt. McDonell and his wife and servants behaves in the most disagreeable manner – I can assure you with truth that I am in the belief that they do me more injury and that they are more against my interest than what rent I paid you. I think every month I have of the Capt. a year – the particulars of which I can tell after this.[58]

Æneas Mackintosh's sympathy was with Donald. He observed to Campbell Mackintosh: 'You perceive by Donald's letter that he has nearly made a sheep fank [fold] which should not have been the case was the one built by Ens McDonell, late of Tulloch, of any use to the farm he leaves Whitsunday first.' He added that he was unwilling to pay any allowance to Angus MacDonell 'for a thing . . . of no use to incoming tenants'.[59]

More strenuous efforts were in hand to get rid of the Captain, to give him the title that Æneas Mackintosh avoided using. Donald showed what was in the wind when he wrote to Campbell Mackintosh on 14 February 1799.

> The sooner you send Sheriff Officer to Execute the warnings against Capt. McDonell . . . the better. It will be a very great loss and disappointment to me if I do not get possession of the farm of Tulloch at Whitsunday next because the Stock that I have on hand requires the whole farm I have also engaged the necessary servants that shall be

required and a manager that is to have a certain share and his residence to be there at Tulloch This great undertaking requires a manager since I could not always be at home Servants cannot be depended upon always.[60]

He was soon beginning to agitate. On 26 February 1799 he said to Campbell Mackintosh, 'I shall expect the officer soon'.[61] On 5 March he was still waiting. 'As the Sheriff Officer has not yet appeared with the warnings for Capt. Tulloch I beg leave to put you in mind of Despatching him.'[62]

Yet another decreet of removing was obtained in the spring of 1799. This was against 'Ensign Angus McDonell alias McDonald of the Invalids, Fort William, pretended tenant [etc.] of the Town and lands of Tulloch or part thereof'.[63] That seems to have been effective: on 17 May 1799, Donald wrote to Campbell Mackintosh to tell him that Captain Angus was busy moving his furniture out and arranging to get all his possessions off the farm. Donald said, perhaps magnanimously, that 'if he requires it I may indulge him for a few days for his Black cattle if Letterfinlay does not leave Clagan'.[64] Captain Angus was moving to Claggan on the Duke of Gordon's land, and the last point was a reference to George MacMartin, otherwise known as Cameron of Letterfinlay, who had been tacksman of Claggan and was being removed from there by the Duke.[65]

The Winter of 1798/99

Bad weather also brought problems at Tulloch. The winter of 1798/99 was a particularly severe one. Three weeks of very hard frost were followed by a great storm on the night of Friday 8 February. Donald said that several hundred sheep in Brae Lochaber had been killed under the snow, and some horses and black cattle had also been lost. Most houses lost all the thatch from their roofs. He had been busy 'endeavouring as I could with all the hands about me relieving my sheep out of great Wreaths of snow and by which means I safed at least three hundred. I only found dead as yet about a dozen but no man can be sure of his losses till the snow is exhausted.'[66] Matters could, however, have been worse, and by 10 May he was seeing a brighter side. 'I had several letters from England and South of Scotland lately and our losses of cattle particularly sheep is nothing to theirs. I understand there shall be a demand for cattle how soon there is appearance of good grass.'[67] (In the usage then, 'cattle' included sheep.)

The Tenants of Murlagan and Achluachrach

The tenants of the adjacent farms of Murlagan and Achluachrach refused to accept that they could no longer graze their animals on the portion of Tulloch previously held by the sub-tenants who had been displaced when Donald got the whole of Tulloch.[68 69] Early in 1798 Donald resorted to the Sheriff Court in Inverness, and on 25 June Daniel Clark, sheriff officer,

warned them about their behaviour. That may not have been effective. On 22 November 1798 Donald obtained a decreet against nine tenants at Murlagan, nine at Achluachrach, and a herdsman. They were required to pay him £300 for his losses and expenses. They had allowed their sheep to trespass on the land which now belonged to Donald, and tups had been allowed among Donald's ewes at the wrong time, with the result that 300 of his ewes were with lambs five weeks early.[70] On a high farm like Tulloch, such early births would lead to heavy losses.

In a long letter to Æneas Mackintosh on 30 December 1798, Donald wrote: 'I am generally distressed by bad neighbours. They do me so much injury that I am afraid of being ruined the fear of which discourages me in so many respects – The Murlagan and Achluachrach tenants I told you formerly about them.'[71]

Æneas Mackintosh sent Donald's letter to Campbell Mackintosh, adding

> In my name tell the Tenants of Murlaggan and Achluachrach that I am creditably informed that they are bad neighbours and if they do not alter their conduct immediately they shall be marked in the Black Book – as proper objects to be removed at a future set and in general, observe to the Tennents that they keep too many Cattle and Sheep on their pasture, therefore a general souming should take place[72]

The tenants were now roused and descended on Æneas Mackintosh at Moyhall 'like mad people through frost and snow when they are not wished for'.[73] Donald returned to his complaints on 10 May, saying that the tenants were oversouming and kept hunting his sheep when searching for their own. He described them as ignorant people.[74] On 12 June he said that his sheep were being killed and he wished that Æneas Mackintosh would write to the tenants.[75]

So far the Sheriff Court and the threat of the Black Book had been insufficient to deter the tenants of Murlagan and Achluachrach. At the end of August 1799, Æneas Mackintosh arranged to visit Lochaber himself, apparently a fairly rare event. He sent elaborate instructions to Campbell Mackintosh in Inverness.

> I shall, God willing, be in Town Monday morning to Breakfast, to get the necessary Cash and everything proper for the Journey – I shall be obliged to you to write to Mr. Ross [Rev. Thomas Ross] as soon as you can & that we will use the Hospitality of his House on Wednesday – I hope we can be provided with corn – It would be obliging if Mr. Ross would enquire if Barnet [McBarnett at Achneich?] could sell some oats for our horses . . . It would be right that Dond. McDond. and the other his neighbours should have intimation of our plans that all may be at hand to have the dispute settled.[76]

That was still not the end of the matter. On 25 December 1799, Donald wrote another letter to Campbell Mackintosh. 'The tenants of Murlagan has also got notice that the Diference betwixt them and me is to be settled on the Second day being Thursday 9 January next. The witnesses to be examined on my part is my own servants.'[77] On 24 March 1800, Donald reported continuing trouble with the 'Murlagan people', but for the first time admitted that there was some disputed ground, Lagnaerifearn, between Tulloch and Murlagan.[78]

At this point Rev. Thomas Ross was asked to attempt to resolve the differences. He wrote to Campbell Mackintosh on 16 April 1800, outlining his intended approach.

> [It] will be necessary to take a particular Precognition and cooly to determine the marches betwixt McDonald and his neighbours without any regard to the Deeds and heat of either party – I am very far from thinking that the Murlagan tenants are the aggressors in this, any more than I think them in other particulars that are laid to their charge, and I am persuaded that when these matters are investigated to the bottom that they will be found less culpable than they are represented at present, it will certainly operate very much against their opponent that he is in continual broils and that he could never agree with his own brothers this last part of his conduct is notorious. – When I remark these circumstances I can say they do not arise from prejudice but from Facts, for I never was an enemy of his but on the contrary one of the firmest friends he ever has as far as I could serve him – but I acknowledge his conduct of late towards these poor people has much disgusted me, for his drift is perfectly clear to me. If McIntosh and you perceived this part of his conduct, as I do, you would never blame the people and far less would you be disgusted at them.[79]

Rev. Thomas Ross may not have had a completely open mind, but Æneas Mackintosh had a poor opinion of his Lochaber tenants and was generally supportive towards Donald. On one occasion he had said 'I see it proves almost impossible to check the itch my tenants have to steal in Lochaber',[80] and on another, 'the inhabitants of that country [Lochaber] are a sad set'.[81] Donald was certainly convinced by now that he was acting in accordance with Mackintosh's wishes.

Donald knew that Mr Ross was going to Inverness on 30 April, perhaps to report his findings, so he sent yet another letter listing in detail the damages he had suffered.[82] Perhaps Mr Ross had some success, but the situation still rankled with Donald and he tried to get control into his own hands by making an offer to lease Murlagan and Achluachrach himself. In June 1802 he offered £340 for 'the Murlagan Lott . . . and for the farm of Bohinie' to add to Tulloch.[83] When he heard that it might go to Alexander

Mackintosh at the next sett, he added to his offer.[84] Nevertheless it did go to Alexander Mackintosh, and subsequently to Alexander Macdonald of Glencoe (see Chapter 8).

Cranachan and Blarnahanin (1795–1804)

The tack of Blarnahanin which had been given to Donald McDonald's father and brother John in 1786 was not renewed in 1795, and it was again joined with Cranachan. The new tack was given to Dr Donald McDonald, the son of John MacDonell of Cranachan, for nine years at a rent of £90.[85] Although he farmed in various locations over the years, Dr McDonald practised medicine at Fort Augustus. His namesake at Tulloch once called him, rather charmingly, 'Dr. Cranachan'.[86]

Dr McDonald wrote on 25 March 1796 from Fort Augustus to Campbell Mackintosh saying that he had signed the tack 'in the presence of Mr. Dond. McDonald, Drover in Tulloch and Peter Mason, Boatman in Fort Augustus'.[87] Dr McDonald was not shouldering the burden of Cranachan and Blarnahanin alone. His cousins Alexander and Angus McDonald were his co-partners. Angus Kennedy, tacksman of Leanachan, was their cautioner.[88] [89]

Keppoch and the Inverroys (1795–1804)

By 1795, Æneas Mackintosh was ready to close in on the Keppoch family and bring to an end their traditional tenure of Keppoch and the adjacent Inverroys and Boline. Alexander Mackintosh, the merchant in Maryburgh, was to be used to replace them.

In the early months of 1795 there was a flurry of confused and confusing letters to Æneas Mackintosh and his factor about the renewal of young Keppoch's tack. On 25 January 1795, Alexander MacDonell, the uncle of Keppoch and one of his curators, sent an offer of £140 'in name of my nephew'.[90] On 19 February, Keppoch's mother, who had remarried and was now Mrs Campbell, wrote to Æneas Mackintosh from Edinburgh to say that 'none but myself and Coll McDonald of Dalness [writer to the signet in Edinburgh] has any power to transact business in my son's name, his uncle's behaviour having been sufficient to destroy our former confidence in him'.[91] Uncle Alexander wrote again on 16 March in some desperation, as he was about to go off once more to the army. He said that he

> would agree to McIntosh giving the farms in Trust to anyone to look after the young man's interest – Campbell and his wife excepted – their present narrow circumstances makes them willing to take advantage of the poor young man and to appropriate to themselves whatever part of his subject they can lay hands on . . . I beg leave to refer . . . to a person

> who is well known to have McIntosh's interest at heart & cannot be accused of too great a particularity for me – I mean Mr. Donald McDonald at Tulloch. Whatever engagements he may enter into with McIntosh in my name and for my nephew's interest I shall without hesitation abide by, in case I shall be at too great a distance.[92]

That was an interesting comment on Donald McDonald from someone who had certainly not been well disposed towards him. He may not have been liked, but he was apparently trusted. On 26 June, Alexander Campbell, Keppoch's stepfather, entered the fray to disown steps which he said Coll Macdonald of Dalness had taken, apparently on the instructions of the curators, to try to keep Keppoch for the current year.[93]

None of those claiming to look after young Keppoch's interests could have known that preparations were already well advanced to put Alexander Mackintosh, the merchant from Maryburgh, into Keppoch and the Inverroys.[94] On 1 June 1795 letters of horning were issued against young Alexander.[95] Trouble was anticipated in getting the family out of Keppoch, and letters of ejection had been obtained from Edinburgh.[96] Campbell Mackintosh passed them to Alexander Mackintosh, who had then to present himself at Keppoch as the new tenant. On 13 June 1796 he reported (from Fort William) that the Keppoch family had gone quietly, although still declaring defiantly that they could have kept possession if they had wanted to. Alexander Mackintosh got entry without even using the letters of ejection.[97]

In his memorial of 1834, written thirty-eight years later, Alexander Mackintosh gave a much more exciting account of his entry to Keppoch.

> [I] was pitched upon to take possession of Keppoch, two Inverroys & Bolyne, as occupied by the Keppoch family, which he did principally to serve his Chief, and was bound to reside with his family at that place at risque of his life . . . the Macdonalds who with their following and adherents principally possessed the country around [the] Memorialist . . . were mortified, provoked, and aggrieved to see another man, particularly that of Mackintosh, a name they hated above all, sit down in their chief's place after possessing it so many centuries.

He went on to explain how he resolved to 'sell his life as dear as possible' and always carried with him two brace of pistols.[98] No doubt Alexander Mackintosh did have his problems to overcome, and it would be unfair to dismiss them lightly. His contemporary account is, however, probably nearer to the truth. Donald McDonald had to run a similar gauntlet at Tulloch, and his well-documented experience, described above, was troublesome and unpleasant, but it had none of the tense drama of Mackintosh's account of 1834.

Alexander Mackintosh's tack did not expire until 1804, but his days at Keppoch and the Inverroys came to a premature end in 1802. In January 1802, Æneas Mackintosh said that he had received 'offers for the farm of the Inverroys'.[99] In 1834 Alexander Mackintosh put his own distinctive slant on what had happened.

> At length when peace reigned [at Keppoch] . . . several gentlemen in the country wished to have these lands for a residence who would never have ventured upon them before, while in the hands of the Macdonalds – Memor^{t} [Alexander Mackintosh] observing that they would give more than double the rent . . . he proposed to Sir Æneas to give them to the highest bidder which he agreed – and through Glengarry's interest with him, Glencoe was preferred, who gave twice the former rent, with some gratuities, consequently in order to do justice to his Chief – Memorialist submitted, so as to raise the value of his lands.[100]

That was a selfless and unworldly approach, but perhaps it was not quite like that.

Alexander Mackintosh was perfectly correct when he said that the rent was doubled. He had been paying £150 per annum,[101] and Macdonald of Glencoe paid £300.[102] The cryptic reference to Glengarry is interesting. Alexander Macdonald signed his tack of Keppoch, the Inverroys and Boline at Glengarry House on 8 May 1802, and Glengarry himself witnessed his signature.[103] There are indictions of an increasing locus for Glengarry in Alexander Macdonald's affairs. In 1796 he was offered 'some land from Glengarry but that land is only a mole hill of hill grass'.[104] However, in 1802 he got a large tack on Glengarry's land at Kinlochnevis in Knoydart (see Chapter 10).

In 1803 the 4th Inverness Volunteer Battalion was formed under the command of Glengarry, who insisted on having the rank of colonel. Alexander Macdonald of Glencoe was his second in command, with the rank of lieutenant colonel. He resigned in 1808.[105]

In 1807 Alexander Macdonald was party to a joint arrangement with Glengarry to take over from Alexander McTurk responsibility for the latter's tack of Killiean, Laddymore and other lands which ran until 1814. This involved a commitment to take over sheep valued at £1762 7*s.* 11*d.*, to relieve McTurk of responsibility for his rent and pay him £200 a year as benefit or profit. The wisdom of entering a joint responsibilty with Glengarry is surely questionable, but an interesting aspect of this agreement is that when it was signed at Invergarry on 21 October 1807 Donald McDonald, tacksman of Tulloch, was there and witnessed the signatures.[106]

There is also reason to suspect that around 1809–1811 Alexander Macdonald of Glencoe was responsible to Glengarry for the rents and stock of

Garrygualoch lying on the south of the River Garry.[107] Finally, in 1814, when Alexander Macdonald was in arrears with his Mackintosh rent, he said, 'I understand Sir Æneas has been complaining to Glengarry'.[108] Why should Sir Æneas have thought it appropriate to do that?

Alexander Macdonald took immediate possession of Keppoch, the Inverroys and Boline. Arrangements for the two years of the old tack from 1802 until the new tack would begin in 1804 were left to him and Alexander Mackintosh to resolve informally. Alexander Mackintosh had second thoughts and asked Glencoe to let him have Inverroymore during the forthcoming nineteen-year lease. Glencoe agreed, but the ubiquitous Donald McDonald reported that 'when his friends heard of it they disapproved of his setting Inverroymore because that would be a bar in his other plans of management'. Glencoe compensated Mackintosh with a payment of £20,[109] but who were 'his friends'? Was Donald indicating in an unusually discreet fashion that he had a hand in this? Was Alexander Macdonald of Glencoe coming under the influence of Glengarry, and perhaps of others?

A comprisement and valuation of Keppoch was undertaken by two wrights, Archibald Wilson and Alexander Taylor, which showed that the house was worth £447 6*s.* 2*d.* A report was also prepared on the fruit and other trees. It was later asserted that Alexander Macdonald of Glencoe did not make any payment of melioration money to Alexander Mackintosh as the outgoing tenant.[110] That was to be an issue in the wrangling between the Glencoe trustees and Æneas Mackintosh's successor many years later when the tack was coming to an end in 1823.

As a merchant, Alexander Mackintosh had other interests. In 1802, when about to vacate Keppoch, he undertook to purchase and extract oakwood from the land of Simon Fraser at Borlum, on the banks of the River Ness, to the south of Inverness. For this privilege he paid £320.[111] It would not, however, be long before he was a tenant on his chief's land in Lochaber again, to be displaced once more by Alexander Macdonald of Glencoe (see Chapter 8).

Settling the Marches

Settling the marches, or boundaries, between neighbouring proprietors and neighbouring tenants was a frequent issue at this time. Perhaps the reason was simply that with population and livestock and the value of land all increasing the location of boundaries assumed greater importance. Any estate maps or plans which existed were not good enough to be of much help. The process of settling consisted of each side producing as many witnesses as they could, the older the better, to say where they believed the marches to be. Representatives of the proprietors or tenants had then to

try to agree. Donald McDonald has left an account of how he did his best to safeguard Æneas Mackintosh's interests.

On 28 August 1796 Donald wrote to Campbell Mackintosh to tell him about the marches at Urchar, to the east of Tulloch, and at Inverroy, to the west of Roy Bridge. Mr Tod, the Duke of Gordon's factor, had been in Lochaber earlier in the month.

> I attended when the Urchar Marches was in dispute and called those that knew the marches. McIntosh lost part of what was Claimed by me more than I thought he should Do if the proof was examined but I cannot say it was very material. Captain Moy [Alexander MacDonell, the Duke of Gordon's tenant in Moy] caused his people to Show more than they could swear to which I took notice of at the time and beged of Mr. Ross [Rev. Thomas Ross] to insist on their Oaths before credit was given them. I attended at Inverroy upon Another day that was fixt upon to settle the marches there and had a great many witnesses with me but the day was raining & Mr. Tod left [it] for Mr. Ross & Inch [Angus MacDonell of Inch or one of his sons] to settle it.[112]

In spite of the attention to the Inverroy marches, or perhaps because Mr Tod had preferred not to stay out in the rain, there was still a problem there a couple of years later. On 30 December 1798, Donald McDonald told Æneas Mackintosh that

> the March at Inverroy is neglected . . . a dyke should be built . . . it is against your interest to delay settling the Marches there as there is a number of old men that can not live long that can prove what is claimed on your behalf. I could not like the manner in which the march at Urchar was settled . . . they had some pretended witnesses that Lieut. McDonald [Alexander MacDonell at Moy again, demoted from captain on this occasion] brought there indeed people that never saw anything about it he brought to make a show believing they would not be put on oath.[113]

Donald McDonald's Influence with Æneas Mackintosh

The extent to which Æneas Mackintosh supported Donald McDonald is remarkable. Donald's letter to him of 30 December 1798 was long and touched on a variety of matters other than the marches.[114] Æneas Mackintosh forwarded it to Campbell Mackintosh, with a covering letter dated 5 January 1799, in which he accepted most of what Donald had put forward.[115]

A new public house had been built at Achaderry, but Donald complained that 'a great number of your Tenants did not in proper time attend to carry their respective proportions of timber and sclate . . . and in the end I was obliged to look out for people to do it for them rather than stop

the work. I have a state of what was done by other men that I paid for.' Æneas Mackintosh instructed Campbell Mackintosh '[to] take the money out of the rents . . . you will please settle with Donald for the expense he has been at according to his own letter paying labourers and tradesmen for carrying materials over and above his own proportion'.

Donald had been building a dyke 'and so far as is done I can say with great truth that there is not a better Stone Dyke in the Highlands than what I built. I had masons from Argyleshire that is really good. I began in the March between you and the Duke at Urchar.' What could Æneas say to that? He instructed that Donald was to be paid a total of £40 for 1,200 yards.

There was also a problem about 'several transgressions done to the wood in several parts', and Donald was intending to apply to Rev. Thomas Ross, as a justice of the peace, 'to give me a warrant of search and how soon I get it I shall go to every suspected place on the Duke's property'. He added: 'Duncan McIntosh, the forester has not the resolution enough to act properly as a forester. I can get more information from other people than from him.' Æneas Mackintosh was torn between the suspicion that his own tenants were guilty and the hope 'that the Duke's people are to blame, and I wish Dond. may be successful in tracing it to that quarter'. He accepted the criticism of the poor forester. 'Dond. is right for I am sensible from Observation that our namesake is too soft for his office and I must soon change him for some person who has more of the Devil about them that is more active and resolute and knows the value of Wood – for better it should be sold than stolen.' References to Duncan McIntosh, forester, in 1805 and 1814 suggest that Æneas relented.[116] [117] Nevertheless, he gave Donald sufficient backing to make him powerful in Lochaber, and far from universally popular.

Problems over Religious Observances

Around 1798 or early 1799, Donald's occasional instinct to help those in difficulty led him to ask Campbell Mackintosh to act for John Cameron, one of his servants at Tulloch, who, like him, was a Roman Catholic. Cameron had fallen foul of the insistence in the Parish of Kilmanivaig that Catholics should pay the dues of baptism and recording in the parish register.[118] Campbell Mackintosh was apparently successful on John Cameron's behalf. On 10 May 1799 Donald wrote to him:

> In regard to the Roman Catholick's process which you carried on please know that I spoke to the priest and people Sunday last at which time both joined in prayer for you, on account of your attention – they agree to pay you thankfully . . . Mr. Ross [the parish minister] appeared displeased when I told him how matters stand, particularly so at me for

> interfering – But I could not help taking a concern for my servant John Cameron when they wanted to impose on him. Mr. Ross says positively he shall bring it before the Court of Session . . . but for my part I shall no longer be at the head of the business and I shall collect the amount due for you for your expenses . . . I was witness to the Roman Catholicks finishing a cask of whisky lately drinking your health on this account.[119]

There is no record of what Campbell Mackintosh thought of that.

In 1801 Donald was under threat himself, ostensibly about his failure to comply with a royal proclamation requiring the King's subjects in North Britain to observe a fast day on 12 February 1801, on account of the critical situation of his kingdoms. Some disaffected neighbour probably saw a chance to make trouble for him. Information was lodged with Neil McIntyre, the procurator fiscal at Fort William, that Donald McDonald, tacksman of Tulloch, 'had taken upon him that very day . . . to come from his own farm to the foot of the River Roy at Achaderry and there, as tenant of Mackintosh of Mackintosh oblige the Smith to warm his forge and work for him, and likewise got down his grain to the mill and obliged the miller to grind it, which he was induced to do by threats of his master's displeasure'. The fiscal said that he was 'duty bound to take notice of any derogation from his Majesty's proclamation' and proposed that Donald should be fined £300 with £50 costs.

Donald's response was vigorous and informative. 'He flatters himself that his conduct . . . has been uniformly that of a loyal subject warmly attached to his King. . . . The Respondent is one of His Majesty's Roman Catholick subjects of Scotland, and as such was one of the first of that description in this Country who Qualified to Government in the year 1793 . . . he subscribed £5 10*s.* 0*d.* towards the subscription for carrying on the war.'

He said that because of the remoteness of the area they did not get early notice of 'affairs going on in the Public World'. He had left his own house on Monday 9 February with his servants and horses and with corn to be dried and milled at Achaderry. He remained there with his servants attending to this until Wednesday morning [11 February], when he went to Bohuntine to procure some potatoes for himself, his servants and his workmen. He returned to Achaderry on Thursday morning [12 February], 'with an intention of hearing Sermon at the Meeting house of Achluachrach'. First he called at the smithy, where he found the smith busy at work. He told him that he would send him a horse which he wanted shod that evening because he was leaving the country [district] next morning.

Then he went off to Divine Service. 'It was upon hearing the Priest's

discourse, and from the tenor of it, and not until then, that the Respondent understood the occasion of the sermon to be a National Fast.' Donald then played his trump card. 'No notice thereof whatever was given in the Parish Church of the Parish in which the respondent resides, nor does he believe it came to hand on the preceding Sunday, and when he found at Bohuntine foresaid on the day of Thursday the 12 February the person who is a schoolmaster . . . busily employed teaching his school the same as on ordinary days.'

When the case came before the sheriff substitute, John Campbell, at Fort William on 24 March, he allowed time for Donald's answers to be seen by the Fiscal. On 18 November, he renewed the diet. When the case came back on 1 December 1801, he dismissed it on the grounds that the fiscal had failed to bring forward his proof. Donald was awarded costs, a miserly ten shillings, which was probably an inadequate recompense for the anxiety which he must have suffered.[120]

It is difficult to believe that this episode was anything other than an attempt to cause trouble for Donald, and it is impossible to tell if the fiscal ever intended to proceed to the limit. If he had, Donald would no doubt have made trouble for the Rev. Mr Ross, the parish minister, who seemed to have been equally neglectful of the King's wishes, and perhaps with less excuse. The episode does, however, provide an interesting account of how Donald and a retinue of servants spent four whole days going about their business with the miller and the smith at Achaderry, buying potatoes at Bohuntine, and going to church at Achluachrach.

Developments on Mackintosh's Lochaber lands during the long tacks of 1804–1823, when Alexander Macdonald of Glencoe and subsequently his trustees held much of that land, will be described in Chapter 8. Donald McDonald's other interests, along with some of those of his brother Alexander Dhu, will be described in Chapter 9.

5

Corpach, Glenure, Auchteraw and Cullachy

While the Mackintosh lands in Lochaber were being developed, Alexander Macdonald of Glencoe was expanding on the lands of other proprietors. His marriage in 1786 and his father's death in 1787 seem to have encouraged him to find other openings to add to the tenancy of Fersit which he had held since 1778. The tocher of 'six thousand merks scots money' from Ewen Cameron of Fassifern under the Marriage Contract of 1787 would have been worth about £330. Along with the loan of £450 from his brother-in-law, the Rev. John Kennedy, he must have had nearly £800 at his disposal. He lost no time. In 1787 he became tenant of Corpach, on Locheil's land, and immediately set about obtaining the tenancy of Glenure and Barnamuck, on the land of Campbell of Barcaldine.

While Alexander Macdonald of Glencoe was spreading out in these directions, his brother-in-law, the Rev. John Kennedy, was a sheep-farmer on Lovat's land near Fort Augustus. In the first instance he was in partnership with Lieutenant Evan Macpherson, his brother-in-law by his second marriage, but they soon separated. Kennedy was then identified with Auchteraw, and Macpherson with Cullachy. Both had careers worth relating, and their Badenoch origins introduce some insights into that district. In due course Alexander Macdonald was drawn deeply into Kennedy's affairs and marginally into those of Macpherson.

Corpach

Alexander Macdonald of Glencoe probably became tenant of Corpach at Whitsunday 1787. It is apparent from the Locheil rentals that this was a substantial property for which he paid a relatively high rental, well above the average for the Locheil estate.

The Locheil estates had been forfeited and annexed after the 1745 Rising. In 1784 they were returned to Donald Cameron, 9th of Locheil, who was then a minor. Ewen Cameron of Fassifern was one of the trustees who managed his estate.[1] In the early months of 1786 he arranged to let Corpach to his future son-in-law, Alexander Macdonald of Glencoe. In April, Donald Cameron, Ewen's cousin in London, expressed approval of the terms on which he had 'let the Farm of Corpach to Mr. McDonald of Glencoe who I understand to be a man of good character'. He added, 'I suppose you

have reserved the House at Corpach subject to Locheil's use in case he needs it.'[2]

On 16 May 1787 Alexander Macdonald of Glencoe wrote from Corpach to Alexander Campbell of Barcaldine.[3] That suggests he had by then taken up the tenancy arranged by his father-in-law. The Locheil rentals for 1788 show that he was paying rent of £120.[4]

Locheil married in 1795, and soon after that he returned to Lochaber.[5] Achnacarry had still to be rebuilt and he and his wife lived in a house in Corpach said to have been built for Henry Butter, the factor during the annexation. This was known as 'Corpach House'[6] or the 'Grey House'.[7] Judging from his correspondence, Alexander Macdonald was still living at Corpach on 13 May 1796[8] but moved to Inverscaddle shortly after. He probably gave up the house at Corpach to make way for Locheil.

In May 1804, Duncan Cameron WS, son of Ewen Cameron of Fassifern, revealed details of proposed changes in the disposition of Locheil's land. Much of it was being offered for rent to the highest bidders, but he said that 'Corpach is to be crofted'.[9] The crofts were intended for small tenants being displaced from other parts of the Locheil estate. This coincided with work becoming available to them on the construction of the Caledonian Canal. Alexander Macdonald's interest in Corpach must have ceased by the early 1800s.

Inverscaddle

Inverscaddle probably served Alexander Macdonald as a home for the few years after he ceased living at Corpach and before he settled at Keppoch. Thomas Garnett travelled along the east side of Loch Linnhe in 1798 and remarked that 'on the opposite side of the water, we saw Inverscadle house, the present residence of Macdonald of Glencoe'.[10] Alexander Macdonald never seemed to use Glencoe as his permanent home.

Inverscaddle was at times part of the Ardgour estate, but in the eighteenth century it was owned by the Camerons of Fassifern. In 1742 John Cameron of Fassifern was the 'Heritable Proprietor'.[11] The 1751 Valuation Roll for Argyll shows him in possession of 'Inverscadil, Conaglen, Achaphuil'.[12] John, the eldest son of Ewen Cameron of Fassifern, was born at Inverscaddle in 1771.[13] In 1792 there was a reference to a vein of lead at 'Inverscaddle belonging to Mr. Cameron of Fassfern'.[14] The Fassifern connection probably explains why Inverscaddle was available to Alexander Macdonald when he had to vacate Corpach and needed a residence.

Glenure and Barnamuck

From 1788 Alexander Macdonald of Glencoe leased land in the upper part of Glencreran from Alexander Campbell of Barcaldine. The lease must have been renewed several times. It was the longest of Alexander

Macdonald's leases and continued until relinquished by his trustees in 1817.

Alexander Campbell inherited the estate of Barcaldine and Glenure from his father Duncan in 1784.[15] His sister Lucy (or Louisa) was the wife of Ewen Cameron of Fassifern[16] and the mother of Mary Cameron, who married Alexander Macdonald of Glencoe in 1786. Alexander Campbell of Barcaldine was therefore Mary Cameron's uncle.

Alexander Macdonald of Glencoe wrote from Corpach on 16 May 1787 to Alexander Campbell about leasing Glenure and Barnamuck. After that letters were exchanged regularly. Many have survived in the Campbell of Barcaldine Papers in the National Archives of Scotland, but some are damaged and cannot be read in full.[17]

The land concerned was in the upper end of Glencreran, not far from Glencoe by the direct routes over the hills which would have been used in the eighteenth century. The full description was 'Glenure, Barnamuck, Diraloch, and Bottom of Elarick'. This was usually shortened to 'Glenure, Barnamuck, etc', or even to one or other of these places alone. It was good land. Barnamuck, 'the highest farm in the parish in that direction has always been noted for the excellence of its pasture'.[18]

The first record of a payment of rent is in a letter of 30 November 1789 from Campbell of Barcaldine. He acknowledged receipt by the hand of James McDonald, of £280 in part payment for the sheep stock at Glenure and £100 in part payment of the rent of Glenure, Barnamuck, etc. for the year from Whitsunday 1789 to Whitsunday 1790. Barcaldine promised to give a regular receipt on stamped paper when required to do so.

On 23 November 1790 Alexander Macdonald told Campbell of Barcaldine that he had sent him £427 11*s.* 0*d.* Of this sum £327 10*s.* 10*d.* was to be applied to the value of the Glenure stock and £100 to the Martinmas rent. These payments show that the land was already under sheep and that Alexander Macdonald was purchasing the stock.

On 14 June and 15 December 1791 and 18 June 1792, Campbell of Barcaldine wrote short letters acknowledging receipt of various payments towards rents due. There is no further mention of payment for sheep, and the purchase may have been completed in the first two years.

On 16 June 1793 Alexander Macdonald wrote saying 'I am very sorry I could not send you rents . . . owing to the [shortage?] of money at the market but I send you the enclosed draught on the Stirling Bank which I am hopeful will serve you equally well'. On 4 June 1794 he said that he was 'sending you these few lines at present to [??] a particular favour of you, to delay the rents of Glenure untill the beginning of August if you can conveniently do it . . . The reason I am asking is that I have. . . .' Evidently Barcaldine declined, because on 13 June Alexander Macdonald wrote again. Only a few words are decipherable. 'I was favoured with your . . . I am sorry you could not . . . your money . . . and I understand some of

them are not willing to pay it . . . if I could ask the favour of you to examine these . . . the bearer bringing them'. The sense of this may be that Alexander Macdonald was claiming that could not pay because he was having difficulty in collecting payments due to him. That was quite a common problem in 1794.

It is not clear how the matter was resolved that year. However, on 5 June 1795 Alexander Macdonald wrote saying, 'I dar say youl be longing to hear from me and I have sent you . . . one hundred pounds'. He finishes by saying 'I am afraid youl think me a bad payer . . . but . . . .' There was a totally unreadable letter from him on 9 August 1796, and on 3 December that year he said, 'I dar say youl be thinking it time to send you the Glenure rents now youl please receive £80 10*s.* 0*d.* per bearer'. Any further letters in this correspondence do not appear to have survived.[19]

Alexander Macdonald placed two brothers, John and Duncan Rankin, as his sub-tenants in Glenure and Barnamuck. They were steelbow tenants:[20] that is to say the landlord provided the stock as part of the rental arrangements, and the tenant was obliged to return the equivalent stock at the end of the tenancy. This seems to have been the only occasion when Alexander Macdonald granted a tenure of this kind. Dr I.F. Grant said that this was a Lowland form of tenure but she had found allusions to it in Ross-shire, Central Perthshire and Kintyre.[21] Other examples in the Highlands can, however, be cited. In 1742 John Cameron of Fassifern gave to Allan Cameron a steelbow tack of 'all and whole the Two Merklands of Inverscadell and Conaglen'. This document includes a very detailed list of all the livestock in the tack.[22] There are references to steelbow tenancies later in the eighteenth century – to 'Ranald MacDonald commonly designated of Meoble and Steelbow tenant of the lands of Drumindarroch and Guisdale';[23] to Peter MacNab and sons, tenants of Allan Macdonald of Gallovie;[24] and to James McCaul as tenant of Loch Treig.[25] There were occasions, perhaps not surprisingly, when landlords had difficulty recovering the stock due to them at the end of such tenancies.

There are two stones placed by these Rankins on Eilean Munde in Loch Leven. One has the inscription: 'This stone is placed here by John Rankin Tacksman of Glenuire and Barnamuck and his spouse Christian Rankin, 24 September 1801.' The other reads: 'This stone is placed here by Duncan Rankin, Tacksman of Glenuire and Barnamuck and his spouse Mary Rankin.'[26] As the Rankins were sub-tenants, the use of the term 'tacksman' was perhaps slightly pretentious.

Alexander Campbell of Barcaldine died in 1800 and was succeeded by his eldest son Duncan. He was then a boy of fourteen at school in England. Later he pursued a military career.[27] Alexander Macdonald continued to lease Glenure and Barnamuck, but there are almost no further surviving records until his Trustees took over after his death at the end of 1814. The

only scrap of information was ominous, being a foretaste of the troubles that the trustees were to have. In 1808 Alexander Macdonald took John Rankin to the Sheriff Court in Fort William to recover '£100 stg. being the balance of rent due by him at Martinmas last for the said farm of Barnamuck which he holds in sub-sett'.[28] It is slightly curious that although Barnamuck is in Argyllshire this case was heard in Fort William and not in Inveraray.

After Alexander Macdonald died, the Glencoe trustees had to continue to collect rent as best they could from the two Rankins until the tack ended. An account of how they fared will be found in Chapter 12, along with descriptions of other dealings that the trustees had with Rankins.

The Rankins

Rankin is a name that occurs with modest frequency in Lochaber and north Argyll and is well known in Glencoe. Several Rankins became involved in sheep-farming, sometimes in association with Alexander Macdonald of Glencoe or with Adam Macdonald of Achtriachtan, and sometimes on their own account. The relationships between various Rankins are not always clear. Angus Rankin, tacksman of Dalness and thus a tenant of Coll Macdonald WS of Dalness, was a senior figure among them and tended to be involved in most of their affairs. His wife, Helen, was a Cameron of Kinlochleven and may have been a sister of Major Angus Cameron of Kinlochleven.[29] [30] Angus Rankin and his relatives were great pluralists, taking up opportunities as they presented in different locations and pursuing them simultaneously. It is not clear how their many ventures were funded, but when sheep-farming slumped most of them were exposed in very vulnerable positions.

Angus Rankin, his brother Donald, and the John Rankin at Barnamuck were deeply involved in the machinations over Fersit which will be described in Chapter 12. There was also a John Rankin associated with Angus and Donald at Achnacon who will be mentioned in Chapter 6, but he may have been a different man. A Horning at Inveraray Sheriff Court shows that in 1802 Angus and a John Rankin were tacksmen of Sallachil. This lies close to Barnamuck but on the opposite side of the River Creran on the land of the Stewarts of Fasnacloich.[31] [32] Angus can be identified with confidence because the messenger from the court delivered the charge for him to a servant at Dalness.[33] It is not clear which John Rankin was involved, but this illustrates the great extent of Angus Rankin's commitments.

Auchteraw and Cullachy

Auchteraw and Cullachy are near Fort Augustus in the parish of Boleskine and Abertarff where the Rev. John Kennedy and Lieutenant

Evan Macpherson of the Ovie family, brother of Kennedy's second wife, Christina Macpherson, became sheep-farmers in 1785. In 1786 Christina died, and in 1787 Kennedy married Jean Macdonald, sister of Alexander Macdonald of Glencoe. The careers of Kennedy and Macpherson show that marriage was important in creating partnerships, but some tantalising questions about the source of Kennedy's funds remain unanswered. The later fortunes of Kennedy and Macpherson diverged in quite dramatic fashion. Kennedy remained solvent; Macpherson became bankrupt with huge debts that affected others who had supported him.

The Reverend Mr John Kennedy

John Kennedy was the son of Alexander Kennedy at Knappach in Ruthven.[34] From November 1767 until May 1776 he was the schoolmaster at Ruthven in Badenoch, where the parish school of Kingussie was located. He was an Arts student at King's College, Aberdeen, where he graduated in March 1770.[35] Although he used the title 'Reverend' he did not obtain a charge in Scotland and had to be content with poorly paid employment as a schoolmaster. (His successor, James MacLean, had to exist on an annual salary of £5 19*s.* 10*d.*[36 37]) In 1774, when the parish of Laggan became vacant, William Tod wrote from Badenoch to his superior James Ross, the Duke of Gordon's cashier. 'I shall enclose a letter from one of the Candidates – I should like that His Grace did something for Mr. Kennedy because he is a Tenant's Son and has few else to depend on – but it should not be in his *own* country'.[38] Tod apparently viewed Kennedy with favour, but not for this appointment. The successful candidate was the Rev. James Grant, who in 1779 married Anne MacVicar. As Mrs Grant of Laggan she was to become well known as an author.

Soon after his failure to obtain that appointment Kennedy went to East Florida, then a British possession, where he settled at St. Mark's.[39] By then he had married his first wife, Elizabeth, daughter of Evan McPherson, tacksman of Laggan.[40] While in Florida he was apparently caught up in the American War of Independence because in 1780 he was deputy Chaplain to the 71st Regiment of Foot serving in America.[41] In 1783 Florida was ceded to Spain, and around that time Kennedy returned to Scotland. In 1784 there is a record of the 'Rev. John Kennedy of East Florida presently residing at Knappach'.[42] His wife, Elizabeth McPherson, must have died because on 6 November 1782 he had entered into a Contract of Marriage with Miss Christina Macpherson, second daughter of Hugh Macpherson of Ovie.[43] Ovie is on the Duke of Gordon's land in Badenoch, a little to the west of Newtonmore on the road towards Laggan. Hugh Macpherson had left Ovie in 1773[44] but, as often happened, the territorial designation remained with him and his family. He had undertaken to pay Kennedy a tocher of £100 but failed to produce

the money. In 1784 Kennedy took action in the Sheriff Court in Inverness to enforce payment.[45]

Kennedy was also the defender in a case in Inverness Sheriff Court in 1784, which provides more information about his new in-laws. In the last few months of 1783, Captain John MacPherson of Ballachroan

> had a communing with the Reverend Mr. John Kennedy at Knappach respecting Æneas Macpherson of Flichity and some difficulties which he laboured under when the said Mr. John Kennedy who is married to a Daughter of a Brother of the said Æneas Macpherson said that he would contribute £5 stg. to the relief and assistance of the said Æneas Macpherson if he was in cash and the complainer [Ballachroan] answered that he would advance the money on condition that the said Mr. John Kennedy would pay it to him.

As Kennedy was now married to a daughter of Hugh Macpherson of Ovie, that statement seems to establish that Hugh Macpherson of Ovie and Æneas Macpherson of Flichity in Strathnairn were brothers.

Kennedy denied having made any such promise and refused to pay the £5 to Ballachroan, who raised the action in 1784 to recover it from him. Ballachroan proposed to prove his case by calling two witnesses: Mrs Christina Macpherson, spouse to the Reverend John Kennedy, and Lieutenant Evan Macpherson of the Sixteenth Regiment of Foot, Christina's brother, who was then residing at Ballachroan. Kennedy's lawyer objected to Mrs Kennedy as a witness on two grounds. First, as Ballachroan's sister-in-law she ought not be a witness for him; second, as Kennedy's wife she could not be a witness against him. He also objected to Lieutenant Macpherson because he was a brother-in-law of Ballachroan, adding that he 'stands in the same degree of connection to the defender [Kennedy]'. Without these witnesses Ballachroan could not prove his case and in the end was found liable for expenses.[46]

These statements imply that Evan Macpherson and Ballachroan's wife were, like Kennedy's wife, children of Hugh Macpherson of Ovie. This is credible. Evan Macpherson's position as one of the Ovie family is well established.[47] [48] Ann, the eldest daughter of Hugh Macpherson of Ovie, married John Macpherson of Inverhall, near Ruthven, in the 1760s and was widowed towards the end of 1770.[49] [50] She would then have been free to marry again. Ballachroan is said to have married 'Ann Macpherson' in 1777.[51] She died in Edinburgh on 29 December 1796 'in the 51st year of her age'.[52] In 1784 Evan Macpherson was married at age 24 or thereby,[53] which would make him some fourteen or fifteen years younger than his sister Ann.

On 6 November 1784 Kennedy and his family moved from Knappach

to 'the Town of Perth'.[54] Although he cannot have remained there for long, there are recurring references to an unexplained connection with Perth.[55] [56] [57]

Kennedy and Macpherson as Joint Tacksmen

In 1785, a few months after Kennedy left Knappach, the trustees of General Simon Fraser granted to his brother-in-law, Lieutenant Evan Macpherson, a tack for nineteen years of Little and Meikle Auchindarrochs, Achteramore, Achterabegg, Balmain, Carngoddy, Inchnacardoch and other lands and grazings, all near Fort Augustus. The Rev. John Kennedy 'had an interest and concern in terms of an agreement entered into between him and the said Evan McPherson'.[58] This probably means that although Kennedy's name was not on the tack as a tenant he had put up some of the money. This is the first indication that John Kennedy, once a poorly paid schoolmaster, had somehow obtained some capital. His first wife, Elizabeth, had been entitled to about £120 plus interest by her parents' marriage settlement and by her father's will. In 1780 Kennedy tried to recover about £45 due to him by bills indorsed to him by his father.[59] These were not, however, very large sums, and he must have managed to increase his capital because he was soon able to acquire a large stock of sheep and to advance money to others from time to time. He appears never to have borrowed money himself, and he was solvent when he died.

Very soon after this tack had been granted, Archibald Fraser of Lovat bought these lands from the trustees of General Simon Fraser. Lovat was obliged to issue a corroborative tack. What he did was quite complicated. He gave to Kennedy and Macpherson jointly a new tack that was to run for nineteen years from 1787, i.e. two years beyond the original one. Lovat apparently wanted to get Inchnacardoch back into his own hands. As it had been included in the original tack, he let it remain in the new one but made it a condition that Kennedy and Macpherson would sub-set it back to him. However, he added to the new tack Cullachy, Kyltrea and other lands which were then held by Alexander Fraser of Dell. Briefly, the new tack consisted of the old one minus Inchnacardoch plus Cullachy etc., and had Kennedy's name added to it. The rent (including public burdens) was to be £339 4*s.* 0*d.*[60]

The Ovie Macphersons and Cullachy

The Ovie Macphersons had an intermittent connection with Cullachy which began in 1773 and ended with Evan Macpherson's failure and departure in 1794. Several authors have referred to the Ovie family at Cullachy. In 1897 Charles Fraser-Mackintosh said, 'some time before the close of last century several Macphersons who had been dispossessed in Badenoch came to Boleskin Parish, and amongst others Lieutenant Evan

Macpherson of the Ovie family took Cullachy'.[61] That must refer to Lieutenant Evan's share in the tacks of the later 1780s, but these were not the family's first connection with Cullachy.

The Macphersons of Ovie left that place at Whitsunday 1773, when Hugh Macpherson did not get his tack renewed.[62] He must then have obtained the tenancy of Cullachy. There are references to the family at Cullachy around this period in a statement prepared some time after 1815 on behalf of John MacDonell, then 'a man past 60 years', who had worked at Cullachy for much of his life. 'In the early part of his life he was servant to a most active and industrious farmer Mr. Macpherson of Uvie [Ovie] who resided at Cullachy.'[63] The early part of John MacDonell's working life would have been in the 1770s.

Hugh Macpherson's correspondence shows that he was still at Cullachy in January 1779,[64] but in 1780 'Macpherson of Uvie was ejected from Cullachy'.[65] In 1782 he was apparently at Etterish in Badenoch.[66] In September 1783, when the Circuit Court was sitting in Inverness, 'John McPherson, Younger of Uvie at Carrochinlee' was one of those present 'to pass upon the Assize'.[67] Carrochinlee has not been identified.

John MacDonell later 'became a tenant in Kyltrea under Mr. Fraser of Dell' and remained when 'Lieut. Macpherson obtained a lease from the late Mr. Fraser of Lovat of the lands of Cullachy and Kyltrea [etc]'.[68] Fraser of Dell must therefore have taken over the lease that the Ovie family had held. That is entirely consistent with the narrative of the tack of 1786.[69]

In 1773, the future Mrs Grant of Laggan arrived in Fort Augustus. She was then Anne MacVicar, daughter of the newly appointed Barrack Master there. She left in 1779 to go to Laggan as the wife of the Rev. James Grant. Her stay at Fort Augustus therefore coincided closely with that of Hugh Macpherson at Cullachy, and she came to know the Ovie family well. In her writings she mentioned them several times. Additional footnotes inserted by her son, J.P. Grant, in the sixth edition of her *Letters from the Mountains*, published in 1845, are helpful in interpreting her references to them.

In a letter of 24 May 1774, Mrs Grant wrote in lavishly complimentary terms about her friend Christina Macpherson, identified in a footnote to a later letter as 'a dear friend of the author' residing at Cullachy.[70] Several of Mrs Grant's letters in 1778 refer to the recent marriage of Christina's brother, John Macpherson Younger of Ovie, to Isabella Macpherson, daughter of the Rev. John Macpherson, Minister of Sleat. Isabella's brother was Sir John Macpherson, Governor-General of India in 1785–6.[71] [72] [73]

Eight years later, on 1 March 1786, Mrs Grant said, 'I feel for the death of my friend Christina Macpherson. Her departure was very sudden; she was nursing her second son . . . and died the second day – you can't think how I was affected by her loss, though already lost to me in an unequal

marriage.' A footnote in the sixth edition of *Letters from the Mountains* adds that Christina 'had some time before married Mr. Kennedy'.[74] Christina did have two sons, James and Hugh,[75] and must have died around this time because by 6 February 1787 Kennedy had negotiated a Contract of Marriage with Jean Macdonald of Glencoe, his third wife.[76]

After a further twenty-two years Mrs Grant referred again in 1808 to Christina. In a letter of 3 February she said: 'many years ago, when I lived at Fort Augustus, I had a friend whose brother, in consequence of my intimacy with her, was well known to me. . . . His father was out with the Prince, and his uncle, Macpherson of Fleigherty [Flichity] marched a company with him to Derby.' She goes on to say that this gentleman's affairs became embarrassed, and he joined the army.[77] As will be shown shortly, Lieutenant Evan Macpherson became bankrupt in 1794 and joined the army.[78] Mrs Grant was writing about Lieutenant Evan Macpherson and his sister Christina at Cullachy during the earlier period when it was held by their father.

Kennedy does not emerge well from these references to him. Mrs Grant leaves an impression that she did not approve of him. She never mentions him by name, but refers to Christina having been lost to her 'in an unequal marriage'. She also fails to take any notice of his subsequent marriage to Jean Macdonald of Glencoe, who must have been one of 'the young ladies of Glencoe' with whom she had claimed a remote connection and great friendship in her letter of 17 May 1773.[79] Is it possible that she resented Kennedy's attempt to secure the parish of Laggan in 1774 when her future husband was appointed to it? Or did she resent Kennedy's intrusion into her friendship with Christina Macpherson? Kennedy, who remained solvent, may have failed to secure Mrs Grant's approval, but she did like his financially reckless brother-in-law, Lieutenant Evan Macpherson, with his fund of Jacobite stories.

Kennedy and Macpherson Part Company

The partnership between Kennedy and Macpherson ended in 1791, on the initiative of Kennedy. An explanation from Lovat sources gives credit to Kennedy's business acumen. 'Mr. Kennedy's sagacity having foreseen a Bankruptcy in the Lieutenant's affairs dissolved the co-partnery by a division on the lands.'[80] Broadly, the Lovat lands were divided between them on the basis that Kennedy took Auchteraw and other lands west of the River Oich and Macpherson took Cullachy and other lands east of the River Oich. However, much more was involved than the Lovat lands. Macpherson retained Crathiecroy and the grazings of Kylarchill in upper Strathspey, which belonged to the Duke of Gordon and Cluny Macpherson respectively.[81] How these came into the holding of the partnership in the first place is not known. It meant that their joint holding of land had

straddled the formidable Corrieyairack pass, which was still an important thoroughfare. Under the terms of the agreement dissolving the partnership, Lieutenant Macpherson undertook to pay a yearly rent of £247 5*s.* 0*d.* to Lovat, and the appropriate rents to the Duke of Gordon and Cluny Macpherson. For this purpose he named Lieutenant Colonel Duncan McPherson of Bleaton (near Blairgowrie), residing at Catlag, Hugh McPherson of Inverhall [Invertromie], and Captain Charles McPherson at Gordonhall, all in Badenoch, as his cautioners.[82] Kennedy seemed particularly anxious to protect himself against liability under the original joint tack for anything that Macpherson might owe to Lovat. Time was to show that he was right to be anxious.

There was some overlap between the subsequent affairs of the Rev. John Kennedy and those of Lieutenant Evan Macpherson, but this was limited and it would be sensible to consider them separately.

The Rev. John Kennedy, Tacksman of Auchteraw

The break up of the partnership in 1791 left Kennedy in possession of a substantial amount of land. It was described as 'the Lands of Little and Muckle Achindarroch, Auchterawmore and Auchterawbeg, Balmain and Carngoddy and two Auchindarrochs' on which there were forty-one subtenants.[83] Relatively little is known about Kennedy's affairs over the next few years, but they probably progressed fairly well. He had a fall from his horse on his way to Invergarry in the Spring or early summer of 1792, and Dr Donald McDonald of the Cranachan family, who practised in Fort Augustus, was called to treat him.[84]

In 1798 Kennedy wanted to visit Florida. He therefore arranged to sublet his lands to Angus McDonald, residing at Fersit 'for the unexpired term of the lease from Lovat'. He reserved to himself the possession of the dwelling house at Auchteraw and a quarter of the garden with access to peat, etc., and the right to melioration money of £200 8*s.* 0*d.* due at the expiry of the lease.[85] A deed drawn up in 1798 showed that the tack was sub-let to 'said Angus McDonald as principal with him Alexander Macdonald of Glenco as cautioner'. The rent was £135 a year, and there was an initial payment of £500 for sheep, with the balance to follow. Alexander Macdonell, Writer in Inverness, was anxious about the position of his father, who happened to be one of Kennedy's sub-tenants. In correspondence with him, Kennedy put his own construction on the arrangement, explaining that 'among other things I have Subsett this farm to Glenco and a ffolanen of his'.[86] The meaning of 'ffolanen' is not clear, but in this context it probably meant 'follower' or something to that effect. It was shown in Chapter 2 that when the Duke of Gordon had Alexander Macdonald of Glencoe removed from Fersit in 1798, there was indeed an Angus McDonald at Fersitriach.

Kennedy must have spent a very short time in Florida. By 1801 he

was back in Scotland and living in Glasgow. His name makes a fleeting appearance in the Post Office Directory for Glasgow for 1801–2 as 'Kennedy, Rev. John. Adelphi St. Hutchesontown'. He made a Trust Disposition and Settlement, signed on 2 October 1801, in which he was described as 'late of East Florida, now residing in Glasgow'.[87] He appointed as his trustees Alexander Macdonald Esq. of Glencoe, Major Charles McPherson, assistant Barrack Master General for Scotland, and Coll Macdonald Esq., WS. He must have had money invested because he instructed his trustees not to sell any of his stock in the funds of government for less than £70 for each £100 sooner than five years after his death. He made provisions for Elizabeth, his daughter by his first marriage; James and Hugh, his sons by his second marriage; and Jane, his daughter by his third marriage. There was no mention of Jean Macdonald, his third wife, which suggests that she had predeceased him.

John Kennedy died very soon after making that Trust Disposition, and his trustees then took over the management of his affairs. Lovat and his son had real or imagined troubles over this tack throughout the 1790s and into the early 1800s.[88] In 1803, Lovat's son, who had often acted for him, died, and by 1805 the old man resolved to end the tack, now of course administered by Kennedy's trustees. Robert Dundas, who handled Lovat's legal affairs in Edinburgh, had reservations about this. Heirs and sub-tenants were specifically mentioned in the lease. Dundas therefore argued that the present possessor, strictly speaking Angus McDonald formerly at Fersit, could claim the right to possession, having been a sub-tenant of Kennedy.[89] In spite of this, Lovat went ahead. In May 1806 Alexander Macdonell, Writer in Inverness, arranged on his behalf that John MacKay, Messenger at Fort Augustus, would execute letters of ejection against Angus McDonald and the forty-one sub-tenants. Because of the numbers involved and the scattered nature of the settlement, MacKay had to employ two Sheriff Officers along with their witnesses and other associates. While this was being done, a suspension was presented at the instance of Alexander Macdonald of Glencoe, Coll Macdonald WS and others.[90] Robert Dundas' views suggest that they may have had good grounds for challenging what was being done.

Lovat had found a new tenant, Thomas Clark, described as 'late tacksman of Drimandrochit'.[91] His wife was a daughter of John MacKay who served the ejection notices.[92] However, Lovat had still to get Kennedy's representatives out, and they were not disposed to go quietly or quickly. On 20 May 1806 Clark presented himself at Auchteraw with the melioration money in his hand, ready to take possession. The money was refused by the person then in possession, presumably Angus McDonald. No doubt he was doing as he had been instructed. Lovat then argued that Kennedy's heirs were 'keeping violent possession of Auchteraw after Mr. Clark's

term of entry'.[93] Clark had managed to agree terms with Angus McDonald concerning the stock which were acceptable to Kennedy's representatives.[94] Perhaps if they had been left to get on with it a settlement would have been achieved, but with Lovat and Alexander Macdonald of Glencoe in the background that would have been too much to expect. Complications were bound to emerge.

Clark raised a process of damages against Lovat in the Sheriff Court in Inverness, and this was later transferred by advocation to Edinburgh. The action was slept for a time, and in May 1809 a very puzzled Coll Macdonald WS, one of the Kennedy trustees, wrote to his colleague Alexander Macdonell in Inverness. He claimed that in spite of the agreement which Thomas Clark and Angus McDonald had reached, Lovat had insisted that Clark should bring this action against him. He believed that Lovat had done this 'to vex Kennedy's representatives and give a pretence for keeping the melioration money in his hands'. Macdonell had acted for Lovat during these proceedings, but Coll Macdonald hoped: 'I may now without impropriety ask what you know of this transaction and the motive of Lovat in insisting for it, for I perceive the papers are not wrote by you in the style you would have done for vindicating the cause of your client if you had not had some instruction to that effect.'[95] Unfortunately, Alexander Macdonell's reply has not been found.

The issue was probably not resolved when Alexander Macdonald of Glencoe died at the end of 1814. Lovat died a year later, at the end of 1815. James S. Robertson WS of Edinburgh, who had been acting for Thomas Clark, then wrote to Alexander Macdonell in Inverness, saying: 'Lovat's death will I presume put an end to these foolish actions which he raised. . . . I do not think Mr. Clark seems disposed to proceed further in the process with Kennedy's trustees.'[96] That seems to confirm that Clark had not been acting on his own initiative, and that Lovat was not only a difficult landlord but also a difficult client whose legal advisers must have found him a sore trial. With Glencoe and Lovat both out of the way, sensible men would no doubt reach a settlement quickly. Thomas Clark was still tacksman of Auchteraw when he died in 1851 at the age of 87.[97]

John Kennedy left four children from his three marriages. It is not certain what became of them. There are some clues in a Decreet of Removal which Lovat obtained at Inverness Sheriff Court in 1806. The Decreet was aimed at the trustees of John Kennedy and at Angus McDonald, who was refusing to budge from Auchteraw, but it was no doubt thought prudent to mention every individual who might have any standing in the matter. And so John Kennedy's sons Hugh and James were both said to be 'now or lately in the Island of Jamaica'. His daughter Jane was 'now at Duthil in the County of Elgin'.[98] His daughter Elizabeth is not mentioned, although he had made provision for her in his Settlement of 1801.[99]

Lieutenant Evan Macpherson, Tacksman of Cullachy

Lieutenant Macpherson's tenure of Cullachy lasted only four years after the partnership with the Rev. John Kennedy was dissolved in 1791. The story was one of developing financial problems ending in bankruptcy, but there were twists and turns involving many people.

Evan Macpherson was heavily dependent on borrowed money. In 1789 he obtained a bond of credit from the British Linen Company, allowing him to borrow up to £500 at any one time at the usual rate of interest. His cautioners for this were Colonel Duncan McPherson of Bleaton and Captain John MacPherson of Ballachroan.[100] There was perhaps a warning of difficulties as early as 1790. A note dated 5 April 1790, apparently intended for Cluny Macpherson, read: 'Lieut. Evan Macpherson of Cullochie wishes to have the accommodation of £500 sterling for three or four months, and proposes Colonel Duncan McPherson of Bleaton and Captain John MacPherson of Balchroan to join him in a Bill for that Amount.'[101]

By the middle of 1793, The British Linen Company had had enough of Evan Macpherson. On 3 August the manager, Walter Hog, wrote to Campbell Mackintosh, Writer in Inverness, saying, 'Captain Evan MacPherson has already had too much indulgence – I beg the Hornings may be returned Executed as soon as possible'.[102] By the following year the bond of credit was overdrawn at £546 14*s.* 2*d.*, which included £25 16*s.* 11*d.* unpaid interest.[103]

Evan Macpherson also had a bond of credit for the same amount, £500, from the Royal Bank of Scotland, using the same cautioners. By the early months of 1794 he was similarly overdrawn on that credit and William Simpson, the cashier of the Royal Bank of Scotland, moved to obtain an inhibition.[104] Haldane said that the years following the founding of the Royal Bank of Scotland in 1727 'saw a rapid expansion of the cash credit system by means of which any reputable person with two guarantors could get credit'.[105] Perhaps it became too easy to get credit.

The British Linen Company and the Royal Bank of Scotland were far from being the only creditors. Among others, Evan Macpherson had borrowed £1,000 in 1791 from the children of the deceased Major Alexander Duff, 89th Regiment of Foot. Interest on that was overdue, and the usual penalty provisions were invoked.[106]

All of his creditors were beginning to close in on him, but the most immediate pressure was from Lovat for the rent of Cullachy. Kennedy was drawn into this because although he and Evan Macpherson had split their partnership, their lands were still held on the joint tack of 1786 on which both names appeared. However, in splitting the partnership Evan Macpherson had named three cautioners for his rent to Lovat, and Kennedy could look to them for some protection. So at the beginning of

May 1794 the unfortunate Captain Charles McPherson was summoned by John Kennedy as one of these cautioners. He had to make the journey from his home at Gordonhall near Kingussie to Auchteraw. From there he wrote to Campbell Mackintosh on 3 May.

> Mr. Kennedy informs me that you have a Caption against him and Captain Evan upon a Draft for rents due to Lovat, which to my knowledge is the Debt of the latter Gentleman guaranteed to the other by his cautioners. I have this day sent an express to Evan urging him in the strongest manner to pay Lovat's rents and before he left home he directed his overseer in the King's Park at Edinr. to sell from his Sheep Stocking immediately as much as would discharge the rents . . . if you can defer putting Diligence in execution against Mr. Kennedy until the other gentleman has a little time to sell the sheep and settle the matter with Lovat it will oblige me and the other cautioners very much.[107]

This letter reveals that Evan Macpherson's sheep-farming interests extended to Edinburgh where, as will become clear, he had a substantial lease in King's Park which still had some years to run. Campbell Mackintosh responded promptly with letters on 6 and 14 May. Charles McPherson had gone back home to Gordonhall, and he wrote from there on 18 May acknowledging gratefully the respite which Campbell Mackintosh had secured from Lovat. He hoped that Mr Kennedy would cease threatening Captain Evan's cautioners for Lovat's past rents.[108]

Within a few days, Charles McPherson was back at Auchteraw about another debt, and he wrote again to Campbell Mackintosh on 23 May 1794.

> A similar business to that which brought me to this place about two weeks ago laid me under the necessity of crossing Corryarick yesterday – you are not unacquainted with my being one of Capt. McPherson Cullachy's cautioners upon a Bill due to Mr. Kennedy and Glenco £271 principal – which I am told with the Diligence and Interest amounts to £290 – Capt. McPherson being particularly desirous to get this debt paid as soon as possible, sent from Glasgow a joint power to Mr. Stewart, Barrack Master of Fort Augustus, and me to dispose of as many of his three year old wedders as would discharge it and we accordingly made a bargain this day with Mr. Mitchell Tullochrom, whose missive I enclose. . . . I shall write to Baillie Inglis [Agent of the British Linen Company in Inverness] from Gordonhall to-morrow requesting him to discount Mr. Mitchell's Bill . . . the sheep are to be immediately forwarded by Mr. Mitchell to an English market.[109]

Evan Macpherson had a printed statement prepared, dated 2 June 1794, to be placed before his creditors at a meeting in the Old Exchange Coffee

House, Edinburgh at 2 p.m. on 19 June. He explained that he held two leases, which he described in glowing terms: one at Cullachy, which had twelve years to run, and the other in the King's Park, Edinburgh, which had six years to run. His debts amounted to £7,614 16*s.* 4*d.*, and the value of his stock along with some debts due to him was £3,518 5*s.* 5*d.*[110]

On the day before his creditors were due to consider his affairs, Evan Macpherson arranged with his overseer to sell twelve score of his sheep to an Edinburgh butcher called Allen. The lawyers involved were uncertain about the status of a bargain made by an insolvent person when sequestration was pending, and some of his creditors moved for immediate sequestration. This caused Captain McPherson at Gordonhall to review anxiously the recent transaction in which he had been involved when, with Mr Stewart at Fort Augustus, he had sold some of Evan Macpherson's wedders. However, he was able to persuade himself that at that time Captain Evan 'had the strongest hopes of getting matters arranged without coming to the unfortunate pitch they have since done'. He was also able to recall that

> some months ago Capt. Evan was in terms with Glenco for as many of his sheep stocking as would pay up the Bill due to him and Kennedy, and in order to accommodate Evan, Glenco wrote to his Sheep Drover in the South to go to Cullachy to make the purchase, but other engagements interfered which put it out of his Drover's power to comply with the request.[111]

Captain Charles McPherson seems to have been the only one of Evan Macpherson's several cautioners to shoulder his problems. One of his cautioners, Captain John MacPherson of Ballachroan, had troubles of his own. Like his brother-in-law Evan, he also had a bond of credit with the British Linen Company for £500, and like Evan he was overdrawn.[112] When the Duff family tried to move against him concerning their loan of £1,000 to Evan Macpherson, they found that he had 'fled or absconded for his own personal safety'.[113]

It became impossible to avoid a roup of Evan Macpherson's sheep and of his sub-lease of Cullachy, Kyltrea, and various adjacent grazings. This was scheduled for 17 October 1794 in the Old Exchange Coffee House, Edinburgh.[114] The sheep on these farms were valued at £1,383 13*s.* 6*d.*, of which about £70-worth had already been sold.[115] Evan had not yet given up hope, and on 13 October he offered a composition to his creditors of six shillings in the pound. As his cautioners for this he produced John Archibald, whose address is given variously as Braidland or Braidlandhill, apparently near Paisley, and Archibald's father and brother. The creditors declined this offer because they did not have evidence of the sufficiency of the security offered. However, they agreed to postpone the sale and called a general meeting for 17 November. At this meeting the offer was again

rejected and it was agreed that the sale should proceed on the following day, 18 November.

The only offer, of the upset price of £2,000, came from Evan Macpherson and his new associate John Archibald. This was accepted, subject to the purchasers finding adequate caution within twenty days. Archibald promised that he would produce an attestation from Mr Hog, Cashier of the Paisley Bank, in whose area he lived. However, it was noted ominously that he remained in Edinburgh instead of returning to Paisley. The time limit expired, but on 15 December Mr Archibald submitted a bond of caution, signed by himself, James Archibald at Knockandow, and James Archibald in Cumbrae (apparently his father and brother), and a John Ewing. Mr Hog of the Paisley Bank seemed unwilling to give the kind of endorsement needed, and Archibald tried various other possible cautioners without achieving an adequate level of confidence. His next tactic was to try to block attempts by the trustee to reopen the sale, on the grounds that there was already a commitment to sell to him.[116] That did not succeed, but it took up some time.

In the spring of 1795 the possibility of another bid emerged. The potential bidders were two brothers, Thomas and Robert Oliver at Langburnshiels, and Thomas Stavert at Coliforthill, about seven miles and two miles respectively to the south of Hawick.[117] Robert Oliver and Thomas Stavert came to see Cullachy. They were also interested in Scothouse in Knoydart which was then on the market to rent[118] but, as will be shown in Chapter 10, Thomas Gillespie took Scothouse and placed his brother John there. The Olivers and Stavert decided to take Cullachy and paid £2,280 for the sheepstock and for the right to the remainder of the lease, also agreeing to pay a rent of £280 per annum. They were in occupation by Whitsunday 1795.[119]

The lands which Lieutenant Macpherson held from the Duke of Gordon and Cluny Macpherson in upper Strathspey appeared to be entirely occupied by sub-tenants. There is no indication that he farmed there himself. In May 1794 the Duke of Gordon obtained a Decreet of Removal against 'Lieutenant Evan McPherson at Cullachie, Principal Tacksman of Crathiecroy' and against numerous tenants there.[120] Later in the decade, Alexander MacDonald at nearby Garva was tacksman of both Crathiecroy and Kylarchill.[121] Finally, there must also have been action over Evan Macpherson's lease in King's Park, Edinburgh. That falls well outside the scope of this study and has not been pursued.

Thomas Stavert died shortly after the transfer of the lease,[122] but Robert Oliver was to be associated with Cullachy for many years. In 1802 Lovat granted a new tack to the two Oliver brothers, this time at a rent of £450. Thomas Oliver was then tenant of Shankend in the County of Roxburgh, a couple of miles or so to the north of their previous location at Lang-

burnshiels.[123] Thomas Gillespie was their cautioner. In 1809 Thomas Oliver, who was still at Shankend, renounced his interest in Cullachy in favour of his brother Robert. The relevant document was prepared by a lawyer in Hawick.[124] Thus, for fourteen years the Oliver brothers had kept a sheep-farm in the Borders as well as one in the Highlands, before splitting their interests.

Also in 1809 James MacBarnett, Tenant in Torlundy, owed £220 10*s.* 0*d.* arrears of rent to the Duke of Gordon. The Duke petitioned for sequestration of his effects to recover the arrears.[125] Robert offered a rent of £150 for MacBarnett's farm of Torlundy with the hill grazings of Inverlochy, and his offer was accepted.[126]

Lieutenant Macpherson joined the Gordon Highlanders, then the 100th Regiment, later renumbered the 92nd.[127] According to Stewart of Garth, who calls him Ewan rather than Evan, he joined in 1794 in his old rank of Lieutenant. In 1799 he transferred to the 17th Regiment, then to the Veterans, and retired with the rank of Lieut. Colonel.[128] References to him in various legal processes which continued over the years seem to confirm Stewart's information.[129] [130]

6

Glencoe in the Early Nineteenth Century

When the nineteenth century began, the Achtriachtan property was encumbered with the debts left by Angus Macdonald, which were detailed in Chapter 1. It was probably impossible for his son Adam to retain possession. The break-up and final disposal of the Achtriachtan lands was, however, to take about twenty years and was complicated by an unwise involvement in Strathconon in Ross-shire. Alexander Macdonald of Glencoe began the new century in a better state than his neighbour, but in the prevailing boom conditions he rashly committed himself to long leases at high rents. This placed increasing burdens upon his Glencoe estate as security for loans, and created severe problems for his trustees after his death in December 1814. These problems will be described in later chapters.

Adam Macdonald of Achtriachtan

Adam Macdonald inherited Achtriachtan from his father because his two older brothers had died. His mental condition complicated the management and eventual disposal of the estate. Charles Fraser-Mackintosh said that he was 'quite unfit to manage his patrimonial estate' and attributed the following quotation to someone who knew him: 'Mr. Macdonald has shown from his youth the most flexible, facile, and unresisting disposition, a mind the most unsuspicious, weak, and pliable, and a habit of life inconsistent with the ways of prudent men. In short, he showed himself the easy victim of designing men.'[1]

On two occasions, trustees were appointed to manage his affairs, and the second time Interdictors were also appointed.

In 1811, reference was made to the first set of trustees who had been appointed some years before.

> Adam Macdonald Esq., proprietor of the lands of Achtriachtan and Leckintuim, and Tacksman of the lands of Strone, from the state of his affairs, and a conviction of his own facility of disposition, some years ago divested himself of his property by a trust-deed in favour of Colonel John Cameron residing at Achnasaul, Patrick Cameron Esq. of Glenevis, Coll Macdonald Esq. of Dalness, WS, and Colin Mackenzie Esq. of

> Kilcoy. These Gentlemen accordingly assumed the management of his affairs and appointed Mr. Flyter, Writer at Fort William, as their factor.[2]

The words 'his own facility of disposition' point to doubt about his mental capacity.

Adam Macdonald's wife was Helen Cameron, eldest daughter of Ewen Cameron of Glennevis, and three of the four trustees were related to her. John Cameron was her uncle and Patrick Cameron was her eldest brother.[3] Colin Mackenzie of Kilcoy in the Black Isle was the husband of her sister Isabella.[4] Coll Macdonald owned the adjacent estate of Dalness and was a writer to the signet in Edinburgh. There was no representative of the Achtriachtan family among these trustees.

The trustees did not maintain control for long. By 1811 Adam Macdonald was managing his own affairs again, although Coll Macdonald WS remained influential as his legal adviser. In 1816 Adam's finances were in a chaotic state and two steps were then taken.

First, Adam executed a voluntary bond of interdiction in favour of Colin Mackenzie, James Murray Grant and James Grant WS. That imposed on him a legal constraint 'provided for those who, from weakness or profusion, are liable to imposition'. The interdicted person 'obliges himself to do no deed which may affect his estate, without the consent of certain persons . . . called interdictors'.[5] Their function was to protect him from his own actions if these might harm his interests.

The role of the interdictors was solely protective, and they were not responsible for the management of Adam's affairs. For that purpose he executed a trust deed nominating as his trustees Alexander Cameron Esq., Colonel of the 95th Regiment of the Rifle Corps, Colin Mackenzie Esq. of Kilcoy, James Murray Grant Esq. of Glenmoriston, and James Grant, WS. These trustees were given the usual authority over Adam Macdonald's affairs, including power to sell his property.[6] So Adam had three interdictors to prevent him from damaging his own interests and four trustees, who included his three interdictors, to manage his affairs.

Colin Mackenzie of Kilcoy was the only previous trustee who was appointed on the second occasion. James Murray Grant of Glenmoriston was married to Henrietta Cameron, another sister of Adam's wife, Helen Cameron of Glennevis. James Grant WS was the uncle of James Murray Grant and thus also a member of the Glenmoriston family.[7] Colonel Alexander Cameron, who was a trustee but not an interdictor, was described as a cousin of Adam Macdonald.[8] In 1814 he had taken the tack of Erracht after it had been given up by Alan Cameron, the founder of the 79th Cameron Highlanders.[9] It seems probable that he was the son of Donald Cameron born at Murlaggan and Helen, a daughter of Alexander Mac-

donald of Achtriachtan. Colonel Cameron later became a Major-General and was knighted.[10]

Thus control of Adam Macdonald's affairs was taken out of his own hands twice because of doubt about his capacity to manage them competently. There must surely be uncertainty about how fit the poor man was to look after them during the periods when they were in his own hands.

The Gradual Loss of Achtriachtan

When Adam Macdonald inherited Achtriachtan, the three loans his father had obtained in 1791, 1794 and 1796, amounting in all to £5,444 10*s.* 0*d.*, were still outstanding.

Adam's older brother Æneas left him with another problem. He had been factor for the Marquis of Tweeddale on his estates in Appin. After his death the Marquis pursued the unfortunate Adam, as the heir or representative of his brother, for large amounts of unaccounted rent – £68 2*s.* 8½*d.* for 1796, £250 5*s.* 2*d.* for 1797 and £1,337 8*s.* 11*d.* for 1798.[11] Proceedings in the Court of Session began in 1801 and dragged on for many years.[12] After arbitration, the claim was eventually settled for a more modest sum in 1820.[13]

Adam Macdonald also had a debt of his own. From his time in the West Indies he apparently owed a substantial sum to John Cameron of Berbice. This claim was also settled eventually, although not in full, in the Decreet of Preference of 1820.[14] [15]

In 1801 Adam Macdonald repaid the loan of £1,000 that his father had obtained from John Cameron of Camisky in 1794.[16] After that modestly hopeful beginning, he laid the foundations for more trouble in the future. In 1803 he became cautioner for a John MacCallum and his son Duncan who took a large tack in Strathconon in Ross-shire from Mackenzie of Fairburn at a rent of £450. It is impossible to discern why Adam Macdonald did this. Fraser-Mackintosh's anonymous commentator referred to 'debt, arising from his cautionary obligations, into which he had been artfully inveigled'.[17] Alexander Macdonald of Glencoe became involved in another of Fairburn's tacks in the same year. This also proved troublesome for him and his trustees. An account of these distant commitments and the complications that arose from them will be found in Chapter 11. For the moment it will suffice to note that Adam Macdonald was left with responsibility for the MacCallums' tack and a debt of £700 or £800.[18]

On 3 January 1809 Adam borrowed £900 from Roderick Mackenzie Esq. of Flowerburn, using Achtriachtan and Leacantuim as security. Coll Macdonald WS of Dalness was a party to this transaction and bound himself jointly with Adam Macdonald to pay the interest.[19] He may therefore have become personally involved in Adam Macdonald's affairs while

acting as his law agent. Mackenzie of Flowerburn was married to Anne, second daughter of John Grant of Glenmoriston.[20] [21]

In February 1809 the loan of £2,444 10*s.* 0*d.* that Adam's father and his brother Æneas had obtained in 1791 from the trustees and executors of Lieutenant Colonel Elphinstone was repaid. That was a substantial sum, but Coll Macdonald's name was again associated with the transaction, adding to the impression that he now had some stake in Adam's affairs.[22]

During these years, Achtriachtan and Leacantuim, which Adam Macdonald owned, and Strone, which he held from the Appin estate, were leased to Angus Rankin, tacksman of Dalness on the estate of Coll Macdonald WS of Dalness. Angus Rankin was frequently involved with both Alexander Macdonald of Glencoe and Adam Macdonald of Achtriachtan, but he also had extensive interests of his own.

In 1811 Angus Rankin explained to the Court of Session that some years earlier Adam Macdonald's trustees had given him a lease of Achtriachtan, Leacantuim and Strone. When the lease was about to expire he had been surprised to receive notice from Adam Macdonald himself telling him that it would not be renewed. To make matters worse it was to be terminated

> in a manner quite unusual in that part of the country, and which would go far to ruin the petitioner. By the custom of the country the flock of the outgoing tenant is uniformly taken off his hands, at a valuation, by the landlord or the incoming tenant. But an intention was avowed of depriving the petitioner of this accommodation. He was to be turned out without any pasture to put his stocking on, and without the means of disposing of it without great disadvantage.

Angus Rankin wanted the whole proceeding declared invalid because Adam Macdonald's affairs had been put into the hands of trustees. His petition was refused by the court and in 1812 John Stewart is identified as the tacksman of Achtriachtan.[23] [24] Perhaps Angus Rankin did not come out of it in the end as badly as he had feared. In 1812 he obtained upwards of £400 at the Killin market, chiefly from John Stewart, Achtriachtan.[25] That might have been payment or part payment for his stock.

In 1812 Coll Macdonald of Dalness bought almost half of Achtriachtan from Adam Macdonald, thus extending his own adjacent estate. The price was to be based on a division of the rent, and that in turn was to be decided by arbitration.[26] Fraser-Mackintosh was highly critical of Coll Macdonald's behaviour.[27] Perhaps it was difficult to combine the role of law agent with that of neighbouring proprietor.

In 1813 Adam Macdonald repaid to his brother-in-law, John MacLachlan of Aryhoulan, the £2,000 his father had borrowed in 1796.[28] That completed the repayment of the whole of the debt of £5,444 10*s.* 0*d.* on Achtriachtan

left by Angus Macdonald at his death. Adam had, however, borrowed £900 himself and sold almost half of Achtriachtan.

However, Adam Macdonald's affairs were getting into deeper trouble and, as mentioned above, interdictors and fresh trustees were appointed in 1816. These trustees immediately took formal possession of his land[29] and set about disposing of it. Robert Downie, a former merchant in Calcutta who purchased the Marquis of Tweeddale's lands in Appin in 1817,[30] [31] was willing to buy the remaining Achtriachtan property. The trustees agreed to sell it to him for £6,000 and were later accused of letting it go for too low a price.[32]

Meanwhile, in 1817 the trustees borrowed money to settle some pressing claims. The sum of £300 was borrowed from 'Mary Macdonald, relict of Allan Stewart sometime Tacksman of Kilisnacon [Caolasnacoan], and Robert Stewart late Captain Royal Marines their son'.[33] In another context Allan Stewart is described as tenant of Shuna, which helps to identify Mary Macdonald as one of Adam's sisters.[34] That loan was probably a noble effort. Allan Stewart had died in 1799 'in embarrassing circumstances' and left sixteen children. His wife took over his affairs, and in 1802 she had 200 ewes, eight cows and two horses belonging to her sub-tenants at Caolasnacoan sequestrated for non-payment of rent.[35] A further £1,000 was borrowed in 1817 from Colonel Alexander Cameron, one of the trustees.[36]

In 1818, the £900 borrowed by Adam Macdonald from Robert Mackenzie of Flowerburn in 1809 was repaid to his daughter Elizabeth, as he had died in 1812.[37] [38] The money borrowed by the trustees in 1817 was also repaid in 1818.[39] [40] These actions were criticised by Walter Ferrier WS, common agent for the creditors, because they gave preference to creditors favoured by the trustees, to the detriment of others.[41]

The sale of the Achtriachtan property ran into trouble because of Adam Macdonald's many debts. By 1820 Robert Downie had paid two instalments of £2,000 each. The final instalment of £2,000, on which some interest had by then accrued, was still in his hands. At this point Adam Macdonald's creditors obtained arrestment orders against the money he was holding. The whole tangled affair had to be disentangled by a process of multiplepoinding in the Court of Session.[42]

Adam Macdonald's extensive debts fell into three groups. The first and largest arose from his interests in Glencoe and thereabouts. His trustees appointed Thomas MacDonald, writer in Fort William, as their factor to deal with these interests. He was one of the Drimintorran family, who will be the subject of Chapter 7. A second group of debts involved Adam Macdonald's interests on the lands of Mackenzie of Fairburn in Strathconon. The trustees appointed Robert Logan, the agent of the British Linen Company in Inverness, as their factor to deal with these interests.

They will be described more fully Chapter 11. Third, there was Adam's own debt to John Cameron of Berbice.[43]

The trustees inserted a notice in the *Inverness Journal* of 6 March 1818 inviting Adam Macdonald's creditors to lodge their claims with Thomas MacDonald in Fort William. This brought forth an extensive list of claims, some of which had been outstanding for many years. There were relatively small amounts due to shopkeepers, the ferrymen at Ballachulish and Corran, John Taylor the Surgeon in Fort William, and so on. There were some more significant claims, such as the £500 tocher promised by Adam in 1807 when his sister Jessy married John Stevenson Esq. of Glenfeochan. Major Cameron of Kinlochleven claimed £55. The Glencoe trustees wanted £28 6*s.* 10*d.*, the balance due to them from John Ban MacDonald at Corrychurochan, for whom Adam Macdonald had been cautioner. John Ban had previously been in Knoydart; his position will be explained in Chapter 10. The executors of the Marquis of Tweeddale were still trying to recover money due to them from Adam's deceased brother Æneas. Arbitration reduced this considerably from the initial sum claimed.[44]

Some claims illustrate the way in which the Achtriachtan family had been living. Mrs Ann Campbell, who had been governess to the family and now lived at Loch Awe, claimed £5 10*s.* 0*d.* unpaid wages. The Rev. Donald MacNaughton, who had been tutor to the children, claimed wages of £15 per annum for five years from March 1813. He also claimed repayment of £180 17*s.* 8*d.* that he said that he had lent to Adam Macdonald.[45]

It fell to Thomas MacDonald as a factor for the trustees to deal with the family's living expenses. His accounts show that on 4 June 1817 he paid £386 8*s.* 3½*d.* rent arrears for the farm of Ardsheal where the Achtriachtan family had lived for some years. He also had to pay £100 18*s.* 10½*d.* for 'Taxes due & for which stock was poinded'. Two servant maids, Mary Munro and Isabella McLean, were paid wages of £2 each, the gardener got £1, and Alexander McIntyre, dancing master, received £5 2*s.* 0*d.*[46]

In 1817 the Achtriachtan family moved from Ardsheal to Leacantuim, which Lieutenant Donald Campbell had vacated.[47] A modest quantity of livestock was sold to the incoming tenant at Ardsheal, a Captain Cameron. On 1 June 1818 Thomas MacDonald 'Paid Mr. Stevenson, Merchant in Fort William, for furnishings to the family at Leckentuim – £48 18*s.* 5½*d.*'. This must have been a short-term move. On 31 March 1819, he 'Paid Ewen Cameron, master of the *Lady Ann*, for Removing Leckentuim furniture to Glennevis – £2 0*s.* 0*d.*'. He also noted receipt of £88 2*s.* 0*d.* from the Leacantuim roup, and listed those who had not yet paid. Glennevis, where the furniture went, was of course the home of Adam's wife's family.[48]

By the end of 1820 the Court had decided that the money still in Mr Downie's hands, less some expenses, was to be paid to John Cameron of

Berbice who had said that he would be willing to accept fifteen shillings in the pound to get the matter settled. In 1821, Downie finally became the owner of Leacantuim and that part of Achtriachtan which Coll Macdonald had not purchased.[49] The dismemberment of the Achtriachtan property was completed by Coll Macdonald, who acquired the salmon fishing rights 'upon the Water of Co and the Lake of Lochtriachtan'.[50]

Alexander Macdonald's Increasing Ambitions

In the first few years of the nineteenth century there was little change in Alexander Macdonald's position in Glencoe itself. The main thrust of his activities was elsewhere. Between 1802 and 1807 he took long leases on lands belonging to Mackintosh of Mackintosh, MacDonell of Glengarry, Cameron of Locheil, Mackenzie of Fairburn and Fraser of Lovat. These will be described in later chapters.

In 1804 he prepared a trust disposition, naming as his trustees his wife Mary, her father Ewen Cameron of Fassifern, her brothers Major John Cameron of the 92nd Regiment, Duncan Cameron WS, and Peter Cameron, and the husbands of her sisters Jean and Catherine, Roderick McNeil of Barra and Colonel Duncan Macpherson of Cluny. Along with this collection of Camerons of Fassifern and their in-laws there was one Macdonald of Glencoe, Donald, his father's brother who was then Lieutenant of Invalids at Fort George and later lived at Invercoe. Finally, there were two trustees not apparently related to either family: Major Angus Cameron of Kinlochleven, and Colonel Alexander Macdonell of Glengarry.[51] Duncan Cameron WS was to be the dominant figure responsible for the work of the trustees after Alexander Macdonald's death.

The great expansion on which he had just embarked no doubt influenced the intentions expressed in his trust disposition in 1804. These were based on the assumption, which was to prove totally unrealistic, that he would be worth 'about £10,000 at my death'. His trustees were to keep this sum 'intact until they can invest in land as near to the present estate as possible'. His ambition was therefore to add more land to his small estate of Glencoe. Presumably he believed that this could be financed out of the profits from sheep-farming. The state of the market then might have seemed to justify that belief. Wool prices were going up and up and sheep-farmers were vying with one another to get more land at almost any price. He had just leased several more farms, and the large investment in sheep-farming in Sutherland was about to begin. In this heady atmosphere, could he have been expected to foresee that such a hectic boom was about to be replaced by a correspondingly deep slump?

The figure of £10,000 was probably in line with the expectations of others who, like him, had invested in sheep-farming. Allan Cameron, tacksman of Inverscaddle on the west side of Loch Linnhe, said shortly before

his death in 1810 that he would be worth at least £6,000 after his debts were paid.[52] His trustees were able to lend £5,000 to Alexander MacLean of Ardgour, which suggests that his prediction had been fairly accurate.[53]

Alexander Macdonald's Increased Borrowing

Borrowed money had always featured in Alexander Macdonald's affairs. The loan of £450 from his brother-in-law, the Rev. John Kennedy of Auchteraw, in the late 1780s has been mentioned in Chapters 1 and 5. When his cousin John MacDonald of Drimintorran married, probably about 1798, he lodged £1,000 with him. The liferent was to go to John's wife, and the fee to pass to his children.[54] This was in effect money put at the disposal of Alexander Macdonald. More will be said about this money in Chapter 7.

There were several other loans outstanding at the time of his death. Captain Donald Macdonald, his uncle who lived at Invercoe, had lent him £800, and Angus Kennedy of Leanachan, a well-known sheep-farmer in Lochaber, had lent him £500. George MacKenzie of Dundonald had lent £1,000.[55]

In 1809 and 1810 Alexander Macdonald obtained three large loans, suggesting that he was coming under some pressure. In 1809 he got £2,000, his largest single loan, from Cluny Macpherson[56] and £1,000 from Miss Isabella MacLeod of Bernera.[57] Then in 1810 he borrowed £1,000 from the Rev. Alexander Rose, a distinguished minister of the Church of Scotland in Inverness.[58] Finally, a loan of £425 that his brother-in-law, Duncan Cameron WS, arranged for him in 1813 was probably intended to help in meeting immediate difficulties. The lender was Mrs Catherine Cameron, the second wife of Ewen Cameron of Fassifern, and stepmother of Duncan Cameron.[59]

Alexander Macdonald of Glencoe, like his recently deceased neighbour Angus Macdonald of Achtriachtan, was burdening his Glencoe property with heavy debts that would remain to trouble others. The problems faced by his trustees in the years following his death will be described in Chapter 13.

The Family of Alexander Macdonald of Glencoe and Mary Cameron

According to one authority, Alexander Macdonald of Glencoe and Mary Cameron had six sons – John, Ewen, Colin, Alexander, Ronald and Alister – and two daughters, Louisa and Jane.[60] According to another, their family consisted of Ewen, Ronald, John and Jane.[61] Neither author gives sources, but other evidence tends to support the first version. John must have died between 1807 and 1814.[62] [63] The death of Louisa was intimated in the *Inverness Journal* of 9 October 1812, but nothing is known of Alexander and Alister. When Alexander Macdonald died in December

1814 he was survived by three sons – Ewen, who was then his heir, Colin and Ronald – and one daughter, Jane Cameron.

Ewen, the eldest surviving son, was born on 12 July 1788. He joined the service of the East India Company as an assistant surgeon on 24 July 1809 and served in the Bengal Establishment throughout his career.[64] On arrival in India he was posted briefly to the general hospital at Berhampore and then became assistant surgeon with the First Battalion of the Regiment of Artillery. In 1811 attention turned to the Dutch possession of Java because of fear that Napoleon might use it to establish control of the sea routes to China. Ewen was there with his regiment during the campaign to capture the island. By 1813 he had been posted as assistant surgeon with the 24th Regiment of Native Infantry. He remained with them for about ten years.[65] [66]

Colin, the second of the surviving sons, was born on 12 August 1789. He became a cadet in the service of the East India Company in 1804.[67] One of his Fassifern uncles, Major John Cameron, obtained the cadetship for him through 'the means of the Paymaster who discovered that an old acquaintance of his and mine had it in his power and procured it for me'. In June 1805 the major again gave attention to Colin's affairs and was with him in London. On 22 June 1805 he told his father, Ewen Cameron, that he had 'got him passed at the India House and arranged everything for him, leaving him in very good hands, he will sail in about 10 days'. He described Colin as 'an uncommon fine lively spirited boy and I do think he will do extremely well'.[68] Colin reached Bengal on 16 May 1806, having been appointed Ensign just before his arrival. He was posted to the 18th Regiment of Native Infantry, and remained with them throughout his service. He was promoted to lieutenant in April 1809.[69] [70] He may have done well, but his service in India was marked by a series of personal tragedies, culminating in his death at Cuttack on 29 April 1818.[71]

Colin returned to the United Kingdom on furlough in January 1815. He consulted John Taylor, surgeon at Fort William, whose report describes his early misfortunes in great detail. In the storming of Bowanie in 1809, a musket ball had entered his right leg. It remained lodged in the upper part of the leg, giving him acute pain on rapid movement and limiting the function of the leg. Worse was to come. In August 1811, at the storming of the lines of Cornelius on the Island of Java, he was blown up in a redoubt and found among the rubbish totally insensible, severely wounded and with his skull fractured in several places. 'After struggling through a six-month confinement and nearly two years' subsequent suffering, he began slowly to recover, and when his strength was found adequate to the task he was ordered to embark the shattered remains of his constitution for Europe.'

It was John Taylor's decided opinion

> that this officer's health will always remain in the most precarious state, utterly unfit for the action of the climate of India; his hearing also will permanently continue much impaired and he will never recover the function of his left limbs; without taking notice of the very serious inconvenience he suffers from the Bullet remaining lodged in his right leg.[72]

But he did return to India.

The youngest son, Ronald, was born on 31 December 1800.[73] Nothing is known about Ronald up to the time of his father's death in December 1814. Similarly, nothing is known about Jane's life before her father's death. After that both appear in the records of the Glencoe trustees. Their later lives will be considered in Chapter 13.

Glencoe after the Death of Alexander Macdonald

When Alexander Macdonald of Glencoe died at the end of 1814, the trustees whom he had nominated in 1804 had to assume the management of his affairs. They found that he held the leases of Strone and Achnacon that had previously been held by the Achtriachtan family. Presumably he had taken over these leases when Adam Macdonald was in difficulties. Angus Rankin and some of his relatives were still there as tenants. Alexander Macdonald also had the lease of the adjacent farm of Innerigan which must have been held by the Glencoe family for a long time. The Glencoe trustees had then to collect rent from the occupants of all three farms and pay £200 at half-yearly intervals to Robert Downie of Appin.[74] Problems with the tenants of these farms emerged very quickly.

Innerigan had traditionally been occupied by a family of Macdonalds who are invariably mentioned in accounts of the Massacre of 1692. Several members of this family moved away from Glencoe in the early years of the nineteenth century and became sub-tenants of both Alexander Macdonald of Glencoe and Adam Macdonald of Achtriachtan in Glenstrathfarrar, the Forest of Monar and Strathconon. Their misfortunes there will be described in Chapter 11. In March 1815 a man called D. Kennedy renounced the tenancy of Innerigan.[75] That suggests that all the Innerigan Macdonalds had left their traditional location some time before that.

The trustees had to find a new tenant for Innerigan and accepted an offer from Donald McMaster. He was to survive at Innerigan for quite a long time, in spite of recurring difficulties in meeting his commitments. In January 1817 the trustees told Inveraray Sheriff Court that McMaster owed them £72 10*s.* 0*d.*, the rent for a half year, and that they 'were creditably informed . . . that the said Donald McMaster is in bad circumstances'. They got authority to sequestrate his stock.[76] However, in July

1817 the trustees received £46 9*s.* 0*d.* 'by cash from Mr. Allan Cameron, Meoble, being the amount of Innerigan Roup Roll and for rents due by McMaster'. That saved McMaster then, but in February 1818 Allan Cameron had to step in again, making a payment 'to account of Innerigan rent' of £16 12*s.* 0*d.*[77] Cameron of Meoble was an ambitious and thrusting sheep-farmer in South Morar, identified by Charles Fraser-Mackintosh as a MacMartin-Cameron whose immediate family hailed from Glengloy.[78] More will be said about him in Chapters 9 and 10, but his reason for supporting Donald McMaster is not clear. At Whitsunday 1818 the lease held by the Glencoe trustees from the Appin estate ended.[79] Robert Downie then became McMaster's landlord and continued to have problems in getting the rent paid.[80]

There were more problems for the Glencoe trustees at Strone and Achnacon. At their third meeting in January 1817 it was reported that 'Angus Rankin, Tacksman of Dalness, and brothers Donald and John, wish to be relieved at Whitsunday next of the tenement of Achnacon'. Strone was probably included, although not mentioned separately. The wish to be relieved of these farms was not surprising in view of the financial difficulties then facing the Rankins on several fronts. However, the trustees thought that although Mr Downie would probably be willing to accept a renunciation it would not be in the interests of the tenants to leave at the stock prices prevailing then. They noted that Mr Downie would be obliged to receive all of these farms on the expiry of the lease at Whitsunday 1818, and they reached a modified agreement with the Rankins to remain. The concern of the trustees for 'the interests of the tenants' seems touching, but they may have had an ulterior motive. On 20 April 1817 they paid a fee of £5 16*s.* 1*d.* for the legal work involved in 'sequestration agt. Rankins of Barnamuck and Achnacone as charged agt. them'.[81] Perhaps they wanted to encourage the Rankins to keep their stock on the farms to give them a chance to get their hands on it. It was quite common for the stock of tenants in difficulties to disappear before their landlords could seize it. The Rankin involvement in Barnamuck is explained in Chapters 5 and 12.

At their meeting in January 1817, the trustees also 'considered the state of the country' and 'authorised the factor to allow the tenants of Carnoch, Tayfuirst and the Milne Croft a gross deduction of £40'. This was to be divided in proportion to the rent paid.[82] There was great public distress and discontent in the Highlands at this time, and several proprietors were lowering rents by as much as 15 to 20 per cent.[83] [84] The trustees were following the trend.

The trustee accounts show that on 2 February 1817 there was a payment of £33 1*s.* 0*d.* 'by Balachulish to Drimintorran to acct. of rent due by quarriers at Martinmas'.[85] 'Balachulish' was of course Stewart of Ballachulish, who

then owned the slate quarries. 'Drimintorran' was Donald MacDonald of Drimintorran, who was appointed by the trustees as their factor in that year.[86] Many tenants in Glencoe were working at the slate quarries in Ballachulish and their rents were collected in bulk from their employer. Similar collections are recorded in later years.[87]

On 25 November 1820 Duncan Cameron was 'drawing advertisement of farms in Glencoe to be let'.[88] One of those farms was probably Gleann-leac-na-muidhe. Although it was next to the Appin properties of Strone and Achnacon, it seems to have been owned by the Macdonalds of Glencoe. On 6 June 1821 the trustees' factor, Donald MacDonald of Drimintorran, reported 'McPhie put in possession of Glenlecknamuy he declined to give a bill of security for the Rankins' arrears of rent'.[89] Landlords seemed to expect that new tenants would undertake to pay the debts left by their predecessors. Perhaps when times were good they could get away with that, but times were not good in 1821. Donald McPhee was paying a rent of £140 for Gleann-leac-na-muidhe.[90]

On 13 November 1821, Robert Scott, Duncan Cameron's partner who was dealing with most of the day-to-day Glencoe business, wrote a long letter to Ewen Macdonald in India. It was addressed c/o Mackintosh & Co., Agents, Calcutta, to be forwarded to Ewen. Robert Scott said: 'I prepare this for going with your Uncle's Ship, The Balcarres, for Bengal and China, at the beginning of next month.' The uncle was of course Peter Cameron of Fassifern, and the letter was a comprehensive report on the current state of the affairs of the trustees. Robert Scott told Ewen that by November 1821 most of his father's interests outside Glencoe had either been relinquished or would be brought to a conclusion fairly soon as leases ended or arbitration was completed.[91] That must have been a great relief to the surviving trustees and to Ewen. It meant the only estate that had now to be administered was the traditional Macdonald land in Glencoe itself. The massive debts on that estate of course remained; this will be considered in more detail in Chapter 13.

Towards the end of his letter Robert Scott wrote:

> The earlier markets this season for Highland produce were particularly unfavourable and must have a serious effect on all money payments at this term. Glengarry I am told is to give a deduction of 25 per cent to all his tenants. Locheil again gives this year 15 per cent to his crofters; but then in 1816 and 1817 he gave 20 per cent to all and in the same way 10 per cent at Martinmas 1818.

If Robert Scott was trying to prepare Ewen to concede deductions in Glencoe, he may have spoiled his case by adding that 'the later markets have been considerably better'.[92]

Many of the accounts of the trustees for the later years of their

administration of the Glencoe estate have not survived. There is, however, a record in 1823 showing that they had to sequestrate the effects of several tenants in Glencoe. One of those was Malcolm Robertson of Tayfuirst. On 23 November 1820 Duncan Cameron had a meeting with him about his lease, consisting of a house and shop at Tayfuirst and a croft and shealing, at a yearly rent of £36 5*s.* 4*d.*[93] [94] When times were bad, shopkeepers and merchants were particularly vulnerable, and by 1823 Malcolm Robertson had fallen behind with his rent. Because Glencoe was in the county of Argyll the sequestration was handled in the Sheriff Court at Inveraray. The trustee appointed over Malcolm Robertson's sequestrated estate applied to the sheriff at Inveraray to fix diets for the examination of Robertson's affairs. These were to be on 4 July for the first examination and 19 July for the second. The unfortunate Robertson was, however, unable to attend on these dates because of sickness, and a surgeon's certificate was submitted to that effect. In the circumstances the sheriff agreed that the examinations could be carried out locally before a justice of the peace.[95]

Another entry at this period shows that on 25 March 1824 the trustees 'Received from Drimintorran to account of his intromissions, being the content of two Bills by the Manager of the Balahulish Slate Works – £116 15*s.* 7*d*'.[96] This would no doubt be rent money again paid on behalf of Glencoe tenants who worked at the quarries.

Initial moves were made by the trustees in 1828 to transfer the estate into the hands of Ewen Macdonald. The formalities were not completed until 1831. Ewen remained in India until the latter part of 1835, so the estate had still to be administered for him. In April 1834, Donald MacDonald of Drimintorran, who was apparently still acting as factor, reported that he expected that the rent arrears for 1833 would be £234 11*s.* 0*d.*[97] These were substantial arrears in an estate that had a total rental of only £578 13*s.* 4*d.* in 1821/22.[98] Donald MacDonald went on to make a fairly frank comment about Ewen, who was by then in full possession of the estate.

> It is now to be taken into consideration that a prospect of a deduction has always been held out to these poor people for they were distinctly told that when Glencoe came home or when he was heard from they would in this respect be dealt with like their neighbours. In this view of matters Glencoe has actually pocketed more money from his estate than he would have done had the common deductions been given to his tenants.[99]

That was quite a strong comment, and not couched in the respectful language generally used when referring to the affairs of Highland proprietors.

By 1836 Ewen was back from India and the rents were being collected by

James Macgregor, Solicitor in Fort William. Donald MacDonald of Drimintorran must by then have been increasingly involved in his own financial problems, described in Chapter 7, which were to lead him into bankruptcy in 1837. James Macgregor wrote on 7 January 1837 to Ewen Macdonald, who was then in Cheltenham with his uncle, Sir Duncan Cameron. They were said to have gone there for health reasons. He was at pains to bring to Ewen's attention the unfortunate position of the Glencoe tenants. He reported that he had placed money to Ewen's credit at the British Linen Company's Bank in Fort William, going on to say:

> None of the tenants came near me and on enquiry I found that it was impossible to get money from them without going down to the Slate quarry and there effect an arrangement between them and Messrs. Stuart. I accordingly took a ride down last week and received from Messrs. Stuart payment of the rents of all in their service, and in most cases I observed that these good men advanced the rents without being indebted. . . . The people expressed considerable anxiety respecting a message from you delivered by the manservant at Inverco, to the effect that you wished them to perform some work at Draining etc. at Inverco. They stated that they would willingly do so if they could absent themselves from the quarryworks, but that from the great demand for slates this season the Messrs. Stuart cannot spare them a day without taking in new hands in their places, and that if this became done it may not be possible for them to get in again, which would be a great loss to them as well as to you.[100]

James Macgregor was quite diffident in his approach to Ewen Macdonald. Nevertheless, he seemed to feel, like Drimintorran a few years earlier, that Ewen was not fully aware of the subtleties of managing a Highland estate. Perhaps that was understandable: he had spent almost thirty years in India. He may well have felt aggrieved that his father had left him with such a financial mess, and he was quite seriously ill for the last few years of his life. What James Macgregor was trying to tell him was that his rents depended on his tenants earning cash wages at the nearby slate quarries. Around this time the village of Ballachulish had some of the characteristics of a company town and accommodated about three-quarters of the workers.[101] As the owner of the adjacent property, Ewen Macdonald had become, in part at least, a landlord providing housing for most of the remaining workers. For example, Carnoch was no longer a farm, as it had been in his father's time.[102] It was now in multiple occcupation, providing homes for twnety-six tenants. Their rents ranged from £3 16*s.* 0*d.* to £15 4*s.* 0*d.*, but most of them were paying £7 12*s.* 0*d.* Carnoch produced a total rent of £201 8*s.* 0*d.*[103]

Nevertheless, most of the land in Glencoe was still in agricultural use,

and by 1836 the famous John Cameron of Corriechoille in Lochaber was occupying the farms of Croft and Mill Croft. He was proving to be a little difficult because he was refusing to pay the arrears of a previous tenant, but James Macgregor hoped that 'Drimintorran may have some document to show that he is liable'.[104] Charles Fraser-Mackintosh described Corriechoille as 'a shrewd and careful shepherd', better known initially as a drover, who was able to profit from the downfall of 'great monopolists' such as Alexander Macdonald of Glencoe and become 'one of the most noted men of his day'.[105] There is a good account of him by Joseph Mitchell, who said that he was 'at the height of his influence about 1840'.[106]

Ewen Macdonald's tenure of the estate of Glencoe ceased with his death on 19 August 1840.[107]

7

The MacDonalds of Drimintorran, of the Family of Glencoe

The MacDonalds of Drimintorran, near Strontian on Loch Sunart, became commercial sheep-farmers on a large scale. They were undoubtedly related to the Glencoe family, and treated Alexander Macdonald of Glencoe as a trusted senior relation to whom they turned at critical times. That trust was reciprocated. In 1817 Colin Macdonald, Alexander Macdonald's second surviving son, recommended to the Glencoe trustees that they should appoint Donald MacDonald, tacksman of Drimintorran, as their factor.

Relationship to Glencoe Family

The founder of the family was Donald MacDonald, described by his descendants as 'Donald MacDonald of Drimintorran, of the family of Glencoe'. He was born in 1730, and was with the Glencoe men in the 1745 Rising. After a few years in France, he returned to Scotland in 1752, and died in 1802.[1] [2]

Donald's son John, by his first marriage, was stated in Court of Session proceedings to be a cousin german of Alexander Macdonald of Glencoe.[3] That suggests that Donald was a brother of John Macdonald of Glencoe, and therefore a son of the Alexander of Glencoe who led his clansmen in 1715 and 1745. Alexander, however, had a son Donald born about 1738 who lived latterly at Invercoe and died there in 1821.[4] Could he have been the father of another Donald, born a few years earlier?

When Ewen Macdonald of Glencoe made his disposition and deed of entail in August 1837 he named his daughter as his heir. As alternatives, if that succession failed, he named his surviving brother Ronald and then the descendants of the Donald Macdonald who had died at Invercoe in 1821. Failing all of these, the succession was to go to his mother's family. He did not mention the Drimintorran MacDonalds.[5] If their relationship had been a legitimate one through the male line they would surely have ranked somewhere ahead of his mother's relations. The explanation for their omission may be that Donald MacDonald at Drimintorran had been an illegitimate son of the Alexander Macdonald of the 1715 and 1745 Risings.

In December 1776 Angus Macdonald of Achtriachtan had referred to an uncle of young Glencoe called Donald who was about to leave a farm at

Kingairloch.[6] That raises an interesting possibility, mentioned briefly in Chapter 2, that this might have been Donald MacDonald who was later at Drimintorran. That must remain speculative. Much of the later history of the family can, however, be pieced together with greater certainty. The original Donald MacDonald at Drimintorran was married twice and had two families. His first wife, Flora, is believed to have been distantly related to the Macdonalds of Glencoe. There was a large family from that marriage,[7] but only a few were identified during the present study. Two sons from that first family, John and James, were both involved in farming during their father's lifetime.[8] Donald's second wife, also Flora, was a sister of Duncan Campbell, factor on the Ardgour estate.[9] [10] They were married in 1788 and there were three children by that marriage, Donald, Thomas and Jane.[11]

The Death of John MacDonald at Drimintorran (1800)

John MacDonald, a son of Donald by his first marriage, died in 1800. John and his father had equal shares in Drimintorran. John and his brother James had equal shares in Ockle on the north coast of the Ardnamurchan peninsula, some distance from Drimintorran. After John died, James bought his half share in the stock at Ockle for £230.[12] The total stock there must therefore have been worth £460. In 1802 the stock at Drimintorran consisted of 1460 ewes, 113 wedders, 65 tups and 315 hogs.[13] Half the stock there had belonged to John. These few facts show that the family was prospering and its members had spread their wings beyond Drimintorran itself. There are other indications that they were in a comfortable situation, and indeed it was claimed later that John had been 'a man of considerable substance'.[14]

John's wife was Elizabeth, a sister of Adam Macdonald of Achtriachtan.[15] She was still unmarried in 1796,[16] and as John died in 1800 the marriage must have been of short duration. They had a son, Angus, born about 1799. At the time of his marriage, John lodged £1,000 with Alexander Macdonald of Glencoe, the liferent to go to his wife and the fee to pass to his children.[17] That confirms John's wealth and shows the trust that the Drimintorran family placed in Alexander Macdonald of Glencoe. The £1,000 was no doubt used to finance Alexander Macdonald's sheep-farming. After he died, his trustees paid interest regularly to John's widow until they eventually repaid the capital sum in 1822.[18] The few surviving papers belonging to John include an invoice in May 1800 from a merchant in Greenock for two dozen bottles of port, one dozen of sherry and five dozen London porter. This adds to the impression of a family that was prospering. For John's funeral, his father paid £12 for whisky and bought six stones each of oatmeal and flour.[19] John left no will, and his widow, his father and his brother James ('his only Brother then in this Kingdom')

asked Alexander Macdonald of Glencoe, Angus Cameron of Kinlochleven, Adam Macdonald of Achtriachtan, John MacLachlan of Aryhoulan and Alexander MacDonald of Dalilea to be tutors and curators for John's only son, Angus. Alexander Macdonald of Glencoe was John's cousin german,[20] Adam Macdonald of Achtriachtan was the brother of his widow Elizabeth, and John MacLachlan of Aryhoulan was married to his widow's sister, Isabella Macdonald of Achtriachtan.[21] Kinlochleven and Achtriachtan 'being at a considerable distance declined to act'.[22] Alexander MacDonald of Dalilea was an agent for the Leith Bank and lived in Callander, which placed him at a disadvantage, but he did not withdraw.[23] Alexander Macdonald of Glencoe was thus able to dominate the management of the estate, with MacLachlan of Aryhoulan the only other tutor and curator in the vicinity. He set about the task in his usual fashion, failing completely to maintain any useful records. That was to be troublesome for Dalilea, who explained later that 'it was his [Dalilea's] wish that all the actings of the Trustees should be entered into a book kept for that purpose. . . . But he found it impossible to carry into execution this plan, his colleagues having no idea of keeping regular accounts.' He was quite specific about Alexander Macdonald – 'Glenco who had much of the subject in his hands . . . and who I could never get to a regular settlement'.[24] No wonder Dalilea the banker showed signs of despair, but that was the style of many of that generation of Highlanders.

John MacDonald at Drimintorran had behaved similarly. Although he had bought and sold livestock extensively, it was found after his death that he had kept no regular accounts and very few vouchers. 'In short, his affairs appeared to be in the utmost state of confusion.' It was said that 'every poor man and woman in the County [placed] their money in his hands with the utmost confidence in his integrity without any written acknowledgement'.[25]

Stewart of Garth confirmed that business was commonly done in the Highlands in this casual way. He observed that 'In the common transactions of the people, written obligations were seldom required, and . . . there were few instances of a failure in, or denial of their engagements.' He regretted that 'new regulations, new views . . . and the novel practice of letting land to the highest bidder, regardless of the fidelity and punctual payment of old occupiers' had brought about 'a melancholy change'.[26] Stricter accounting was now needed. When Stewart took control of the debt-ridden Garth estate from his brother in 1820 he said, 'my brother kept no books – not even a Rent roll'. Large sums had remained uncollected for long periods, and Stewart referred bluntly to his brother's conduct as 'mismanagement'.[27]

The Death of Donald MacDonald of Drimintorran (1802)

Old Donald died in 1802, a couple of years after his son John. Alexander Macdonald of Glencoe was again brought into the affairs of the Drimintorran Family. This time, he and Duncan Campbell, brother of Donald's second wife Flora, were appointed tutors and curators for the three children of the second marriage.[28] They were all quite young, Donald having been born in 1791, Thomas in 1792 and Jane in 1795.[29]

The children's mother, Flora Campbell, and their stepbrother James managed Drimintorran, which carried a substantial stock at the time of old Donald's death.[30] They got into difficulties that were later blamed, perhaps conveniently, on James.[31] In 1808 the landlord, Sir James Riddell and his trustees, petitioned the Sheriff Court at Inveraray for the sequestration of sufficient stock at Drimintorran to cover unpaid rent of £332. They included the customary statement that there was reason to believe that the tenants were in bad circumstances.[32] At this time young Donald was employed in Fort William at the office of Mr Robert Flyter, writer and sheriff substitute. His duties took him as a witness to the Sheriff Court in Inverness in 1807, where he was described as 'Clerk to Mr. Robert Flyter, aged about 15 years'.[33] In 1809, as Donald explained later, 'the Trustees of the late John MacDonald to whom half the stock belonged (although managed by James) gave up their share which by the advice of Glencoe and Mr. Campbell who were the tutors appointed by my father, I took and accordingly . . . entered upon the management'.[34] Donald was silent about the position of John's widow and young son Angus. When Angus reached adult years he was to make an issue of this.

While Donald took over the family concerns, his younger brother Thomas remained with the law. He became a respected writer in Fort William and agent there for the National Bank of Scotland. After Alexander Macdonald of Glencoe died, the Glencoe trustees entrusted to him much of the legal work which had to be done for them in and around Lochaber.[35] In 1819 he was offered the post of Collector of Customs at Fort William, but declined to accept it.[36]

Donald MacDonald of Drimintorran, Farmer and Wool Merchant

Several problems surfaced when young Donald took over Drimintorran. Some of the children of his father's first family made a claim against his estate. Not much is known about this. An action was raised in the Court of Session against Alexander Macdonald of Glencoe and Duncan Campbell, as curators of the second family, by Ronald MacDonald, lately residing in Glasgow, and Flora MacDonald, spouse of John Stevenson, Merchant in Oban. The case was delayed for a time because of the illness of the Lord Ordinary, but details are missing.[37] Donald said many years later that

some payment did have to be made to them, which reduced the amount available to him and the others in the second family. That was not all. His maternal uncle, Duncan Campbell, failed and the family lost 'a considerable sum of securities we were in for Mr. Campbell'. Finally, arrears of rent incurred from 1803 until 1809, while his mother and stepbrother James were managing Drimintorran, had to be paid.[38] However, young Donald managed to survive these difficulties and become a highly successful sheep-farmer and wool merchant for almost thirty years.

In 1814 Donald married, but his wife lived for only a short time. Her identity is not known, but he and a cousin of hers succeeded to her property, and after some legal dispute Donald benefited to the extent of £400. He was advised to purchase wool on commission, which he did, reckoning that he made about £400 profit by 1822.[39] In 1817, when he was proving himself to be a capable farmer and man of business, the Glencoe trustees appointed him as their factor. They did this on the recommendation of Colin Macdonald, the second surviving son of Alexander Macdonald of Glencoe, who was home from India on a long furlough.[40]

On 9 October 1820 Donald married again. His second wife was Ann, the daughter of Adam Cummings of Pallinsburn, close to Flodden in the northern tip of Northumberland. Over the years they had a family of eight children.[41]

Following his second marriage, Donald expanded his farming interests. At Whitsunday 1821 he took over the farm of Ariundle, about a mile further up and on the other side of the Strontian River from Drimintorran. The two farms were managed jointly. Inversanda on the shores of Loch Linnhe followed, and was managed by what he called the 'flying stock principle'.[42]

The Claim of Angus MacDonald

About this time, John's son Angus, now a young man in his early twenties, was becoming restive about his father's assets, claiming that insufficient had come his way. His main target was Alexander MacDonald of Dalilea, one of those who had in 1800 accepted responsibility as tutors and curators for young Angus, who was then an infant. As already mentioned, some of those nominated had declined to act. Others, such as Alexander Macdonald of Glencoe, were by now dead. Angus was studying law in Edinburgh and was an apprentice with the firm of Cameron, Scott and Arnott, who were handling the business of the Glencoe trustees. His action against Alexander MacDonald of Dalilea in the Court of Session appears to have commenced in 1825.[43]

Angus fastened on items such as the stock at both Drimintorran and Ockle. Half of that stock had belonged to his father at the time of his death. He was also able to say that his uncle James had intromitted with

his father's affairs and produced no accounts. He also mentioned in his claim the £1,000 which his father had placed in the hands of Alexander Macdonald of Glencoe at the time of his marriage.[44] The nature of Angus' complaint about that is not clear. The accounts of the Glencoe trustees show that the £1,000 had been repaid in 1822 to 'Mr. Thomas MacDonald, Writer, Fort William, Assignee of Mrs. John and Mr. Angus MacDonald, Drimintorran'.[45]

Alexander MacDonald of Dalilea was in a difficult situation. He had accepted the position of tutor and curator but had been too far away to be actively involved. Alexander Macdonald of Glencoe had attended to almost everything and had resisted Dalilea's pleas to keep a record of what he was doing. Dalilea started a frantic search for any trace of written material. On 10 December 1825 he wrote to Thomas MacDonald saying that he was having a search carried out at Aryhoulan and asking if Thomas had seen anything in 'Glenco's jottings', or if Mr Flyter or he could remember anything or find any paper about James' affairs when he failed in 1808. He wrote in similar vein to Duncan Cameron, who had of course been deeply involved in the work of the Glencoe trustees. However, Duncan Cameron was doubtful whether any paper existed. Dalilea's comments about Glencoe's failure to keep records became increasingly critical.[46]

The Failure of Donald MacDonald

While this case was rumbling on unsatisfactorily, Donald MacDonald was finding renown as a farmer. In 1829 he received a silver cup from the Highland and Agricultural Society.[47] In 1833 and 1834 he added Ardmollach at the mouth of the River Moidart and part of Kinlochmoidart to the land that he already farmed. These were followed in 1835 by Ardslignish on the north shore of Loch Sunart, a couple of miles or so beyond Glenborrodale Castle.[48] He was described in the account of the Parish of Ardnamurchan in the *New Statistical Account of Scotland* written in 1838 as 'the late enterprising tenant of Drimantarran'. The farm offices that he had built were particularly commended.[49] These words were written a year after 'Donald MacDonald, General Merchant and Dealer in Wool, Meal, and other Articles of Merchandise residing at Drimintorran near Strontian, Argyll' had become bankrupt.[50] He was probably described thus, rather than as a farmer, to take advantage of the legal provisions for mercantile sequestration which were not available to tenants of land unless they were gaining a material part of their living dealing in animals or grain not produced on their land. It was quite usual then for farmers who became bankrupt to be described in such terms.

Donald MacDonald's assets were substantial and worth in total almost £7,000, of which sheep accounted for over £4,000 and other livestock almost £600. His creditors included several well-known tacksmen in and

around Lochaber. It is not really clear from the information available why he was precipitated into bankruptcy at a time when, according to Evander Maciver, sheep and wool prices were beginning to rise.[51] Donald MacDonald should have been well poised to take advantage of that. Perhaps he had been persuaded by improving conditions to expand too quickly and found himself overextended.

Donald's effects were auctioned,[52] and in 1840 he and his family left for New Zealand, where he died in 1849.[53] The Highlands lost an enterprising farmer and man of business. This is amply supported by the impressive testimonials provided by prominent citizens in North Argyll when he left. These reveal that he had been a magistrate for twenty-two years and a deputy-lieutenant for the county of Argyll. He was said to have been unfortunate in his wool speculations in 1835 and 1837. The writers included two of Donald's landlords, Colonel Alexander MacLean of Ardgour, from whom he held Inversanda, and Colonel Robertson Macdonald, who had given him a lease of the farm of Kinlochmoidart on his wife's estate in 1833.[54]

Thomas MacDonald, Writer in Fort William

Donald's brother Thomas, who had done all that he could to help him, continued his career as a highly respected writer in Fort William. He held the office of procurator fiscal and was agent for the National Bank of Scotland. Donald's experience did not deter him from personal involvement in sheep-farming. By the 1840s he had replaced Captain Moses Campbell as tacksman of the farm of Achnadaul. In 1847 he took a leading part in refuting the critical comments of Robert Somers about Lord Abinger, who had purchased part of the Duke of Gordon's Lordship of Lochaber.[55]

When the 1851 census was taken, Thomas MacDonald was living in Fort William, where his legal and banking work was centred. Many of those whom he employed at Achnadaul had been born in the parish of Ardnamurchan or, more specifically, in Strontian. Duncan McPherson, the farm manager aged forty-five, and Thomas Rankin, a shepherd aged thirty-five, had both been born somewhere in the parish of Ardnamurchan, while Ewen McPherson, another shepherd aged thirty-eight, had been born in Strontian. It was quite usual for the tenant of a farm to bring in men from his own locality, but in this case these may have been people who had been with Thomas's brother Donald in his prosperous days at Drimintorran and elsewhere in Ardnamurchan.

Thomas MacDonald died in Fort William on 30 June 1856 at the age of sixty-two.

8

The Mackintosh Lands in Lochaber, 1804–1823

The development of the Mackintosh lands in Lochaber up to 1804 was described in Chapters 3 and 4. In 1802 Alexander Macdonald of Glencoe had replaced Alexander Mackintosh, the merchant from Fort William, as tacksman of Keppoch and the Inverroys. Most of the new tacks given by Æneas Mackintosh 1804 were for nineteen years, lasting until 1823. The rents were high, reflecting the boom conditions at that time. Unfortunately, livestock and wool prices fell substantially during these long leases, leaving tenants committed to high rents. This was to cause much difficulty and distress.

Three of the main tenants were gone long before the tacks ended in 1823. Alexander Mackintosh, the merchant who took Bohinie, Murlagan and Achluachrach, was removed in 1807 and again replaced by Alexander Macdonald of Glencoe. He withdrew, hurt and resentful, to Fort William. Donald McDonald at Tulloch died in 1808. Finally, Alexander Macdonald of Glencoe died in December 1814, leaving his trustees to struggle with his substantial commitment to Æneas Mackintosh until his tacks finally ran out in 1823.

Preparations for the Tacks of 1804

When arrangements were being made for the sett of 1804, the attitude towards sheep-farming was euphoric. On 4 February 1802, Donald McDonald at Tulloch made the exuberant comment that 'people that pretends to have superior knowledge of sheep farming particularly white faced south country sheep thinks that Sheep farming is only at its infancy in the Highlands'. He proposed that Æneas Mackintosh should try to lease additional land from the Duke of Gordon. He could then add to the tacks that he could offer and make them suitable for rearing 'the English White faced Sheep with great safety'.[1] Nothing came of that.

The tenancy of Keppoch, the Inverroys and Boline had been settled in 1802. Alexander Macdonald of Glencoe had taken over the last two years of Alexander Mackintosh's tack of 1795, and got a tack in his own name to follow from 1804. The other new tacks had however to be prepared, and drafts were ready towards the end of 1802. Donald McDonald sent lengthy and critical comments to Campbell Mackintosh on 14 January

1803. In spite of his optimism about sheep-farming he expressed a reservation about the level of rents proposed. 'Both the Lairds and Tenants should pray for continuance to the present times. I am afraid the rent is above the proper mark when some of your lands exceed five shillings for sheep's grass it cannot stand & I see the whole above four shillings per sheep.'[2]

Donald wrote a memorandum about the new proposals for his own tack of 'Tulloch, Dalnaderg, Tollie, & Urchar, also Clachaig and Corybeg'. He was to be required to build a new house to a prescribed plan for which he would be allowed £150 melioration money. He wanted £180. He was also to be allowed £130 for stone dykes, sheep fanks, tup parks and a garden. He wanted £200. He was to be required to live at Tulloch for at least six months of the year, but he had other interests and was not happy about that. He was not to be allowed to sub-let, but he wanted to be able to give some ground to a stockman to help him with his stock. If he died, the landlord was to have the option of removing heirs within two years. Donald wanted a stipulation that any incoming tenant would be required to buy the stock at valuation. He also wanted, as in his existing tack, a departure from the usual arrangement of finding cautioners, 'because the Tacksman's good intentions towards the proprietor's interest has long ago rendered him Disagreeable to his Lochaber Friends'.[3]

Donald McDonald got some concessions in the final version of his tack. For example, he was to get £160 for the house and up to £50 for out-houses and other farm buildings. For the most part, however, the landlord got his way, and Donald accepted his new tack at a rent of £320.[4]

Donald also wanted some lower ground to complement Tulloch. He had his eye on Achaderry, because 'I am greatly affraid of my stock in time of storm without such a place as Achdery'.[5] He seemed fairly confident that Æneas Mackintosh would let him have it, but he did not. Achaderry went to Alexander McDonald at Murlagan, but with 'a reservation to Donald McDonald at Tulloch to bring his sheep to Achderry in case of a severe winter'.[6] Perhaps that was a partial success.

The prospects of new tacks in 1804 caused others to try to secure their positions. Dr Donald McDonald's partners at Cranachan and Blarnahanin, his cousins Alexander and Angus McDonald, were in a dilemma. They said, 'the Dr. our Cousin . . . does not appear inclined to let us know his intentions although we asked him several times'. So on 16 June 1802 they put in an offer themselves of 160 guineas for Cranachan and Blarnahanin. They said that was 'as high an offer as any man is safe to give'.[7] The doctor, their cousin, must have been conserving his resources for something more expensive. He and two different partners took a seven-year lease of Scot-house in Knoydart from Whitsunday 1804 at a rent of £650[8] (see Chapter 10). His two cousins back in Glenroy wisely hedged their bets by making an

alternative offer if they should fail to get Cranachan and Blarnahanin. If the farms of Bohuntine and Achvady (or Achvatie) 'as presently possessed by Dr John Mackintosh and others' were to become vacant, they were willing to pay £200 for them. Finally, they asked if they could be told if their offers were not to be accepted 'before the Duke's Sett is over', because there were some farms on offer on the Duke of Gordon's land.[9] They did not get Cranachan and Blarnahanin, but their situation illustrates very well the dilemmas facing tenants when tacks were being given out at these great setts.

Others also put out feelers. On 23 July 1802 Captain Angus MacDonell, late of Tulloch, wrote from Fort William to Campbell Mackintosh claiming that Æneas Mackintosh had promised to consider him for a farm. His company was being reduced and he was to retire from the army.[10] Æneas Mackintosh certainly did not want him but had difficulty in finding a satisfactory way to refuse. In the end he told the unfortunate Campbell Mackintosh that he must 'endeavour to put him off as best you can',[11] which he dutifully did.

Tulloch (1804–1823)

In 1804 Donald McDonald seemed secure in his position at Tulloch. Captain Angus MacDonell of Tulloch was well out of his way. Murlagan and Achluachrach were in the hands of Alexander Mackintosh (see below), and the disruptive tenants there were no longer a problem. Donald could give attention to other matters. In 1802 he had bought property in Fort William to develop into the New Inn. In 1805 he made offers for farms on the Duke of Gordon's land. These expanding interests outside the Mackintosh lands, described in more detail in Chapter 9, all involved additional financial commitments. Æneas Mackintosh expressed a timely concern about what Donald was doing. 'I wish Don^{d} may not have too many irons in the fire – and that he may not end as he began.'[12] Donald's anxiety about whether the prevailing high rents could be sustained was soon to be tested. By 1807 livestock prices had fallen and the harvest was poor.[13]

Several bills were protested against Donald at the Sheriff Court in Fort William in 1806, 1807 and early 1808, suggesting that all was not well with his finances.[14] Towards the end of 1807 he tried to pay some of his Gordon rents by a bill at three months notice, but the Duke's factor told him that he would not take accommodation bills for rents.[15] His brother Alexander, by then tacksman at Rifern in South Morar, later claimed that he had advanced money to help Donald, but he did not specify the amount.[16]

When Campbell Mackintosh visited Lochaber in 1807, Donald 'was at pains in explaining his situation to me and I was induced to help him with the loan of some money'.[17] This discussion probably led to the preparation by Campbell Mackintosh of Donald's trust disposition and

deed of settlement dated 31 August 1807.[18] On 1 February 1808 Donald's property in Fort William was transferred to Campbell Mackintosh, perhaps as security for the loan.[19]

Early in 1808 Donald's health was failing. On 25 March, Alexander Macdonald of Glencoe, by then a fairly close neighbour, wrote to Campbell Mackintosh: 'Poor Donald du is getting weaker every day. I can't think but he will leave a good subject after paying his debts for I saw him lately and he told me a good deal about his affairs.'[20] Donald died within the next few days. Alexander Macdonald of Glencoe wrote again to Campbell Mackintosh on 8 April, expressing concern about the effects of the weather on the stock at Tulloch. 'There is hardly a bit of Black ground to be seen yet as for lambs there will be very few . . . the Tulloch farm will suffer considerably . . . a number of the black cattle will be lost.'[21] There is no doubt about the severity of the winter of 1807/08. Snow covered inland areas as late as the end of April. Fodder became very scarce, and many black cattle did die.[22]

On 9 April Campbell Mackintosh informed Æneas Mackintosh of Donald's death, 'which I realy regret as he was a clever sensible fellow'. He also thought that there would be 'a good deal of Funds after paying all his debts'. He hoped that his property in Fort William, consisting mainly of his New Inn, would sell 'in consequence of the advantages of the [Caledonian] Canal'. He also observed, 'I have reason to think that Gillespie will be a candidate for his farm but the poor man who is dead has done a great deal about the farm of late years'.[23]

Donald had appointed five trustees: Campbell Mackintosh, the Rev. John Macdonald (or Macdonell), minister at Forres, Thomas Gillespie at Ardochy, Ranald Macdonald, tacksman of Strathmashie and son of Allan Macdonald of Gallovie, and Donald Mackintosh, a lawyer in Edinburgh who was a son of Campbell Mackintosh.[24] Thomas Gillespie and Ranald Macdonald, the two farmers among the trustees, lost no time in seeing to the management of Tulloch. By June 1808 they were arranging the sale of some livestock.[25] [26] In January 1812 Gillespie forwarded £470 to Campbell Mackintosh and said that £1,400 was awaited for sheep that had been sold. He added that he and Mr Macdonald, Strathmashie, did not want to put Mr Mackintosh to the trouble of sending them an account of the late D. Macdonald's affairs. This confirms that he was referring to the realisation of Donald's assets.[27] Campbell Mackintosh busied himself, among other things, with Donald's property in Fort William.

In 1802 Æneas Mackintosh had observed that 'Donald is looking out for a wife',[28] but nothing came of that. His trust disposition, which will be considered in more detail in Chapter 9, shows that he had six children by three mothers but none was legitimate. His brothers Alexander Dhu and John claimed to be his legal heirs and attempted to take over the tack of

Tulloch, which still had fifteen years to run. Æneas Mackintosh exercised his right to remove heirs on the second Whitsunday after the tacksman's death.[29] As he was dealing with at least one stubborn character, he took the precaution of getting a decreet of declarator and removing from the Court of Session.[30] Thomas Gillespie began paying the rent for Donald McDonald's tack, and in 1813 Æneas Mackintosh granted a tack for eighteen years to Gillespie and James Greig. They paid the same rent as Donald McDonald, but they were required to pay to Donald's trustees the melioration money, potentially over £300, due to Donald at the end of his tack.[31] [32] Gillespie died in 1824, but Greig was still in possession of Tulloch when he died in 1837.[33] [34]

The exact state of Donald McDonald's complicated affairs when he died is not clear. The year 1807 had been a bad one in which sales were stagnant and livestock fetched low prices. Many tenants in Lochaber were in arrears with their rents.[35] Nevertheless, Donald possessed so much stock on his farms as well as his property in Fort William that Campbell Mackintosh and Alexander Macdonald of Glencoe were almost certainly correct when they said that he would leave a significant amount of money after his debts had been paid. He had come a long way from his precarious existence as a drover some thirty years earlier.

Keppoch, the Inverroys and Boline (1804–1823)

As explained in Chapter 4, Alexander Macdonald of Glencoe was installed in Keppoch, the Inverroys and Boline in 1802 and had signed his lease in that year. It was a condition of that lease that he would 'forthwith after his entry build a complete set of Substantial office houses, and a Kitchen of Stone and Lime Covered with Slates and also Inclose the mains of Keppoch with a substantial Stone Dyke or Sunk fence faced with stone four and a half feet high to the north betwixt the said mains and the Hill'. Subject to comprisement at the expiry of the tack, he would be entitled to payment not exceeding £300.[36] Much of that work was not carried out, and this was eventually the subject of dispute between Æneas Mackintosh's successor and the Glencoe trustees.

Within a few years, Alexander Macdonald of Glencoe was to become even more heavily involved in Mackintosh land, once again in succession to Alexander Mackintosh, the merchant from Fort William.

Bohinie, Murlagan and Achluachrach (1804–1823)

At the 1804 sett, Alexander Mackintosh, the merchant from Fort William, 'took the big lot and paid £2,000 ready cash for the Stocking thereof'.[37] Bohinie, Murlagan and Achluachrach were combined in a single tack known popularly as 'the big lot'. It was indeed a large holding, for which his rent was £340, one of the highest of all the Mackintosh rents in

Lochaber.[38] This move was probably the last part of Æneas Mackintosh's long-term plan to consolidate his Lochaber estate into larger holdings.

The unfortunate Alexander Mackintosh, having again proudly supported his chief, came to grief fairly quickly. By late 1806 or early 1807 he was in some financial difficulty. He claimed later that 'a family relation forged a bill to a considerable amount on [him] & who immediately absconded, which harassed him'. His first reaction was to ask permission to sub-let his tack. Æneas Mackintosh was not going to make it easy for him. 'In Mr. A. Mackintosh's case . . . I gave him a larger Farm and on easier terms than I would have given to any other person, it is not to be supposed I am to transfer that favour to any stranger without a valuable consideration paid to myself.' Alexander Mackintosh would have to pay him a whole year's rent or an additional £20 a year for the remaining sixteen years of the lease.[39] That probably put the option of sub-letting beyond his reach.

Æneas Mackintosh obtained a decreet of removing against him in the Sheriff Court in Inverness on 15 April 1807.[40] Campbell Mackintosh did the rest. Alexander Cameron in Inverguseran, on the west coast of Knoydart, was due some money from Alexander Mackintosh. He wrote to Alexander Macdonell, writer in Inverness, in some alarm on 23 July 1807.

> I am told that Mr. Campbell McIntosh was lately in the Braes of Lochaber & that he took possession of Mr. Alex. McIntosh's stock and farm, that he was turned out of house and hall, some people says that the stock will pay every farthing of his debt, if it is properly managed, but I am afraid it will not be managed as it should. I beg if you did not get a decreet already that you will push everything on to get it.[41]

Alexander Mackintosh, ever loyal to his chief, blamed Campbell Mackintosh for seizing the opportunity to sequestrate his stock while he claimed to have had sufficient funds to cover his liabilities.[42]

Land was still in great demand, and six men – Alexander MacArthur, Bochaskie, Donald MacArthur, Murlagan, Angus Beaton, Achnacoichan, John MacGillivantich and Murdoch McPherson, both from Bohuntine, and Donald MacDonald, late Bohinie – promptly made an offer to pay a rent of £530.[43] However, Alexander Mackintosh's farms were once again given to Alexander Macdonald of Glencoe, at a rent of £500.[44] This was substantially more than Alexander Mackintosh's rent of £340. With hindsight, it may seem that Alexander Macdonald was rash, even although the other offer of £530 shows that he was keeping in step with the market. He may have felt this himself, because in February of the following year he displayed uncharacteristic realism and declined the offer of a farm from Glengarry with the words: 'As so much unexpected Lands Fell into My Hands since Whitsunday last, I beg to inform you that I will have no occasion for the farm of Glenlay.'[45] Hindsight also suggests that it may

have been unwise of Æneas Mackintosh to put another large tack into his hands. Perhaps it had been thought that he was still financially sound, but there had been warning signs. Alexander Mackintosh, for example, had complained in 1805 that even 'Lairds will not pay me the fourth part of the small debts I have against them, I mean Glenco, in place of £400 I was glad to get his Bill for £101 0*s.* 0*d.*'.[46]

Thomas Gillespie and James Greig were called upon to value the stock that Alexander Macdonald of Glencoe was to take over from Alexander Mackintosh. They valued the total at £1,764 8*s.* 7½*d.*, of which sheep accounted for £1,543 11*s.* 10*d.*[47] A first instalment of £600 was paid on 13 October 1808,[48] but the large amount unpaid would have helped Alexander Mackintosh to meet his rent arrears and other debts.

Unlike the position at Keppoch and the Inverroys, where Alexander Macdonald of Glencoe had a formal lease, a simpler procedure was adopted at Bohinie, Murlagan and Achluachrach. As Campbell Mackintosh explained after Glencoe's death:

> Of the lott of Murlagan etc. there was no extended lease only a minute between him [Glencoe] and me in which the sale of Sandy Mackintosh's Stock was made as well as the Sett and latterly Glenco was indifferent about this farm and taking any lease of it considering the rent was very high which it certainly is now.[49]

Glencoe's payments for Bohinie, Murlagan and Achluachrach were invariably shown as being 'in place of Alex. McIntosh'.[50]

By early 1808, Alexander Macdonald was not meeting his commitments. Campbell Mackintosh said that 'Glencoe besides his arrear of rent is due a very large sum for Sandy Mackintosh's stock a great part of which will go to pay Sandy's arrears of rent and woods money'.[51] That was a reference to the stock taken over from Alexander Mackintosh.

In 1811 Campbell Mackintosh warned Æneas Mackintosh that 'Glencoe is again falling behind with the rent'. He had remitted 'only' £300 which included 'a Dft upon a man in the neighbourhood of Perth'.[52] The latter referred to a bill upon a Mr Drummond of Crieff for £150.[53] Alexander Macdonald evidently wrote towards the end of 1811 pleading difficulty because of some commitment as a cautioner. Æneas Mackintosh took a robust line on that. He had difficulty with Glencoe's writing, which was less and less legible as the years passed, but said:

> I understand enough to judge that we shall not long remain good friends, he must either resolve never again to become Cautioner, or I must get another Tennant tho' I once borrowed money to befriend him he must not suppose I am to repeat the same indulgence . . . his doing foolish things is no excuse to me.[54]

In late December 1812 Alexander Macdonald passed through Inverness and left some money with Campbell Mackintosh, but 'still there will be a large balance due when this fund answers'.[55]

On 4 January 1814 Alexander Macdonald told Campbell Mackintosh that he was coming to Inverness and that he would pay £500 of his arrears as well as the rents. He commented that 'it was a very bad bargain for me, but there is no help for it he shall not want a shilling of what I promised'. He may have been referring to the second lease that he had taken from Alexander Mackintosh in 1807. Even so, he was still prepared to add to his commitments. He said that he wanted Colarick in which the Bohuntine tenants were interested and was willing to pay £10 more than their offer of £270.[56] That creates doubt about his insight into the true state of his affairs.

Alexander Macdonald of Glencoe as Cautioner (Cranachan and Blarnahanin)

As explained above, Dr Donald McDonald did not attempt to get a further lease of Cranachan and Blarnahanin. His two cousins failed in their offer. The tack went to (Dr) John Mackintosh at Achvatie (in Glenroy) and John Mackintosh, his son, along with some grazing rights at Bochaskie, for nineteen years at a rent of £200 yearly. Their cautioners were Rev. Thomas Ross of Kilmanivaig and a Mr Allan Cameron.[57]

By 1807 John Mackintosh senior had died[58] and his son John sought permission from Æneas Mackintosh to sub-let the tack. This indulgence was granted 'upon account of the numerous family left to your charge and the respectable . . . '. Æneas Mackintosh introduced this qualification because he was anxious not to set a precedent. He had imposed a prohibitive condition when Alexander Mackintosh made a similar request. John Mackintosh sub-let his tack in two parts, and Alexander Macdonald of Glencoe was cautioner for the new sub-tenants of both.

Blarnahanin was sub-let to Alexander MacDonald, residing at Bohuntine, and Cranachan was sub-let to Angus MacDonell at Munassy (Monessie). John Mackintosh's mother, Isobell Macdonald, was to continue to occupy her house and byre at Cranachan, to have grass for two cows to be herded and cared for by Angus MacDonell (the new sub-tenant), room for her peats in the barn and ground for her potatoes. In both instances the lands were to be held 'immediately under the proprietor' at rents of £145 and £120.[59]

The motive of Alexander Macdonald of Glencoe in undertaking to be cautioner is unclear. He might have had his reasons for wishing to be genuinely helpful, but it did put him at some risk. John Mackintosh had trouble in extracting the rent from Angus MacDonell at Cranachan. In 1808 he went to the Sheriff Court in Fort William to force him to pay his rent of £120 plus £6 public burdens and £2 9*s.* 4½*d.*, the last being half

the fee due to Campbell Mackintosh in Inverness for drawing up the sub-tacks.[60] However, there is no evidence that Glencoe was called upon to make good any deficiencies in either of these sub-tacks.

John Mackintosh was a messenger and surveyor of taxes in Fort William, which may have been his reason for sub-letting his tack.[61]

The Death of Alexander Macdonald of Glencoe

When Alexander Macdonald of Glencoe died in December 1814, he held three leases of Mackintosh land in Lochaber. They had a long time to run and were to trouble his trustees until they expired. It may be helpful to summarise the position.

The first lease consisted of Keppoch, the Inverroys and Boline. Keppoch is at Roy Bridge and included the mansion house where he lived. The Inverroys lie to the west of Roy Bridge and were divided into Inverroymore and Inverroybeg. Each was sub-let to about a dozen tenants.[62] Boline (or Bo-Linn) is an area of hill ground lying to the north of Keppoch and the Inverroys. That lease was due to run until Whitsunday 1823.

The second lease was of Bohinie, Murlagan and Achluachrach, which he had taken over from Alexander Mackintosh in 1807. Strictly speaking, this lease was never in Alexander Macdonald's name, but for all practical purposes it became his. Like his first lease, it was due to run until Whitsunday 1823, and it had tenants or sub-tenants. After Glencoe's death the Glencoe trustees usually referred to this collectively as the lease of Gaelmore.

There was also a third lease, of Park of Culdiavin (or Caoillediavin). This was on the west side of the River Roy, to the north of Roy Bridge, probably next to land held under the first lease. It is not clear how this lease was acquired, but it was a separate one and its duration was exceptional in that the trustee accounts noted that it 'falls in 1821'. There were also some tenants there.[63]

Alexander Macdonald's death on 14 December 1814 was a significant event for the Mackintosh estates. He was their largest tenant in Lochaber, responsible for an annual rent of over £800, and he was in arrears.[64] [65] Within less than three weeks, Campbell Mackintosh wrote four letters about the implications of his death to Sir Æneas Mackintosh, who had been knighted in 1812.[66] On 22 December 1814 Campbell Mackintosh told Sir Æneas:

> I heard of the death of poor Glencoe who died suddenly at Fort William . . . but there can be no loss to you and the security of the lease will be quite sufficient for you, particularly the old lease of Keppoch and the meliorations – even if the other farm was given up the Sub-tenants would pay the rent which they are doing to Glencoe. I mean the great Lot of Murlagan etc.[67]

Sir Æneas was perhaps anxious. On 24 December, Campbell Mackintosh tried to reassure him with an immediate explanation.

> I had the honour of your letter of the 24th with regard to Glenco. We have not been so clear with him for a number of years. He is due three hundred odd pounds Stg which by the receipt given under his hand is fixed and expressed to be due for Keppoch and the Inverroys etc. included in his first Lease. This was done on purpose to secure it in case of accidents and I have often been plagued with the honest man having matters in a loose and imperfect state of partial payments so that I was much gratified not only to have the sum due so reduced but to have it fixed and the payment so secured.[68]

Another letter followed on 28 December. By then, Campbell Mackintosh had heard from Duncan Cameron WS, the late Glencoe's brother-in-law and trustee on whom the management of his estate largely devolved. Duncan Cameron said that 'the sudden death of Glenco and the season of the year would occasion some little delay'. He trusted that 'under such circumstances you will be disposed to grant some indulgence to the Managers and representatives of the deceased'. Campbell Mackintosh took some comfort from the indication that Duncan Cameron was about to take matters in hand.[69] However, it soon became difficult to find any source of comfort in the muddled and overstretched affairs of the late Alexander Macdonald of Glencoe.

On 7 January 1815 Campbell Mackintosh wrote again to Sir Æneas, having had a further letter from Duncan Cameron. First he reported: 'MacArthur the drover who called upon me just now from Lochaber tells me that the eldest son of Glencoe came to the Country a few days ago – he has come from India rather ailing but otherways his arrival is seasonable to assist in looking into his father's affairs.'[70] That information was not quite accurate. It was Colin, the second surviving son, who had arrived. He did try to help (see Chapter 13), but there was a limit to what he could do.

The position at Keppoch and the Inverroys was quite clear, but there was no formal tack of Bohinie, Murlagan and Achluachrach in Glencoe's name, and Campbell Mackintosh said:

> It is of little consequence in my opinion whether the Glenco people give it up or keep it – at present it is sett to a number of sub-tenants many of them the same persons as possessed before Sandy Mackintosh and making up the same rent as payable by Glenco neither more nor less. You know this lot joins Gillespie's farm [Tulloch, taken over after Donald McDonald's death] but it is a question with me whether he would even give the present rent for it.[71]

On 4 January 1815, Duncan Cameron had said that enquiries were already

being made by people anxious to take over Glencoe's leases. He expressed a tentative view that 'if your constituent [Sir Æneas] agrees to it they may have the last lease that Glencoe took from him [Bohinie, Murlagan and Achluarach] but as to the farm of Keppoch you may tell them that it is intended to be kept by Glenco's Representatives'.[72] But Mackintosh did not relieve the Glencoe trustees of responsibility for Bohinie, Murlagan and Achluachrach. Whatever the reason for this decision may have been, it remained with the Glencoe trustees until it had run its course in 1823.

Duncan Cameron was still taking a fairly cheerful view of the task of clearing up the affairs of his late brother-in-law. He thought that 'we shall very soon have it in our power to pay you the greatest part of the rent as our friend Glenco did not collect his rents before his death and I have authorised Mr. Scott [then a writer in Fort William] to collect them immediately if the tenants will pay and to send the money to you'. He was sufficiently realistic to recognise that Glencoe's well known antipathy towards written statements and regular accounting would 'give those who take the management of his affairs more trouble than they would otherwise have'. Nevertheless, 'considering his very great subject my impression is that after paying all debts & Demands there will be a very considerable Reversion besides his Landed property etc.'[73] He was quite right about the trouble from the absence of records, and quite wrong about value of the estate.

The Glencoe trustees were now responsible for the management of the tacks held by Alexander Macdonald of Glencoe. At their first meeting on 18 May 1815, it was reported that there was a 'Surplus of Black Cattle and horses which could not be kept on the Farm of Keppoch'. It was agreed that the surplus should be sold by public roup. On 25 May, Duncan Cameron posted fourteen circular letters 'requesting the attendance of Gentlemen of the country at Keppoch Roup'. The task of organising and managing this sale was given to Thomas MacDonald, writer in Fort William. He was one of the Drimintorran family who, as noted in Chapter 7, was engaged by the trustees on a variety of tasks. Most of the proceeds of this sale had been received by June 1816, but on 16 November Thomas MacDonald posted 'roup bills' to sixteen individuals who had not yet paid. Some payments were delayed until 1817. For example, in March 1817 there was a payment of £6 from Alexander Kennedy, Inverroy, for his purchases. One straggler was Adam Macdonald of Achtriachtan, whose debt was finally settled with his trustees in 1820.[74]

Collecting the rents and finding enough money to make the required payments, including substantial arrears, to Sir Æneas Mackintosh was a major concern for the Glencoe trustees[75] Entries in their accounts show that they received significant amounts of mney from tenants from time to time. For example, in January 1816 they received £271 1*s.* 4½*d.* from the

Inverroy tenants, £358 from the Gaelmore tenants (Bohinie, etc.) and arrears of £20 14*s.* 0*d.* from Culdiavin. In May 1816 there was a £100 more from the Gaelmore tenants, and in November 1816 there was a further £28 from the Culdiavin tenants 'to account of rent'. In February 1817 they received £409 5*s.* 8*d.* from the tenants of all three properties.[76]

On the payment side, in January 1816 the trustees paid to Sir Æneas Mackintosh £625 16*s.* 6*d.*, described as 'balance of rents'. In August 1817 they paid £892 9*s.* 3*d.* 'Balance of arrears of rent and interest'.[77] This substantial amount was paid some ten days after the trustees had received a second loan of £1,000 from Mr John Henderson, a linen draper in Edinburgh.[78] (Details of the trustees' borrowings will be given in Chapter 13.) The Mackintosh Rentals show that at least up to 1818 the trustees managed somehow to pay these very substantial rents in full.[79] In December 1820, Duncan Cameron 'attended upon Mr. Cooper on the subject of the paying up of the Rents to Sir A. McIntosh's executors'. Sir Æneas had died earlier that year.[80] Mr Arthur Cooper was a solicitor in Inverness who appears to have dealt with some Mackintosh estate business after the death of Sir Æneas.

In 1822, after the lease of Culdiavin would have expired, the trustees paid over £900 for 'the rent of Keppoch, Inverroys, etc. for the year to Whitsun 1822' and for 'balance of rents and public burdens'.[81]

The figures available are not sufficient to make an accurate assessment of the balance between income and expenditure on these properties. The impression is, however, that the rents being collected did not match the payments being made. The trustees seemed well aware of this. At their third meeting in January 1817, they discussed difficulties with the tenants at Gaelmore and decided to draw the attention of Sir Æneas 'to the increase in rent paid by them beyond what was paid by Mr. Alex. Macintosh who left the tenement by failure'.[82] The Gaelmore tenants were then allowed an abatement of 20 per cent on the 1816 rent and 10 per cent in 1817.[83]

Keppoch was Alexander Macdonald's own residence and was not sub-let. The trustees decided at their first meeting on 18 May 1815 that 'Possession of Keppoch should be adhered to both on account of the heir and as a kind of family domicile as well as in some measure supporting the Heir's right to a renewal of the lease should he be so inclined'.[84] It took a little time for such well-intentioned optimism to disappear from the minds of the trustees. Keppoch was retained until the lease expired in 1823, although in November 1821 Robert Scott said to Ewen Macdonald, 'even at Whitsunday first your uncle [Duncan Cameron] would readily agree to give up Keppoch providing Mackintosh will take it'.[85] The use of Keppoch as a home by the family must have been quite limited. Ronald was there at times during his school holidays, and Colin used it while home on

furlough from 1815 to 1817. Jane lived there for a short time before she moved to Edinburgh and married, and again for a short time after the death of her husband. Nevertheless, wages and other outlays on the house and farm had to be met. Duncan Cameron corresponded with a firm in Greenock about getting farm implements. Small sums were paid occasionally to various tradesmen for 'work at Keppoch', which could have been at the house or on the farm. At least one longstanding member of the domestic staff, a Miss Stewart, was retained until her death shortly before the lease expired, and in October 1815 the trustees paid an account for 'whiskey at Annabella MacKenzie's funeral'. This payment is listed with payments to members of staff.[86] A gravestone in Kilmanivaig churchyard shows that an Annabella MacKenzie from Glengloy died on 14 February 1815 at the age of thirty years.[87] In May 1817 the trustees settled with a John MacKenzie 'balance due on account between the deceased and the late Annabella MacKenzie – £2 11*s.* 0*d.*'.[88] She had probably been a servant at Keppoch when she died.

Josephine MacDonell has given an account of some work that Alexander Macdonald of Glencoe carried out at Keppoch. 'He built the wing at the back; and he built up the windows of the basement rooms to half, and filled up between the sloping banks and the windows making it level with the lawns, and rendering these basement rooms useless except for servant's offices.'[89] He failed, however, to build the 'Dyke or Sunk fence' required by the terms of his lease, and it was claimed that the offices he built were not slated as they should have been.[90] When the lease expired in 1823, the house was said to be in a poor state of repair and requiring considerable expenditure.[91] This is certainly borne out by a report on the house in 1823 by Alexander Simpson and Donald Cameron from Inverness, who said, 'the whole of the Joists [are] so much decayed and sunk in the centre as to render the House entirely insufficient'. It also talked about decay in the roof and the need for urgent action 'to prevent the immediate falling in'.[92]

With the expiry of the leases of Keppoch, the Inverroys and Gaelmore at Whitsunday 1823, the concern of the Glencoe trustees in the Mackintosh lands ought to have ended. However, there was the irritating business of the melioration money. The trustees not only claimed the full £300 mentioned in the tack, they got it by deducting it from the final payment of rent in November 1823. Mackintosh and his representatives responded with vigour. They pointed out that the maximum of £300 was linked to a list of works, including a stone dyke or sunken fence which Alexander Macdonald had simply not built. They asserted that the whole claim fell, but counsel advised that only the value of the non-existent dyke was at issue.[93] Arbitration followed, and the matter dragged on for some time.

9

Some Wider Interests: Donald and Alexander Dhu McDonald

This chapter has two purposes. The first is to give an account of Donald McDonald's interests outside Mackintosh lands. Brief references have been made to these, particularly in Chapter 8. They dated from about 1802 and were mostly on the Duke of Gordon's property. The second is to explain what Donald's brother Alexander Dhu did between his removal from Tulloch in 1796 (described in Chapter 4) and his arrival in Knoydart in 1815. The Glencoe interests in Knoydart, and Alexander Dhu's later involvement in them, will be explained in Chapter 10.

Donald McDonald's Property in Fort William

In August 1802, Donald McDonald, tacksman of Tulloch, bought a 'Tenement of Houses with Garden ground behind the same in the south side of High Street, Gordonsburgh'[1] (Gordonsburgh was one of the names by which Fort William was known then). The tenement measured 93 feet 3 inches in front from east to west, and 133 feet backwards. It was bounded on the north by High Street, on the south by gardens possessed by Duncan McIntyre, merchant and Paul McPhail, Surveyor of Customs, on the east by 'the road or lane leading to the Cow Hill betwixt the said Tenement and the Burn commonly known as the Spout Burn', and on the west by the property of William Wright, Cooper and the garden ground of Paul MacPhail.[2] The property had once belonged to Allan McLean, a vintner in Fort William, who had left it to his daughter Mary McLean alias Fraser. Her sons, Major Simon Fraser of the Fraser Fencibles and John Fraser, writer in Edinburgh, sold it to Donald McDonald in 1802.[3]

Allan McLean had used the premises as an inn. This was presumably where Donald McDonald's creditors met in May 1784 when he was alleged to be in financial difficulties, and where Alexander MacPherson uttered some of his defamatory remarks about him in 1787 (see Chapter 3). According to Donald, the property was in a ruinous condition when he bought it in 1802. He decided to restore it to its former use and so he got 'an elegant plan for a proper Inn and Stables'. A coffee house was to be included, and the cost was 'no less than £1,500'. In 1805 he found that he wanted to add 'a feu of a small garden presently occupied by Paul McPhail which interrupts his plan'. He therefore submitted a memorial to the

Duke of Gordon requesting the addition of this small garden.[4] This caused some agitation; perhaps Donald had already helped himself to the garden. On 20 May 1807 the Duke's recently appointed factor, the Rev. J. Anderson, wrote to James Robertson WS in Edinburgh, 'relative to Donald (Du) MacDonald's feu in Gordonsburgh, and intimating my suspicion that he had endeavoured to cabbage [appropriate surreptitiously] a Garden, in return for his Good Deeds'.[5] The final phrase was probably sarcastic. The reverend gentleman never did appreciate fully all of Donald's virtues.

By 1805 Donald's plans for the inn were well advanced. On 29 September, James Macpherson of Belleville wrote to one of the Duke's staff at Fochabers recommending a John Campbell for appointment as the innkeeper.

> I trouble you with this to beg you to speak to His Grace of Gordon in favour of a young man named John Campbell who wishes to take an Inn newly built at Fort William by Mr. Macdonald on His Grace's property. . . . The young man served me in London and here for two years . . . behaved entirely to my satisfaction . . . he has since been a waiter at Arbroath and Aberdeen and is now in that situation at Macdougall's at Perth.[6]

Trading in Meal

Establishing the New Inn was not Donald McDonald's only entrepreneurial venture outside farming. In a letter of 24 April 1803 to Campbell Mackintosh, he said: 'It is with great regret I inform you that the [??] with my meal [from] Lossymouth was wrecked near Thurso and the Cargo Totally Lost. The Capt. lost his two sons and another sailor. It was very lucky for me that I had all insured to the amount of what it would fetch at Fort William.'[7] Donald was evidently trading in meal on his own account. It was not unknown for farmers and drovers to turn to trading in meal, but it was probably a risky business. John MacDonell near Fort Augustus was a drover for many years, but in 1813 'he thought it proper to give up the droving business and . . . betake himself to what he did not understand, the meal trade'. He bought meal in Caithness and lost money.[8]

Offers for Claggan and Annat

In 1805, notice was given of 'Extensive and valuable Farms and Grazings in the County of Inverness To be Let and entered to at Whitsunday 1806' on the Duke of Gordon's Lochaber lands. At the end of the list there was a note that 'Mr. Donald MacDonald at Tulloch in Glenspean will point out the Boundaries of the several Lotts as now arranged'.[9] Somehow Donald was now making himself useful to the Duke of Gordon's estate, but he was interested in two of the farms himself.

On 10 August 1805 Donald McDonald offered £185 per annum for the farm of Annat in Braeroy. This farm had once been occupied by his uncle, Donald Cameron (see Chapter 3). It was described as 'The Farm of Annat as at present occupied by William Mitchell [probably of Achnadaul], together with the Shealing of Coryroy formerly annexed to Ratlichbeg [Rattullochbeg]'.[10] Donald explained that his offer was 'no less than 4/– per sheep's Grass, notwithstanding that this ffarm is the very highest and stormiest in Lochaber. At the same time it lays very convenient for me & [I] could attend the management thereof with very little trouble . . . no other place would answer my purpose so well.' Then Donald added a piece of gratuitous information, couched in his best ingratiating manner.

> Several strangers has been viewing your Lordship's property that intends to appear at the sett. I have shown them the boundaries with as much justice to your Grace as in my power and hope it will have the desired effect.
>
> I beg your Lordship's leave to mention that some of the present tenants are very Indeavouring to Discourage strangers from coming forward by saying that Strangers will be only used as Toalls [tools] to assist in making the rent known and get nothing in the End but ill-will from the present Tenants. All I could say was that I saw nothing of that kind last year in Badenoch and that I could not believe that Docteriegn [doctrine]. Some of your Lordship's Tenants who used to get great favours are the most Dangerous people for Discouraging the present Sett.[11]

On 10 September 1805, Donald wrote to Sir George Abercromby, who was then involved in managing the Duke's affairs. He offered £100 per annum for the farm of Claggan, just outside Fort William. He said that he wanted it to complement the New Inn.

> I make this offer to secure the Farm for behalf of the Tenant that shall Occupy the New Inn which I built at Gordonsburgh and shall be ready at Whitsun next for a proper tenant. I shall not Settle with any Tenant for the same without asking the Approbation of His Grace the Duke of Gordon. That Innkeeper cannot accommodate the publick unless he will have such as this farm near him and some ground – I ask no profit by this.[12]

It is an interesting coincidence that the tenant then in possession of Claggan was Angus MacDonell of the old Tulloch family. He had gone there in 1799 when Donald got effective possession of his ancestral lands of Tulloch. (The difficulties in getting him out of Tulloch were described in Chapter 4.) He now wanted to remain at Claggan, but his offer was only £70.[13] Robert Flyter, the writer in Fort William who was sheriff substitute there, was also interested. His initial offer was £80.[14]

Ensign (or Captain) MacDonell was pessimistic about his prospects of holding on to Claggan, predicting gloomily on 28 September that 'America must be my fate'.[15] Robert Flyter took a more positive approach. He wrote on 23 December 1805 saying that he understood that Mr Donald McDonald was pressing hard for Claggan, and that he too would be willing to pay £100.[16] Flyter won and got Claggan for seven years from 1806.[17] Perhaps his appointment from that year as the Duke's baron baillie in Lochaber gave him an edge over Donald.[18] [19] Angus MacDonell seems to have been given Park of Claggan, which must have been a small piece of ground at a rent of only £10.[20] At the end of 1807 he wrote to Campbell Mackintosh pleading for ground on the Mackintosh estate. He referred to previous offers he had made for Tulloch (his traditional place), Bohuntin, Cranachan or Blarnahanin.[21] [22] His plea fell on deaf ears.

Donald never wasted an opportunity. His letter of 10 September to Sir George Abercromby covering his offer of 10 August for Claggan gave him the chance to repeat his offer for Annat, and to return to his theme of the strangers and what he could do for them.

> I beg leave to send this letter by Express to you and I take the opportunity of enclosing two offers one for Annat and the other for Claggan. . . . My Chieff reason for Sending an Express is that I had a letter yesterday per Express from two gentlemen from Belford near Kelso dated 31 August saying they were to be at Gordonsburgh this night and wished me to meet them there in order to show them His Grace the Duke of Gordon's sheep farm. The Express was one of their old shepherds that came before them riding a very fine horse and the Gentlemen are coming in their carriage. As they are strangers and respectable men I mean to attend and go with them myself not trusting to a guide in case the country people might be willing to discourage them. I understand by the express they sent (who is their competent judge) that they are inclined for lands suitable for the white faced sheep, and I shall, after conversing with them, show them the fittest for that purpose. My motive for sending this by express is to learn how long do you wish to continue showing the ground and to give in offers as I am in the Dark whether his Grace may Approve or Disapprove of going on as formerly Since your meetings in Gordonsburgh lately over.
>
> I shall give them every information I can and show them such farms as ought to answer their purpose.
>
> They may perhaps be the means of opening our eyes to know the returns that white faced sheep shall make. Your honour told me you was to be in Inverness to-morrow where I expect the express to return and

> send this letter by post in case of not finding you then there.
>
> I trust that your honour will please have the goodness to Remember my case when convenient to Answer my two offers here enclosed.
>
> N.B. I trust you have good news to give me of his Grace the Duke of Gordon's health.[23]

The Duke had been told about the impressive strangers by John Rutherford at Edgerston near Jedburgh. In a letter of 24 August 1805 he said that Mr Robson and Mr Thomson, two of the best and most considerable farmers in the Cheviot Hills, would be in Fort William about 10 September to look at the Duke's property in Lochaber. Robson 'with his brother farms to the amount of about £3,000 a year in this County and in Northumberland'. Both were young men who wanted 'to extend their operations in hopes of finding a provision afterwards for some of their families'. Finally, Rutherford asked the Duke to arrange for someone to be available to show them the farms.[24]

There was a lack of decisiveness in handling the Duke's affairs at this time, and Rutherford's final request had been overlooked. William Tod, who had been the Duke's factor in Badenoch and Lochaber, was replaced in the autumn of 1806 by the Rev. J. Anderson, parish minister of Kingussie. There was a hint in a letter from Anderson to Tod's son in London that Tod may have been in some personal difficulty.[25] The Abercrombys of Elgin appear to have served the Gordons as Commissioners[26] and Sir George, then head of that family, was perhaps helping by visiting Lochaber in 1805.

The Robson family were certainly well established at Belford which, as Donald said, is quite near to Kelso. In the 1770s Belford was leased by James Robson, perhaps the father of the young man who visited Lochaber. It was described as part of the ridge of Cheviot, and was high with rough, coarse and cold ground. However, the Robson of that day

> is in possession of several other store farms for which he pays £1,000 yearly . . . there is probably not another person in the world more thoroughly skilled in managing and breeding sheep. By great industry, and by frequently crossing the breed, he has at last perfected his flock to his wish, with respect not only to shape and wool, but chiefly with respect to hardiness.[27]

In 1802 Donald McDonald had expressed interest in 'white faced south country sheep' and urged Æneas Mackintosh to obtain more land to make his tacks suitable for them.[28] He would have been referring to Cheviot sheep, which were still relatively uncommon in his area. In 1803 James Hogg said that 'Gillespie hath one farm completely stocked with the finest Cheviot breed, which thrive remarkably well'.[29] In 1807 the Rev. Thomas

Ross in Kilmanivaig sued Ann MacDonald, the owner of a dog that had killed four of his Cheviot sheep and maimed several others.[30] He must therefore have had some, but progress in introducing the new breed was slow. In 1814 the Linton, or Blackfaced, breed was still the most prevalent in Inverness-shire, although Cheviots were gaining ground.[31] Donald was probably right to be excited about what he could learn from the visitors.

Donald McDonald increased his offer for Annat. In a summary of the bids it was noted against Lot 1 [Annat] that 'Present tenant Mr. Mitchell seems preferable but Donald Dow McDonald offered first £180, then £200, and would pay more if asked'.[32] Donald was beginning to throw caution to the wind and was offering the equivalent of a blank cheque. He got Annat for seven years at a rent of £250.[33]

Alexander Dhu McDonald Obtains Rifern

Meantime, Donald's brother Alexander, who had been ejected from Tulloch in 1796 and accommodated on Locheil's land at Achintore through the 'intercession of Mr. MacDonald of Glencoe', was making his own way in the world. Donald helped him from time to time by writing letters for him, and some of these show what he was doing. Alexander's enthusiasm for litigation also created some useful records.

By the end of 1800 Alexander was copying his brother and many others in leasing extensive lands for sheep-farming. Donald said to Alexander Macdonell, writer in Inverness, on 12 December 1800, 'Sandy my Brother Desires me write you to take good care that the Farm of Refearn shall be leased from the present possessors at Whitsunday next the best way is to warn them in proper time '.[34] He was referring to the farm of Rifern in south Morar, which was linked with the adjacent Clachaig and Torrary, all lying south of Loch Morar and west of the River Meoble. Sure enough, on 4 April 1801 a Decreet of Removing was obtained at Inverness Sheriff Court – 'William Fraser Esq. of Queen's Square, Bloomsbury, London, heritable proprietor . . . and . . . Alexander MacDonell, Writer in Inverness, his factor, against Malcolm Gillies residing in Riefern pretended tenant . . . of the lands of Riefern and Donald Gillies, Father of the said Malcolm Gillies'.[35] Unfortunately, there is nothing to show how Alexander found the money that he must have needed to assume the tenancy of this farm.

True to form, he resorted to the courts in a dispute with a family of MacGregors living at Rifern over the use of timber for building purposes. Angus McDonald at Kinchreggan, younger brother of the better known Alexander of Dalilea,[36] wrote on 18 January 1803 to Alexander Macdonell, writer in Inverness, on behalf of the MacGregors. He was uncomplimentary about the litigious Alexander Dhu McDonald, but his letter tells us about the state of the land at that time.

The bearer, Donald MacGregor informs me that he and his brother

> Angus have been summoned at Capt. Fraser's instance, and yours as Factor of the land of Clachaig to compear in a Sheriff Court to be held at Inverness on the 28 inst. anent several charges prepared against them by Alexr. MacDonald Principal Tacksman of the lands of Clachaig, Riefern & Torrary with regard to the wood part of them, it consists with my knowledge that the MacGregors used no improper liberties further than that Donald MacGregor used some of the Birtch growing on the farm of Riefern with which he built a dwelling house of stone and Angus in like manner built a House of the same kind for their wooll. I am perfectly convinced were you, or Captain Fraser to know the reall circumstances of the case that so far from prosecuting the poor people you would thank them for their industry, as there were no houses formerly on these farms but miserable Hutts that could not be used for any purpose besides that the wood on their property consisting entirely of Black Birch of not the best quality could not possibly be better applied for the interest of the proprietor than in making comfortable houses on the property, as there inland situation and the difficulty of access render part of them unfit for any other use. I know MacDonald's whole design is to distress the Honest people as by the change in terms since he gave them those farms he finds that they are making by them – I assure you from what I know myself of his litigious and troublesome disposition (altho a namesake I would not like him as a neighbour) – I hope you will not go according [??] to his report of matters without you having information from better authority.[37]

Nevertheless the action went on, the MacGregors lost and they were ordered to pay £28 to Alexander McDonald. This led to an extraordinary confrontation. Although he was now tacksman of Rifern, Alexander McDonald still lived at Achintore, and on 28 March 1804 Robert Flyter went in his capacity as a writer with Angus MacGregor to his house to pay the money. Alexander McDonald had, however, put the matter into the hands of Peter MacDonell, a Messenger in Fort Augustus, to secure payment. He therefore thought that MacGregor should have gone to MacDonell at Fort Augustus with the money. He lost his temper over this, got hold of a stick or staff and threatened Mr Flyter who prudently left the house. Alexander McDonald followed him brandishing a 'Kybe or potato spade', while his wife and children tried to restrain him. The efforts of his son, Allan, thought by witnesses to be aged fourteen or fifteen, were particularly noted. Alexander McDonald was detained and put on trial on 2 May 1804. He claimed that he had picked up the stick with the intention of beating his children, who were being noisy, and he was bound over in 300 merks for twelve months.[38] Angus MacGregor must have succeeded somehow in handing over the £28, and no more is heard of that.

Removal of Alexander Dhu from Achintore

Colonel Cameron of Locheil wished 'to new model his farms which could not be done without removing many small tenants'. In the spring of 1804 he brought summonses of removal into the Court of Session against his small tenants, including Alexander McDonald at Achintore. He had no trouble in obtaining decreets against all except Alexander McDonald and one other. Alexander entered a defence founded on an Act of Sederunt of 14 December 1756, which authorised removings only before the Judge Ordinary of the Bounds. He therefore argued that the case should have been brought in the Sheriff Court rather than in the Court of Session. The legal arguments were heard in June 1804, and the Court sustained the defence and found Alexander McDonald entitled to expenses.[39] Presumably Locheil then took the obvious step of getting a decreet in the Sheriff Court, because Alexander McDonald soon left Achintore. For a few years he rented a house in Fort William belonging to Donald MacDonald, a mariner there. In about 1810 he left that and lived at Rifern.[40] [41]

Offers for Blarmachfoldach

Alexander Dhu tried to lease even more land. In 1802 he had expressed an interest in getting a farm on Mackintosh's land, but that had been quietly ignored.[42] In 1805 he attempted, along with others, to obtain land from the Duke of Gordon. The notice of farms to be let from Whitsunday 1806 included 'Lott 16 Blarmacphildach as now occupied by John Campbell Younger of Glenmore'.[43] The Duke's factor received 'two equal offers, £300 each, one by Angus McDonald and Donald McPherson, subtenants to Keppoch, and the other by John Rankin, Ewen Boyd, and Alexander McDonald, Brother of Donald Dow, and both have proposed to find security'.[44] John Rankin was in Garvan of Ardgour, and Ewen Boyd was in Blaich of Ardgour. Donald McDonald at Tulloch commended all three, saying: 'there is no risque to any man in securing them. They are very able for the undertaking & careful good managers and will pay every attention to the marches.' He was willing to be their cautioner.[45] There must have been a change of mind about letting Blarmachfoldach. Neither offer was accepted. Instead of letting it as a single farm, it was divided into crofts which were given to people displaced from other parts of the Duke's estate.[46] After he had become factor, Rev. John Anderson said that Blarmachfoldach had been 'most injudiciously crofted'. It was 'at a distance from the Sea and from the great works that are carrying on'. The latter was probably a reference to the building of the Caledonian Canal, where crofters might be expected to find some employment. There was good grass at Blarmachfoldach, and he thought it

would have been suitable for Cheviot sheep if it had been retained as a single farm.[47]

It seems remarkable that, having obtained the tack of Rifern only a few years before, Alexander Dhu McDonald was prepared to be involved in a bid for Blarmachfoldach. It is also surprising that his brother Donald, who was paying a substantial rent to Æneas Mackintosh and was trying to get Claggan and Annat at high rents, felt able to be a cautioner.

Charge of Perjury

A strange misfortune befell Alexander Dhu in 1806. While a case was in progress in the Sheriff Court in Fort William on 12 March 1806, Alexander McDonald was called in by a sheriff officer 'and Examined anent the Alleged perjury & after which ordered to prison'. The details are not entirely clear, but his brother Donald was immediately summoned, arriving in Fort William late on the evening of 13 March. He wrote in great alarm to Alexander Macdonell, writer in Inverness, on 14 March. It was alleged that the signature of Mr Lindsay, a writer in Fort William, on a registered protest, presumably about an unpaid debt, had been forged. It must have been quite an old document, because Donald remarked, 'the Sheriff Clerk's Office has lost all the records they had in Bailly Cameron's time where the protest should be found which will show if Lindsay's name has been forged'.[48] Donald very reasonably questioned how the signature could be said to be a forgery until 'that office Delivers it up'. He also pointed out that even if it was forged, his brother Sandy could not write. Apparently the suspicion of forgery rested upon Mr Lindsay saying he could not remember having signed the protest. Donald let himself go on that one: 'But how could he possibly remember anything about it at this distance of time particularly a man that is well known would be every day Drunk and unfitt for Business.'

Allan Cameron of Meoble and Duncan McIntyre, a merchant in Fort William, were both in court and offered to provide security so that Alexander could be released from custody, but the sheriff refused this. Donald had his own view about the reason for that.

> I see now that Sheriff Campbell has carried his prejudice against Sandy to the highest Degree he could and that because Sandy had been one of those that complained of him as Collector in consequence of which he was Broke for misconduct. Also Sandy happened to join in an offer for a farm that Mr. Campbell had of the Duke's lands and both these things is the cause of the ill-will and malice.[49]

The mention of the offer for the farm seems to identify the sheriff as John Campbell Younger of Glenmore, who had held Blarmachfoldach.

Donald was not satisfied that the proper procedure had been followed.

'My brother has not yet got a Double [copy] of the warrant which I thought he should.' He was also concerned about where his brother was being held. 'Sandy is confined in a Dungeon where there is no Fire or Light allowed him. I send you the Surgeon's certificate to show that his life is in Danger.' There was no prison in Fort William, and the practice was to hold prisoners in the garrison until they could be transferred to the tolbooth in Inverness or Inveraray.[50] Donald's immediate concern was to get bail for his brother, and he was unsure whether this could be arranged in Fort William, or whether it would have to be done in Inverness. If it could be arranged in Fort William, he said, 'Glencoe and I will be his cautioners or Mr. Duncan McIntyre', but if it had to be in Inverness, he trusted that 'you and my friend Mr. Campbell Mackintosh will manage that. I am writing him a few lines.' Donald no doubt felt that he could rely on Messrs Macdonell and Mackintosh to do this. They had given a bond of caution for Alexander's good behaviour in 1800 when he had been found guilty of assaulting John MacBarnett, joint tacksman of Achneich.[51] Presumably he also had reason to be confident that Alexander Macdonald of Glencoe who had secured a place for his brother at Achintore would be willing to help again.

One of Alexander Dhu's sons was sent immediately to Inverness with Donald's letter and all the relevant papers that he could obtain.[52] The effort was in vain. By the latter part of April Alexander was in the tolbooth of Inverness. He was brought before the Circuit Court there on 29 April 1806, charged with 'the Crimes of falsehood and perjury'. At the outset, however, the advocate depute 'Represented that on account of the absence of some material witnesses he was under the necessity of moving his Lordship to desert the diet against the Pannel pro loco et tempore'. Lord Meadowbank agreed, and that was the end of the matter for Alexander.[53] What really lay behind this is not clear. Were material witnesses simply absent, were they 'encouraged' to be absent, or did they never really exist?

A most curious event followed within a few months. At the next sitting of the Circuit Court in Inverness, in September 1806, Archibald Dhu MacDonell, then at Inverie in Knoydart, was charged with perjury at the instance of the same sheriff in Fort William. An account of this will be found in Chapter 10, but it may be noted here that the charge was dropped once again. This was due to 'Certain Circumstances' that came to the notice of the Advocate Depute after arriving in Inverness.[54]

The possibility that Alexander Dhu and Archibald Dhu were related is suggested by an episode in 1813, when Donald Mor McDonald, tenant of Shenachy in Knoydart, set about removing John MacDonald, a brother of Archibald Dhu, from his farm. John sought help simultaneously from Archibald, who was then at Kyles Knoydart, and from Alexander Dhu, then at Rifern on the south side of Loch Morar. Archibald and Alexander

together arranged that Alexander Macdonell, writer in Inverness, would defend John against eviction.[55] The outcome was unsatisfactory, but it may be significant that John approached both men at the same time and they acted jointly on his behalf. Another interesting circumstance is that Alexander Macdonald of Glencoe was involved from time to time in the affairs of both men, and in 1815 they and their sons jointly took over the farms in Knoydart left vacant by his death. How this came about will be explained in Chapter 10.

The Death of Donald McDonald

Donald McDonald's difficulties in meeting his commitments from 1806 onwards were mentioned in Chapter 8. In the early months of 1807 he was in correspondence with the Duke's factor, Rev. J. Anderson, for whom he had obligingly obtained a pony. In thanking him, Anderson said: 'I have retained your Bill on MacDonald and Macpherson – as you desire and I shall urge them to prepare for payment . . . but I have my doubts about their ability they wish to be just, I believe, but the sum is larger than you ought, in prudence, to have trusted them with.'[56] These people were not identified, but this perhaps suggests that Donald was becoming incautious. He had raised another issue, on which Anderson said

> I really feel distressed beyond measure for the poor peasantry whom you plead for; not having it in my power to provide for them this season, owing to the lands having been nearly all locked up before I came into office, but the Duke is more than commonly desirous to patronize and preserve his poor people, I trust during the ensuing summer I shall be able to arrange a place for accommodating a considerable number of them on Lochyside but at present I can come under no engagement – which I request you will have the kindness to convey to them, as it is real cruelty to feed them with false hopes.[57]

The letter is interesting in showing that Anderson was using Donald to communicate with people on the Gordon land in Lochaber. No doubt this was useful because of the extent of the estate that he had to manage from a distance.

By the end of 1807 Donald's difficulties were such that he had failed to pay a whole year's rent of £250 for the farm of Annat. He offered a bill at three months notice, but the factor refused to take this for rents.[58] Many others in Lochaber were in a similar position, but Rev. John Anderson remarked of Donald that 'Saul is also among the Prophets'. Donald complained that the rent of Annat was 'greatly too dear', perhaps forgetting that he had offered to pay as much as the landlord wished to ask. Anderson remarked cynically that 'another is ready to take His Bishopric when he resigns it. It is worth every shilling it pays; and would be of particular

advantage to a Tennant who had it in his power to connect it with a lower farm.'[59] The last point was certainly a valid one. The need to have some lower ground to complement the hill ground had been in Donald's mind in his negotiations about the Mackintosh tacks of 1804. Anderson's tone was noticeably less sympathetic than it had been earlier in the year.

Donald McDonald died at Tulloch at the end of March or beginning of April 1808. In his trust disposition and deed of settlement of 31 August 1807, he had appointed five trustees (see Chapter 8). He was unmarried, but left three sons and three daughters by three different mothers who were all apparently still living. The provisions made for these dependants ensured that the assets remaining after his debts had been paid would be spread thinly. He directed that his trustees were to dispose of his property, heritable and moveable. Out of the proceeds, an annuity of £5 was to be paid to his brother John, and the sum of £100 to the Rev. John Macdonald of Forres 'in token of my regard'. A further £100 was to be divided among the mothers of his children. A modest £10 was left for the poor of the parish of Kilmanivaig, to be distributed by the trustees 'at the sight of the Roman Catholic Priest in the District or the Bishop'. After these bequests had been made, 10 per cent of the residue was to be paid to Campbell Mackintosh 'in regard that the said Campbell Mackintosh has been friendly to me . . . [I] do Esteem and Confide in his taking an active Concern in the management of my affairs after my death'. Finally, the remainder of the residue was to be divided among his six children, the sons to receive double the amounts given to the daughters.[60]

His brothers Alexander and John claimed to be his legal heirs. Æneas Mackintosh dealt summarily with their attempts to take over Tulloch, and they did no better with Donald's possessions on the Gordon estate. As at Tulloch, Thomas Gillespie got Donald's tack of Annat, at a rent of £200, which was £50 less than Donald had recklessly agreed to pay. There was an arrear of £50 that the factor said 'cannot be struck off at present but neither will it be demanded'.[61] [62] He was referring to the custom that an incoming tenant was required to pay off any debts left by his predecessor. Gillespie was being excused that.

Donald's property in Fort William consisted principally of the New Inn, but it was shown in Chapter 8 that this was in Campbell Mackintosh's hands a couple of months or so before his death.[63] Immediately after his death, the trustees leased the New Inn 'as lately occupied by John Campbell' for seven years to another Donald Macdonald who was then at Highbridge.[64] This Donald has been described as one of the Cranachan family who had been keeping the inn at Highbridge since 1806. He left that in the hands of his brother Angus when he got the tenancy of the New Inn at Fort William.[65] The trustees must however have succeeded in selling the property, because in 1813 Alexander MacArthur of Bochaskie in Glenroy

said that he had purchased it a few years earlier. He was then making an offer of £20 rent for a small park that served the inn. Donald McDonald had held it, and then it was occupied by Miss Bell MacLachlan. MacArthur also offered £12 for a croft at Blarmachfoldach.[66] Bell MacLachlan was involved in innkeeping in Fort William. On 4 December 1816, while Colin Macdonald of the Glencoe family was home on furlough, the Glencoe trustees paid Miss Bell MacLachlan, vintner, Fort William, £6 3*s.* 11*d.* 'for stabling of his and other Keppoch horses'.[67]

The trustees had to deal with another of Donald's commitments. At some unknown date, he and 'John MacKenzie residing formerly at Balachelish now at Glengloy and Duncan McIntyre, Junior, Merchant in Fort William' became joint principal tacksmen of the lands and grazings of Easter and Wester Invergloy, on the estate of George MacMartin, otherwise Cameron, of Letterfinlay.[68] By 1816 Campbell Mackintosh was becoming rather desperate to get rid of this responsibility. In a letter to his son Donald in Edinburgh he said:

> When you see your acquaintance Mr. James Saunders [Robertson, WS] will you remind him of writing me in answer to my letter and enquire when the Tack from Letterfinlay expires or at least the breach. . . . This is a matter of consequence as the farm is a very bad concern . . . having no copy of the lease the whole was sent to Mr. Robertson with the papers about the affairs of Donald McDonald, Tullich. I suppose you knew the circumstances.[69]

Perhaps Donald McDonald's judgement had not been quite up to scratch when he got involved in this tack. It is not known how or when the trustees finally got rid of it.

Alexander Dhu McDonald after Donald's Death

Donald's death deprived his brother Alexander of a relative who could write letters for him and occasionally restrain him from his wilder excursions into litigation. Two other people now began to figure in his affairs. One was Allan Cameron of Meoble, a fellow sheep-farmer who was his neighbour just across the River Meoble to the east of Rifern, the other John MacDonald of Borrodale, who was initially his neighbour to the south but later became his landlord when he bought the land on which Rifern stands.

Charles Fraser-Mackintosh identified Allan Cameron, tenant in Meoble of South Morar, as a MacMartin-Cameron whose family hailed from Glengloy.[70] He must have moved to Meoble in the early 1780s. A gravestone in Gairlochy burial ground used by the MacMartins shows that Allan Cameron was at Meoble by 1784.[71] He was an ambitious and determined sheep-farmer. The interest that he took in Innerigan in Glencoe in 1817 was noted in Chapter 6.

John MacDonald of Borrodale was one of Clanranald's tacksmen. He was born about 1754. His family had connections with North America, and by 1780 he was a merchant in Quebec. He returned to Scotland in 1785 and used his business experience to develop trade. He owned a half share of a boat and had dealings with the firm of Macdonald and Ravenscroft in Liverpool. He was always ready to try to help tenants and others in difficulties who appealed to him because of his standing and his knowledge of business matters.[72]

Alexander Dhu had a brush with Allan Cameron of Meoble in 1810 and turned to John MacDonald of Borrodale. He wrote for him to Alexander Macdonell, the writer in Inverness, much as his brother Donald might have done.

> Your namesake, my neighbour, Alexr. McDonald, Tenant of Clachaig etc. [part of the Rifern tack] was with me here yesterday and complains much of his unneighbourly usage from the Meoble people, indeed from what he states they use him rather harshly, it is true that Sandy was never yet blessed with the most flattering reputation, but my own idea of these things is that every man is entitled to proper treatment and good neighbourhood till his own conduct deserves the contrary usages . . . he desired me to inform you that Meoble's son told his son one day in course of conversation that his (McDonald's) Father had lost a good deal by not being on a friendly footing with his own father's (Meoble's) shepherds – the lad asked what had his father lost by it and Meoble's son answered that he had lost from £5 to £10 worth of sheep already by it . . . McDonald after such hints from his neighbour considers his stock not to be free from dangers and wishes to be informed by you what steps he should take for his own security.[73]

Relations between the neighbours did not improve. In 1814 Allan Cameron had a summons issued against Alexander McDonald in the Sheriff Court in Fort William. A squabble that seems to have started over fifteen shillings had escalated. Cameron claimed £150, including even the value of some butter that McDonald's wife was alleged to have taken.[74] Alexander McDonald had an action against Cameron in the Sheriff Court in Inverness, alleging that he had been assaulted without provocation by Cameron near the boundary between their farms. Robert Munro, a messenger in Inverness, and Allan MacDonald, a sheriff officer from Fort William, had been present. Alexander McDonald wanted £500 compensation for his physical injuries and for the damage to his reputation.[75] It is not clear whether either of these actions was ever taken to a conclusion.

In 1813 John MacDonald of Borrodale purchased from Sir William Fraser the land in South Morar that included Rifern, Torrary and Clachaig.[76] Borrodale set about removing Alexander Dhu McDonald.

The summons to appear in court was delivered on 16 January 1815, but the sheriff determined the case on 24 January in the absence of Alexander McDonald. His absence certainly did not mean that he was content to let the matter rest, and he tried to raise various objections.[77] In the end Borrodale got his way and Alexander Dhu had to leave Rifern.

Alexander Dhu then became tenant of Glaschoile in Knoydart and moved there at Whitsunday 1815. Thereafter his interests were concentrated in Knoydart. Within a few months of entering Glaschoile he joined Archibald Dhu MacDonell in taking over the farms vacated by the death of Alexander Macdonald of Glencoe at the end of 1814. These interests in Knoydart and the complications which arose from them will be explained in Chapter 10.

There were difficulties over the money due to Alexander Dhu McDonald for the stock that he left at Rifern. He had offered it to his landlord, John MacDonald of Borrodale, who did not take it and later counted himself fortunate not to have done so. A new tenant, also named John McDonald, appeared and became committed to purchasing the stock. Within about eighteen months of Alexander Dhu's departure, Borrodale was saying that 'Poor John McDonald is ruined . . . lost about 400 of the sheep by poverty owing to the place being overstocked by Sandy'. A decline in prices at that time was mentioned as a factor contributing to these losses. In September 1816 Alexander Dhu had still not been paid for the stock he had left at Rifern and he was pressing for his money.[78] By then he had paid £922 8*s.* 0*d.* in cash towards the value of stock in Knoydart[79] and was no doubt anxious to replenish his funds. However, John McDonald seems to have given up at Rifern and gone to Kinlochailort, where he may have had family connections.[80]

John Mackintosh, messenger and Surveyor of Taxes in Fort William, tried to help. (The sub-letting of his tack of Cranachan and Blarnahanin in Brae Lochaber in 1807 was described in Chapter 8.) For his trouble he found himself involved in a couple of actions in Fort William Sheriff Court. He said that acting in his public capacity he

> had occasion to intermeddle with the McDonalds called Bitags [Alexander Dhu and three of his sons] and the Defender [John McDonald now at Rifern] to adjust a dispute between them . . . [there was] every appearance that they would never settle . . . incited by a spirit of peace and friendship the Pursuer [John Mackintosh] made an overture to the Defender – then the McDonald's debtor – to which both the disputing parties eagerly consented.

In short, John McDonald was to hand over to John Mackintosh all his assets and John Mackintosh would be responsible for paying what was due to Alexander McDonald and his sons.

The assets of John McDonald, the new tacksman of Rifern, were in the hands of John MacDonald of Borrodale, who seems to have held money in this way for his tenants and others. These assets consisted of Borrodale's bill for £446, and £204 in cash. John McDonald then began to complicate matters, apparently with the intention of holding back some of his assets. He uplifted the cash and 'placed the same in the hands of a young gentleman whom none would suspect of having such a sum, as it was improbable he would yet have begun to deal in cash transactions'. The young gentleman was Angus MacDonald, son of John MacDonald of Borrodale. This was an attempt, presumably without the knowledge of Borrodale senior, to draw a veil over the existence of the cash and to deny it to Alexander McDonald. The attempt failed, and a final settlement seems to have taken place eventually. John McDonald who had bought the sheep at Rifern probably came out of it badly.[81] [82] [83]

It is curious that the nickname or soubriquet 'Bitag' used by John Mackintosh has not been noted in any references to Alexander Dhu or his family until about 1816, after the move to Knoydart. Thereafter it was commonly used. In 1818 Robert MacKay, writer in Fort William called one of the family 'young Alister-na-bitaig'.[84] In later years Donald, the youngest son, married and became tacksman of Finiskaig in north Morar, where he remained into the 1870s. His family were known as the Biodag MacDonalds. This is thought to have been a rendering of 'Biodag' (a dirk), which presumably reflects a tendency to use that weapon.[85] In fairness to Donald and his family, it was probably old Alexander Dhu who had displayed such a tendency, as will become clearer in the next chapter.

10

Knoydart and Glendessary

In 1802 Alexander Macdonald of Glencoe leased land in Knoydart from Alexander MacDonell of Glengarry. In 1804 or thereabout he leased Glendessary from Donald Cameron of Locheil. These lands were close geographically and linked by a route over Mamclachard that appears as a road on maps published in the eighteenth and nineteenth centuries.[1] The Glencoe trustees were able to disengage from these leases soon after Alexander Macdonald's death. Disengaging was, however, a troublesome process which left a trail of loss and damage in its wake.

Early Sheep-Farmers in Knoydart

Sheep-farming had been pursued in Knoydart for almost twenty years before Alexander Macdonald came on the scene. In the late eighteenth century most of Knoydart belonged to MacDonell of Glengarry, but a small part was owned by MacDonell of Scothouse. An advertisement in 1790 offering Scothouse to rent revealed in glowing terms that it 'has been occupied as a sheep store farm for six years bygone'. In 1788, upwards of 1,000 wedders had fetched 17s 6d each, and it was claimed that the land could graze 6,000 sheep.[2]

Sheep-farming was also well established on Glengarry's land in Knoydart. In 1793 his 'old' tenants there complained about the effects of sheep-farming in causing removals, increasing rents, curtailing grazings and causing actions for trespass.[3] In the same year it was reported that 'the estate of Scothouse, as also a great part of Glengarry's property . . . are mostly laid out in sheep-walks'. Wool was being shipped to Greenock, Dumbarton and Liverpool.[4]

Loss of population from Knoydart is quite well recorded. There was a large emigration in 1786 and an even larger one in 1802.[5] There may thus have been sufficient movement of people out of Knoydart to make room for others to come into the area. It would not have been the first time that this had happened. In the early 1770s, 'Skye was so short of successful tacksmen that the Chief of Sleat was being forced to bring in clansmen from the mainland to take over abandoned tacks'.[6]

By 1792 Archibald Dhu MacDonell, whose affairs were to become linked to those of Alexander Macdonald of Glencoe, was a sheep-farmer at

Shenachy in Knoydart.[7] On his way back from the Falkirk Tryst in September 1794 he sent a letter from Maryburgh [Fort William] to Alexander Macdonell, writer in Inverness, about the poor prices at the market.[8] It was probably written for him because he could not write, and indeed was not fluent in English.[9] By 1796 Archibald Dhu had moved from Shenachy to the large farm of Carnoch at the head of Lochnevis.[10]

Alexander Cameron, who was to take Glendessary after the death of Alexander Macdonald of Glencoe, also established himself in Knoydart about this time. Initially he 'had the sub-set of Inverguseran from Strone and MacLachlan',[11] but by 1794 he was 'Tacksman of Inverguseran'.[12] He remained there until he died about thirty years later,[13] but his interests were not confined to Inverguseran. In 1802 he was the principal tenant of Inverie,[14] which was about to be taken over by Archibald Dhu MacDonell.[15] In the same year he was also described as a sub-tacksman of Inverskilavulin in Glenloy.[16]

There was some interest from the south. In 1795 Robert Oliver and Thomas Stavert had thoughts about leasing Scothouse, which had been offered to rent.[17] However, they stopped at Fort Augustus where they examined Cullachy and, as explained in Chapter 5, settled there instead. The lease of most of Scothouse was taken by Thomas Gillespie.[18] His brother John was installed there and paid the rent of £354 from 1795.[19] [20] [21] The Scothouse family had substantial debts,[22] and in 1804 they sold their land to Patrick Grant of Glenmoriston.[23] Gillespie's lease was not renewed, and John moved to Kilmuree in Strathaird on the Isle of Skye.[24]

The end of the Gillespie lease also ended southern involvement in sheep-farming in Knoydart at that period. Scothouse was leased for seven years by three Highlanders: Dr Donald McDonald, Fort Augustus, Donald MacDonald in Torgulbin and Donald Grant in Bohuntine. Their cautioners were Alexander MacDonald in Cranachan and Angus McDonald in Fersitriach. The rent was £650.[25] Although he practised medicine in Fort Augustus, Dr McDonald was a member of the Cranachan family in Glenroy and had held Cranachan and Blarnahanin from 1795 until 1804 (see Chapter 4). His partners and cautioners for the Scothouse tack all came from various locations in Glenroy and Glenspean. Donald MacDonald from Torgulbin was entrusted with the management of the farm at Scothouse.[26]

Alexander Macdonald of Glencoe in the Lochnevis Lands

Archibald Dhu MacDonell moved out of Carnoch in 1802, and Alexander Macdonald of Glencoe became tenant of that farm. He also obtained several other farms adjacent to Carnoch. The management of all these, and of the land obtained in Glendessary, was entrusted to John Ban McDonald, who had formerly been at Achtriachtan in Glencoe. By early

1802 he had moved to Knoydart and resided at Carnoch.[27] The Glencoe trustees were to have a good deal of trouble with John Ban and his sons.

Shortage of money does not seem to have been Archibald Dhu's motive in leaving Carnoch. He was said later to have been 'a wealthy man' and 'the deceased Mr. Macdonell [sic] of Glencoe who succeeded Archibald MacDonell in the lands of Carnoch owed to Archibald MacDonell a large sum of money'. Archibald had even lent a sum of money to Glengarry, which he 'uplifted' in 1813.[28]

Archibald Dhu and his son Donald became tenants of the smaller farm of Inverie.[29] [30] Archibald Dhu then had money to spare for investment elsewhere. He attempted to secure part of the farm of Kinlochmorar, which led to a lengthy dispute with Allan Cameron of Meoble in South Morar. Alexander Macdonald of Glencoe was drawn into that dispute as an arbiter.

Archibald Dhu MacDonell and the Kinlochmorar Dispute

Kinlochmorar was part of the Lovat lands in North Morar. It was let to four tenants called Gillies on a lease due to expire at Whitsunday 1804. In 1802, two of these tenants (William and John Gillies) decided that they would go to America when the lease expired. The other two, Donald and Angus Gillies, agreed to lease the whole farm from 1804. Archibald Dhu MacDonell then arranged with William and John to take over the lease of their half of the farm for its last year up to Whitsunday 1804. Presumably he hoped to remain there beyond 1804. He undertook to pay the rent, give them a premium of £40 and buy their few black cattle. He got possession at Whitsunday 1803.[31] To stock the land he bought 500 lambs from Mr Cameron of Clunes at twelve shillings each.[32]

Alexander Cameron of Inverguseran also had his eye on Kinlochmorar, but he suspected that Donald and Angus Gillies were designing to get their neighbour, Allan Cameron of Meoble in South Morar, to join with them. He took a poor view of this. Lovat had given 'a deduction to serve the tenants' and he did not think that they should bring in someone like Cameron of Meoble to share the benefit of that.[33]

Archibald Dhu probably had little prospect of retaining his half of Kinlochmorar in the longer term, and on 1 August 1803 he relinquished it to Donald and Angus Gillies, who were willing to relieve him of it.[34] He was to remove the cattle that he had bought. They would refund the £40 premium that he had paid and be responsible for paying for the lambs that he had brought in.[35]

This agreement was recorded in the presence of the Rev. Mr Ranald MacDonald, Roman Catholic priest in Morar, who kept possession of the written agreement. Donald and Angus Gillies then deposited £40 with the Rev. Mr Charles MacDonald, Roman Catholic priest in Knoydart.

Presumably this money was to be given to Archibald Dhu, but it was later taken back by Angus Gillies.[36] Archibald was soon expressing concern to his lawyer that he had not received it.[37]

It now became apparent that Allan Cameron of Meoble wanted to become 'a sharer in the farm' of Kinlochmorar. Donald and Angus Gillies brought him into the arrangements as cautioner for the price of the lambs, for which they had agreed to be responsible.[38]

For reasons of his own, Allan Cameron insisted that Archibald was still responsible for half the farm of Kinlochmorar and had agreed to let him (Cameron) have it, but was failing to honour that agreement. A meeting to try to resolve this was held at Kinlochmorar, attended by Donald and Angus Gillies, Allan Cameron and Archibald Dhu MacDonell. They were all said to be 'much the worse for liquor'. In spite of the liquor, Archibald had the wit to tell Cameron, quite correctly, that he was no longer responsible for any of Kinlochmorar. Cameron's desire to have a share in the farm was a matter between him and Donald and Angus Gillies.[39] This made no impact on Cameron.

Early in 1804, Archibald agreed with Donald and Angus Gillies to refer to arbitration their failure to repay the premium of £40 to him. The two arbiters chosen were Alexander Macdonald of Glencoe, by then residing at Keppoch, and the Rev. Thomas Ross, minister of the Parish of Kilmanivaig.[40] As was explained in Chapter 2, Alexander Macdonald was a nephew of Mr Ross's wife.

The arbiters concluded that Donald and Angus Gillies should pay Archibald Dhu the £40 with interest and his legal expenses of £6. That was consistent with Archibald's contention that he was no longer responsible for any of Kinlochmorar. Nevertheless, in February 1806 Cameron of Meoble began a civil process against Archibald Dhu before John Campbell Esq., Younger of Glenmore in the Sheriff Court in Fort William. He claimed £100 damages because Archibald had deprived him of the proper use of half of Kinlochmorar in the year to Whitsunday 1804, as the houses on it were still occupied. This had come about because Donald and Angus Gillies had agreed that William and John Gillies could continue to live there until they left for America.[41]

Archibald Dhu was in some difficulty. The record of the agreement with Donald and Angus Gillies to release him from his commitment to take half of Kinlochmorar was still with the Rev. Ranald MacDonald many miles away in Morar. If that paper had been available to the court in Fort William, Allan Cameron's case should have collapsed. The court seems to have ignored that possibility and deliberated on what may have passed between Cameron and Archibald Dhu. As there was no written agreement before the court, the matter had to be decided by what the parties said on oath. Archibald swore that there was no agreement between him and

Cameron. He was then charged with perjury and committed on a warrant to the Garrison at Fort William to await trial at the Circuit Court in Inverness in September 1806. Arrangements for the trial were made and lists of witnesses were prepared. Alexander Macdonald of Glencoe and the Rev. Thomas Ross were among those to be called on Archibald's behalf. They were to speak to various matters including 'Character of Pannel as to correctness and substance'.[42] Archibald must therefore have been known to both of them.

In Edinburgh, Coll Macdonald WS of Dalness set about obtaining letters of exculpation, that is to say a warrant compelling the attendance of the witnesses for Archibald Dhu. On 28 August 1806 he wrote to Alexander Macdonell, Archibald's lawyer in Inverness, enclosing letters of exculpation. He disclosed an odd situation.

> Owing to some Cause or other the Memorandum of the particulars of Archd. Dow's Trial was mislaid in the Justiciary Office. They say they do not always keep them. But if they do not they certainly ought. The enclosed letters of Exculpation were therefore made out from the information I furnished and if there be any Error in the designation given to our client – or anything else the Crown cannot in these circumstances take advantage of it.[43]

When the case came before the Circuit Court in Inverness, the advocate depute stated briefly 'that since his arrival in town he had Received information of Certain Circumstances concerning which he considered it necessary to make enquiry before proceeding to the trial of the Pannel and therefore moved the Court to desert the diet pro loco et tempore'. Lord Cullen agreed.[44] That seems to have been the end of the matter.

This experience was remarkably similar to that of Alexander Dhu McDonald six months earlier, described in Chapter 9. Like Archibald, he had been charged with perjury before Sheriff John Campbell Younger of Glenmore, and committed for trial at the Circuit Court in April 1806. In both instances the advocate depute had immediately moved to desert the diet.

The similarity of these experiences perhaps contributed to confusion on the part of Charles Fraser-Mackintosh between Alexander Dhu McDonald and Archibald Dhu MacDonell. In the Fraser-Mackintosh Collection in the National Archives of Scotland, there is a substantial bundle of papers marked on the outside 'Case against Alexander Dow McDonald for Perjury 1806'.[45] The papers themselves are, however, solely about the charge of perjury against Archibald, and most of the above account has been drawn from them.

Fraser-Mackintosh added to the confusion between the two men in his account, considered below, of the attempt by Glengarry's factor to remove

'Archibald Dhu' from Kinlochnevis in 1817.[46] It will be shown below that it was Alexander Dhu who was the subject of this failed attempt at removal.

Archibald Dhu MacDonell as Tacksman of Kyles Knoydart

After he was rebuffed at Kinlochmorar, Archibald Dhu took a lease of Sandaig to add to the land he held at Inverie. Then, in 1811, he and his sons moved to Kyles Knoydart, with which he was to be identified for many years.[47] His tack extended east to include Torcruin,[48] where he installed his son Donald. To the west, his tack appears to have included the adjacent farm of Brunsaig.[49]

When word of Archibald's intended move got around, the tenants of Kyles and Torcruin were anxious to know whether Glengarry would make it a condition that 'Archy Du' as the incoming tenant would purchase their sheep at valuation. They went to John MacDonald of Glenmeddle, who wrote on their behalf to Alexander MacDonell, writer in Inverness.[50] Archibald did purchase the stock of two of the existing tenants, John and Roderick MacDonald, who decided to go to America. The cash for this was advanced by Mr John Elder of the firm of McDonald and Elder, merchants at Isleornsay in Skye.[51]

The fortunes of Archibald Dhu will be left at this point and taken up again after Alexander Macdonald of Glencoe died in December 1814.

McDonald and Elder, Merchants, Isleornsay

The firm of McDonald and Elder, who figure in some of these transactions, traded as merchants for at least thirty-seven years. In 1790. Mr Donald Smith wrote from Inverness to Mr George Gibson in Rotterdam: 'I am applied by my Two Young Friends Messrs. Alexander McDonald and John Elder to write you on their behalf whose orders you will please execute on the very best terms within your power . . . to strengthen their credit I hereby promise to see you regularly paid.'[52]

In 1798, McDonald and Elder wrote from Camuscross, near Isleornsay in Skye, to Alexander Macdonell, writer in Inverness, to get a bill protested.[53] Most of their business seems to have been as general merchants, with customers in Skye and on the adjacent parts of the mainland. For a time around 1816 and 1817 they had a third partner, Duncan M'Innes.[54] There are indications that besides the usual business of merchants, they were willing to advance sums of money. Their assistance to Archibald Dhu at Kyles Knoydart is one example. In 1819 they took steps to recover £244 5*s.* 0*d.* plus interest from another Archibald MacDonald, the tacksman of Ord in Skye. The amount is large enough to suggest that money had been advanced to him.[55] In 1818 they were interested in obtaining oak bark from the Keppoch wood in Lochaber. They were still in business in Sleat in 1827, and there is a

reference in the same year to 'Mr. Colin Elder sometime merchant at Kyleakin, now residing at Isleornsay'. He was John Elder's son. [56] [57]

Alexander Dhu McDonald in Knoydart

The movements of Alexander Dhu McDonald from Tulloch in Brae Lochaber to Achintore and then to Rifern in South Morar have been traced in Chapters 4 and 9. He was removed from Rifern at Whitsunday 1815, and he leased Glaschoile in Knoydart from the same date. As explained in Chapter 9, he left his sheep at Rifern to be purchased, according to the usual custom, by the incoming tenant. However, he took his furniture and some cattle, perhaps for domestic use, with him to Knoydart. [58] [59]

His move took place very soon after the death of Alexander Macdonald of Glencoe in December 1814. Within a few months of his arrival at Glaschoile, Alexander Dhu joined with Archibald Dhu MacDonell at Kyles Knoydart in taking over the tenancy of the farms at Kinlochnevis, which had been held by Alexander Macdonald of Glencoe. With hindsight, it is clear that this was not a wise move. Perhaps he did not give sufficient weight to warning signs about the future of the wool market. The disastrous entanglement with Glengarry which followed will be explained below.

Glendessary

On 23 March 1804, Duncan Cameron of Fassifern wrote that 'Locheil has had a Sett lately', involving lands on both sides of Loch Arkaig and further west. Macdonald of Glencoe was apparently interested in land on the south side of Glendessary, for which a rent of £610 was being offered. Duncan Cameron believed that 'Glencoe wd. get it if he paid the sum but he says that it is too much'.[60]

Alexander Macdonald of Glencoe must have got Glendessary. In 1805 he obtained a decreet of removing against several sub-tenants there.[61] Little else has, however, been found about how he used Glendessary.

Knoydart after the Death of Alexander Macdonald of Glencoe

The minutes of the first meeting of the Glencoe trustees on 18 May 1815 refer to 'Deceased's possession and stocking of Lochnevis and Glendessary'. It was agreed to 'get clear' of these and that 'Stocking would be delivered to the proprietor of each or his incoming tenant at valuation'. The trustees had, however, some difficulties over the position of John Ban McDonald.[62]

John Ban McDonald

It can be surmised from the records of the Glencoe trustees that John Ban had been a partner with Alexander Macdonald of Glencoe, holding a one-sixth share of the Lochnevis and Glendessary interests and undertaking

the local management of them. The trustees said:

> From the connection Glenco had with John Ban McDonald in his Lochnevis and Glendessary concerns and from his being by the giving up of these thrown out of employment and even residence with a large family, conjoined with the certainty that the small possession of Corychorichan [Corrychurochan] held by Glenco under Locheil could never have been made a source of profit under the management of the Trustees, the previous arrangements with Locheil's concurrence for giving up that possession to John McDonald are approved of.[63]

In other words, Duncan Cameron had already moved him to Corrychurochan, on the east side of Loch Linnhe, a few miles south of Fort William. The trustees corroborated what he had done.

At the third meeting of the trustees in January 1817, the problems with John Ban and his family were exposed in more detail. It is unlikely that Alexander Macdonald had a formal agreement with John Ban, but his position seems to have been sufficiently well established to cause the trustees difficulty in disengaging from him. He claimed one-sixth of the value of the 'concerns of Lochnevis, Glendessary, and Keppoch', amounting to £1,980. The trustees countered with several demands, including five-sixths of 'a boat for Lochnevis delivered to him at Whit 1815'.[64] The boat was probably intended for the export of wool.

The trustees held against John Ban a letter from him to Alexander Macdonald on 2 March 1814 acknowledging a debt of £1,316 8*s.* 0*d.* They also wanted from him the value of the stock at Corrychurochan.[65]

What may have incensed the trustees most was a claim submitted by John Ban's sons Angus and Donald for wages. They thought it 'extraordinary that claims for wages should have been so long over, or that Angus could have purchased packs of sheep when keeping was said to have been allowed in place of money wages'. They said that 'this needs satisfactory evidence and explanation' and instructed the agents to get on with it.[66]

Problems continued. John Ban did not pay the rent of Corrychurochan. Adam Macdonald of Achtriachtan was his cautioner, but he was in difficulties himself.[67] In December 1817 the trustees had to pay £428, John Ban's rent from 1815 to 1817, to Locheil's factor – 'payment of which you are bound', as Duncan Cameron explained to the trustees.[68] John Ban was also being pursued in the Sheriff Court by several creditors for various debts.[69] By 1820 the trustees had had enough of John Ban. It was arranged that Corrychurochan would be let to James Greig, who was probably the nephew of Thomas Gillespie's wife, Christian Greig.[70] [71] The sheriff at Fort William was asked for a decreet to have John Ban removed.[72]

John Ban's sons were another matter. A third son, Allan, had joined Angus and Donald in claiming a total of £420 18*s.* 7½*d.* for 'wages, sheep, wool, etc'. To avoid litigation, the claims were submitted for arbitration to 'Mr. McAlpin, General Merchant and Commercial Agent of respectability at Corpach'.[73] The trustees were losing their taste for expensive litigation.

John Ban's son Donald became a drover and left Corrychurochan about the time his father was removed. He went to Strathconon in Ross-shire, where several MacDonalds from Innerigan in Glencoe had settled (see Chapter 11).

New Tenants for the Lochnevis Lands

Descriptions of the land at the head of Lochnevis leased by Alexander Macdonald of Glencoe are sometimes confusing. The whole is often referred to as Kinlochnevis. Nevertheless, two parts can be identified. One is the farm of Carnoch (or Carnach), lying to the north of the River Carnoch. The other is 'the general tenement of Kenlochnevis'. That is, the land in Knoydart lying to the south of the River Carnoch. It contained several farms. The one most often mentioned is Sourlies (sometimes rendered as Souryess or Sourchaise), where there was a change house or inn. That lies in the south of the area. Some accounts and maps also refer to the farms of Gorten and Achglyne lying between Sourlies and the River Carnoch. In others these appear to be included in Sourlies.

The Glencoe trustees would have been anxious to realise the value of the stock on these lands and get rid of responsibility for the annual rent. There was therefore a pressing need to find new tenants, but the arrangements made to replace Alexander Macdonald of Glencoe were complicated and probably reflected the difficulty in finding suitable tenants at short notice for such a large undertaking as Kinlochnevis.

At Whitsunday 1815 two steps were taken. The first was that the sons of Alexander Dhu McDonald, recently arrived at Glaschoile from Rifern in South Morar, and the sons of Archibald Dhu MacDonell at Kyles Knoydart were given a lease of Sourlies. That apparently meant the whole of Kinlochnevis south of the River Carnoch. The arrangement was confirmed in missive letters from Mr John Hood, Glengarry's factor, on 12 May 1815. The sons were not named individually, but it is interesting that the two families were jointly involved.[74] Although he was not mentioned originally, it emerged later that Archibald Dhu MacDonell at Kyles Knoydart was partly responsible for the rent and purchase of stock on this part of the land.[75] [76]

The second step was that the farm of Carnoch, i.e. the remaining ground held by Alexander Macdonald of Glencoe, was let from Whitsunday 1815 to an Alexander McDonell at Inverie. His identity is not clear, but by July he

had come to the conclusion that he was not going to be able to cope with this and Mr Hood agreed to relieve him of it. He then gave Carnoch, with the exception of the hill grazings, to Alexander Dhu McDonald, recently established at Glaschoile, on a fourteen-year lease backdated to Whitsunday 1815.[77] [78] The hill grazings were taken by Captain Archibald MacDonald of Barrisdale. He was then living at Auchtertyre in Lochalsh (Ross-shire), but his family had held the tenancy of Glenmeddle in Knoydart since his grandfather's time.[79] [80] [81]

As part of the deal, Alexander Dhu was allowed by Mr Hood to leave Glaschoile at Whitsunday 1816, and that farm was then let to John McPhee at a lower rent, but this was on condition that Alexander Dhu would pay to Glengarry the difference between the rent he had agreed to pay and the lower one that McPhee was to pay. According to Mr Hood, all this was 'explained to Sanders Dubh [Alexander Dhu] in the Galic language'.[82] Nevertheless, Alexander Dhu was to claim later that he had been relieved of Glaschoile completely.

The rental for all of the land previously held by Alexander Macdonald of Glencoe was £450. This total was to be divided among those involved so that each would pay 'the Rent proportioned by Mutual people chosen for that purpose'.[83]

All of these arrangements were made by Mr Hood, and at the meeting of the Glencoe trustees on 15 January 1816 Glengarry could only refer vaguely to 'the subdivided manner in which Lochnevis is now possessed'.[84] The scene was set for confusion and disagreements, which soon followed in ample measure.

At Martinmas 1815 the stock on all these farms was valued at £2,441 2*s.* 6½*d.* Of this, £1,457 5*s.* 10½*d.* was the value of the stock at Carnoch and £983 16*s.* 8d was the value of stock in the general tenement of Kinlochnevis.[85] [86]

The price of the stock was to be paid to Mr Thomas MacDonald, writer in Fort William, one of the Drimintorran family, who was acting locally for the Glencoe trustees. Payment of the first half was due on 28 November 1815, but Thomas MacDonald wrote 'that the Lochnevis people had not arrived here [presumably Fort William] to settle for stock'. However, on the next day, 29 November, he noted 'Attending with them and Mr. Hood received from them all or partial settlement of the moiety . . . payable at this term'.[87] Alexander Dhu McDonald at Carnoch made a payment of £922 8*s.* 0*d.*, [88] and Archibald Dhu MacDonell from Kyles Knoydart made a payment of £249 5*s.* 6*d.*[89] Alexander paid cash, but Archibald turned again to Messrs McDonald & Elder, who provided a draft on the British Linen Company in Inverness. Of the £922 8*s.* 0*d.* paid by Alexander, £676 8*s.* 10*d.* was credited to the stock at Carnoch and £245 19*s.* 2*d.* to stock at Kinlochnevis because of 'his entire interest in the Stock of the general

Tenement of Kenlochnevis'.[90] Archibald's smaller payment was credited to the stock at Kinlochnevis.[91]

Something must have troubled Thomas MacDonald, because on 14 December he asked Mr Hood for copies of his calculations of the Lochnevis stock. On 31 May 1816 he noted 'going over the Lochnevis Calculations and correcting them, they having been wrong – engaged some hours'.[92] The mistakes were not specified.

The Disputed Melioration Money

The new arrangements were complicated by a longstanding dispute about melioration money. The incoming tenants were expected to pay the Glencoe trustees, representing the outgoing tenant, for the meliorations. In return, the new tenants would have a claim against Glengarry (or the next tenants) when they gave up the tenancy. This was a common arrangement. However, Archibald Dhu had still not received the melioration money due to him when he had vacated Carnoch in 1802. Glengarry probably expected Alexander Macdonald of Glencoe to have paid to Archibald Dhu the value of the improvements that he had made while he was tenant. Archibald had apparently not insisted on payment from him; instead he tried without success to get Glengarry to pay it to him. Eventually he resorted to the simple expedient of withholding £50 12*s.* 8*d.*, which included interest from 1802, from the rent for his other farms. Glengarry would have none of this, and by 1815 he was insisting in the Sheriff Court in Inverness on payment of the full rent.[93]

It is therefore hardly surprising that the new tenants were unwilling to pay the Glencoe trustees for meliorations. At the time of the second meeting of the trustees in January 1816, the matter had not been settled. The minute says 'agents to seek adjustment'. It was still unresolved several months later, and Duncan Cameron's patience was nearing exhaustion. On 28 October 1816 he was 'writing Mr. McDonald [Thomas MacDonald] fully about getting the Lochnevis meliorations immediately valued'. At the third meeting of the trustees in January 1817, the issue had still not been resolved. The minute said: 'if the endeavours already made for a settlement in this way shall fail upon one other attempt the meeting instruct application to be made to the Sheriff for the apportionment of judicial valuation having in view the previous valuation by men from Glencoe.'[94] The 'men from Glencoe' must surely have been Alexander Dhu and Archibald Dhu.

The matter of the melioration money was probably not taken to a conclusion because larger difficulties arose.

Glengarry's Liability

The Glencoe trustees had insisted that Glengarry as the landlord should also be made liable to them for the price of the stock.[95] According to the usual practice, the second moiety would be payable at Martinmas 1816 (i.e. in November), and the tenants were responsible in the first instance. The trustees must have had doubts about their ability to pay and looked to Glengarry. He was in a parlous financial state. By 1804 he had debts of £11,000, and by 1808 he was resorting to bonds over his estate to raise money. He was waiting hopefully for a large compensation payment from the Canal Commissioners.[96] No payment was made to the trustees for Lochnevis stock until 31 January 1817, when Glengarry gave them a bill for £1,358 0*s.* 4*d.* (including interest) at four months.[97] This is rather more than the 'moiety' paid in November 1815, but perhaps Thomas MacDonald's corrections and the interest account for this.

Glengarry then set about trying to recover this amount from the McDonald and MacDonell families. He had two summonses issued, answerable in the Sheriff Court at Fort William on 8 April 1817. One was against Alexander and his sons Archibald and Alexander at Carnoch for a total payment, including interest, of £832 11*s.* 2½*d.*[98] The other summons was against Angus, Duncan, Allan and Archibald, all said to reside 'in the tenement commonly known by the general name of Kenlochnevis', for a total, including interest, of £321 7*s.* 9*d.*[99]

The total demanded by Glengarry in these two actions was therefore £1,153 18*s.* 11½*d.* A draft for £200 on Archibald Wilson, a sheep drover who purchased surplus stock from Alexander Dhu at the end of the 1816 season, had been paid to Mr Hood after the initial payments in November 1815.[100] [101] If that £200 is taken into account, Glengarry was demanding approximately the amount needed to equal the bill that he gave to the trustees on 31 January 1817.

Sequestration and Removal

The new tenants of Kinlochnevis were already in arrears, and in June 1816 the sheriff at Inverness found them 'due the balance of rent' for the farms and made an order for sequestration of some of their effects. It was alleged that they were 'disputing among themselves as to their proportions of the said tacklands . . . in place of paying to the complainer [Glengarry] the balance of rent'.[102] However, both families had handed over substantial amounts for the purchase of stock in November 1815, and they may have been short of money. Alexander Dhu had also been required to purchase the stock at Glaschoile, and he had done this, apart from quibbles over a few small matters.[103] In addition, as was explained towards the end of Chapter 9, the money due to him for stock left at Rifern had still not been

paid by September 1816. His finances must have been under great strain.

Glengarry's substantial demands for reimbursement of his payments for stock came before Fort William Sheriff Court in April 1817. The position of the tenants must have been impossible, because on 23 May 1817 Alexander Dhu McDonald and his son Archibald agreed in writing and on behalf of the whole family to renounce Carnoch and their share of Sourlies at Whitsunday 1817. All of the sheepstock was to be handed over to Glengarry and valued by Allan Cameron at Meoble and Archibald MacDonell at Glenmeddle. Presumably money due to Glengarry was to be deducted from the value of the sheep.

Glengarry then made a concession which was to prove troublesome. 'As the said Alexr. McDonald and Sons have no other holding whatever Glengarry agrees to give them four cows grazing upon the farm of Sourchaise and the house formerly occupied as the Inn for one year only say till Whitsunday 1818 at such rent as shall be laid on by any two of the valuators. Glengarry also agrees to give them the proportion of the arable land that is attached to the Inn upon the lands of Sourchaise.'[104]

Presumably Archibald Dhu and his sons also renounced their interests in the Kinlochnevis lands. They remained at Kyles Knoydart, but in June 1824 Glengarry obtained an order removing them from these lands.[105] Kyles Knoydart was advertised to let in the *Inverness Courier* of 6 July 1825, with entry set at Whitsunday 1826.

On 19 July 1817, Glengarry wrote to Alexander Macdonell, his lawyer in Inverness. His letter has been found recently among papers which appear to have been in the possession of Fraser-Mackintosh and used by him as source material. They are now in the Highland Council Archive in Inverness. The first part of the letter was social and referred to Alexander Macdonell's wife and family. Then Glengarry said:

> I am bothered with Bitag. I gave him the grass of four cows in Sourchaise for this year <u>by Missive</u>, when he renounced <u>by comprisement</u> the sheep stock of Kinlochnevis, <u>still far short of his debt to me</u>, but he keeps in his sons names or his own <u>four more Milchers</u>, (& I believe a young horse) without <u>authority</u> or <u>right</u> of any kind; can I not seize those in part [payment] of his debt still due to me & remove him off the farm wh. he surrendered to me – (I mean to its extremity <u>Sourchaise</u>, where his sons live), by my own authority, or am I <u>necessarily</u> to have him ejected, or go otherwise more formally to work? When Mr. McLean & the Ground Officer went to move him the other day he ran into the house for a Gun, <u>loaded it in their presence</u>, & cocked it, & then taking out an Old Broad Sword worn by his Grand Father & backed by his sons with <u>oak sticks</u>, they outnumbered & browbeat the Factor & his adherents, & so maintain <u>illegal</u> & unwarrantable possession of my property <u>by violence alone</u>.[106]

The Fraser-Mackintosh Account

Charles Fraser-Mackintosh quoted the above passage in his *Antiquarian Notes* of 1897, with the additional words 'at Culloden' inserted after the mention of the broadsword worn by Bitag's grandfather. In introducing the quotation, Fraser-Mackintosh identified 'Bitag' as Archibald Dhu MacDonald, commonly called Archie-du-na-bitaig, who had been dispossessed from Rifern of South Morar in 1815. He said that 'Bitag' got a share in the large farm of Kinlochnevis 'in consequence of him possessing an uncommon address', and that he had six or seven sons, 'all worthy chips off the old block'.[107]

Several processes in Fort William and Inverness Sheriff Courts show that Fraser-Mackintosh's identification of 'Bitag' was incorrect. Borrodale's removal process in Fort William shows that it was Alexander McDonald, known as Alastair-du-na-bitaig, who had been removed from Rifern in 1815.[108] Other processes show that he went to Kinlochnevis and that he had five sons.[109] [110] [111] The removal orders to clear Sourlies which Glengarry obtained in Inverness in 1818 and 1819, described below, were certainly against Alexander Dhu and not Archibald Dhu.

Another of Glengarry's letters shows that *he* certainly knew that 'Bitag' was Alexander Dhu McDonald.[112] The reference to 'an uncommon address' is in a document used during litigation in 1818. This states that Alexander Dubh na bitaig had four sons: Angus, Allan, Archibald and Alexander.[113] A fifth son, Donald, was said to be 'a young boy' in 1818.[114] [115] It would appear that Fraser-Mackintosh was responsible for the incorrect identification of 'Bitag', for the enhanced number of 'worthy chips off the old block' and for the provenance given to the broadsword by adding the words 'at Culloden'.[116]

Prebble has given a briefer account of the same episode, referring to Archibald Dhu Macdonell at Kinlochnevis, his seven stalwart sons and a broadsword carried by his grandfather at Culloden. He quotes the words 'so maintain illegal and unwarrantable possession of *my* property by violence' which appear in Glengarry's letter of 19 July 1817. He relates that 'By 1817 he [Glengarry] had found a way to clear Kinlochnevis of Archibald Dhu, his seven sons and his broadsword, and to put a Lowland sheepman in their place.'[117] Reference will be made later to the succession at Kinlochnevis.

Curiously, Alexander Dhu, the genuine 'Bitag', had once before wielded a broadsword, in the course of a squabble with Ensign Angus MacDonell at Tulloch in Brae Lochaber in 1783. Alexander Dhu 'ran off drawing his durk to his brother's house'. Ensign MacDonell and others insisted that he come out of the house, and 'he came out accordingly with a Broad Sword and as the said Ensign attempted to be at him Alexander drew the sword to

defend himself but the people interfered between and neither of them was allowed to strike one another'.[118]

Glengarry must have tried the patience of the lawyers who served him. When Campbell Mackintosh ceased to deal with his business in 1793, Æneas Mackintosh said 'I am not surprised . . . for you can have little pleasure in acting for a person who seems to act without a plan, and thinks himself above advice'.[119] Alexander Macdonell's advice in response to Glengarry's letter of 19 July 1817 must have been to proceed more formally.

On 4 April 1818 Glengarry obtained a decreet in the usual form in the Sheriff Court in Inverness for the removal of 'Alexander Dhu McDonald residing in Souryess or Sourchaise and Angus McDonald his son residing there pretended tenants and possessors of the change house public house or Inn at Souryess or Sourchaise with the grazings and pertinents thereto annexed part of the land of Knoydart'.[120] This was drafted on a careless assumption that although it was Alexander Dhu and his sons who had been allowed to remain in Sourlies, Angus as the eldest son was the main tenant.[121] Allan then objected that he had rights as a tenant.[122] The attempted removal must have failed, because a year later, on 3 April 1819, Glengarry obtained another decreet. This one was more widely drawn against Alexander Dhu McDonald, late tacksman of Carnoch, and his sons Angus, Allan, Archibald, Alexander and Donald. The first four sons were said to be 'now residing in Sourges', but Donald was in Arisaig.[123] This order was effective. A month later, in May 1819, Alexander Dhu was described as 'late tenant in Carnoch now residing in Arisaig'.[124]

Glengarry v. Alexander Dhu McDonald

Alexander Dhu McDonald's transactions with Glengarry became the subject of lengthy proceedings in the Sheriff Court in Inverness. On 9 March 1818, Mr McLean told Alexander Macdonell, writer in Inverness, that 'in regard to the balance against Saunders Dubh [Alexander Dhu]' Glengarry wished to proceed before a jury 'against him, his sons and Archy Kyles conjointly and severally as the joint occupants of Kinlochnevis'.[125] On 5 July 1818, Mr Hood reported that 'the Old Baitaig has summonsed Glengarry for a balance due to him in the Carnoch transactions'.[126] However, on 4 December 1818 Alexander Dhu and his son Archibald went to Glengarry House and apparently reached an agreement with Glengarry based upon the appointment of Thomas Gillespie as arbiter between them. Glengarry instructed that all judicial proceedings were to be suspended until Gillespie's determination was known.[127]

The prospect of an amicable settlement did not last. Litigation began on 13 May 1819[128] and was continued for several years after Alexander Dhu's death in 1822 by four of his sons. Each party raised claims against the other. There were three issues. Glengarry claimed the rent of Glaschoile

for 1815/16 and the shortfall in 1816/17; Alexander Dhu rejected this, claiming that he had been relieved of his commitment to Glaschoile when he took Carnoch in 1815. Glengarry claimed that a rent of £200 per annum plus public burdens was due for Carnoch, but Alexander McDonald claimed that the rent due was less than this.[129] [130] Finally, the value of the sheep at Carnoch surrendered to Glengarry by Alexander McDonald at Whitsunday 1817 was disputed.

For a long time little progress was made, but eventually, on 15 June 1822, the sheriff granted at the instance of Alexander Dhu McDonald

> a diligence for summoning Alexander Cameron at Inverguseran, Alexander Cameron at Scothouse, and Archd. MacDonald at Lee to appear personally in court on the 12th day of July to be examined on Oath as to the proportion of rent ascertained by them to offer to the said farm of Carnoch and in like manner . . . grants diligence to the said Alex. Dhu MacDonald for citing John Hood factor or late factor for Glengarry to appear in court in the said 12th of July next to be examined on Oath as to the promise or engagement made by him to relieve the said Alex Dhu MacDond. of the farm of Glaschoile as mentioned in the debate.[131]

Before these examinations could take place, Alexander Dhu McDonald died suddenly in Fort William on or about 28 June 1822. At the request of his son Archibald, John Taylor the surgeon carried out a post-mortem examination to discover the cause of his death. The surgeon then attended before the sheriff, presumably to be precognosed.[132] Unfortunately, the precognition does not seem to have survived, and no record has been found of any prosecution concerning this death.

Alexander Dhu McDonald had become a nuisance and possibly a threat to Glengarry. The time of his sudden death in relation to the legal proceedings and the suspicions apparently surrounding it give rise to the possibility that it was instigated by Glengarry. He was certainly capable of violence. He took part in a duel in 1798 in which his opponent died, he was reputed to have killed a gamekeeper, and he committed several 'bloody assaults'.[133] In 1805 Dr Donald McDonald at Fort Augustus, who had crossed him, 'was severely beaten by Glengary and his "tail"'. In a civil action in the Court of Session, Dr McDonald was awarded £2,000 damages, but there were no criminal proceedings.[134] Glengarry's violence towards Alexander Cameron of Inverguseran will be described shortly. Perhaps Alexander Dhu McDonald died a violent death because of his quarrel with Glengarry.

On 12 July 1822, when the witnesses should have been examined, the entry in the court minute book is simply 'to Avizandum'.[135] In May 1823, proceedings started again, with Alexander Dhu's sons Angus, Allan, Archibald and Alexander, but not Donald, in his place. On 1 December

1823 they were granted a diligence similar to the one granted to their father on 15 June 1822, but Alexander Cameron of Inverguseran was now dead and only Alexander Cameron at Scothouse and John Hood the factor were cited to appear.[136] They were examined on 24 February 1824, but they cannot have been helpful to the McDonald brothers. The sheriff substitute held that Alexander Dhu had been liable for rent for Glaschoile as claimed, and that the rent due on Carnoch had indeed been £200 per annum. He also upheld the value of £1,032 18*s.* 11½*d.* put on the sheep at Carnoch in 1817.[137] (This is to be compared with a value of £1,457 5*s.* 10½*d.* at Martinmas 1815, but prices had fallen and the stock might have been reduced.) A State of Accounts between the parties showed that the McDonalds owed £59 8*s.* 2½*d.*, plus some public burdens to Glengarry.[138]

The financial dealings had been complex and the information available is limited. There seems little doubt that Alexander Dhu McDonald and his family lost a substantial amount of money. There is, however, an intriguing suspicion that Alexander Dhu had some money salted away.

Alexander Dhu McDonald's Hidden Funds

In late 1822 or early 1823, Mr Robert MacKay, writer in Fort William, raised an action in the Sheriff Court there against Allan, Archibald, Angus, Alexander and Donald McDonald, all resident in Fort William, for the recovery of legal fees of £9 18*s.* 3*d.* incurred by their father, Alexander Dhu McDonald, while he was at Rifern and at Carnoch. Under Judicial Examination on 4 April 1823, the five brothers each denied knowledge of 'any money belonging to his late Father in the hands of the Right Reverend Mr. MacDonald, Bishop of Lismore'.[139] Roman Catholic Bishops did not then have territorial titles, but Ranald MacDonald lived at the Catholic college at Kilcheran on Lismore[140]

It is just possible that some of Alexander Dhu's money had found its way to the Bishop. In February 1819, John MacDonald of Borrodale said: 'my old tenant Alexander McDonald holds two Bonds of mine for upwards of £300 stg. One of these was due . . . but he never called for it.'[141] So, while Glengarry was pursuing him for large sums of money, Alexander Dhu McDonald was letting more than £300 lie quietly with his old landlord, Borrodale.

Subsequent transactions, known in part to Robert MacKay, must have reduced this amount to a little over £100.[142] [143] John MacDonald of Borrodale wrote to Bishop Ranald MacDonald on 13 Febuary 1823, proposing to send him £100. That was eight months after the death of Alexander Dhu but before his sons revived the action against Glengarry. The bishop replied on 22 February 1823.[144] He was short of money and said, '£100 is a fortune to me just now'. The most interesting sentence in the bishop's letter was the following: 'By the bye, I do not understand what money that

may be but people were not the most correct in their accounts, but it is all one so that I get so timious a supply.' Clearly the money which Borrodale was offering to send was not his own, but he had not made its source clear to the bishop.

Alexander Dhu McDonald has left several unresolved questions. Why had Alexander Macdonald of Glencoe helped him to get a place at Achintore in 1796? What was his relationship to Archibald Dhu MacDonell, with whom he was so often confused? What did the Glencoe trustees mean when they referred to 'men from Glencoe'? What was meant when Alexander Dhu was said to possess 'an uncommon address'? How did he obtain sufficient funds to establish himself as a sheep-farmer and be worth well over £1,000 when he went to Carnoch? Did Glengarry instigate his death? Was it his money that Borrodale sent to the bishop?

Glendessary after the death of Alexander Macdonald of Glencoe

Arrangements for Glendessary were distinct from those for the Knoydart lands, because it was the property of Cameron of Locheil. It was relinquished by the Glencoe trustees and leased to Alexander Cameron of Inverguseran. He paid substantial sums to the trustees towards the price of the stock. A bill lodged with Sir William Forbes (bankers) on 12 August 1815 produced £420 in cash. A draft on the Commercial Bank produced £500 in cash on 1 December 1815. There was a cash payment of £150 on 1 February 1816. Nevertheless, there was a dispute with the trustees about payment for the stock. The general business account of the trustees shows a charge of £1 10*s.* 0*d.* on 11 November 1817 'to account for correspondence etc. – dispute with Inverguseran as to the stock of Glendessary'.[145] When Robert Scott wrote to Ewen Macdonald in India in November 1821, the dispute was still unresolved. The Court of Session had remitted it for jury trial, but to avoid the expense of this Duncan Cameron had agreed that it should be submitted to Messrs Jameson and Robertson, advocates, presumably for arbitration.[146]

After he took Glendessary, Alexander Cameron had problems with Glengarry, whose behaviour was arbitrary and sometimes violent. In August 1817 Glengarry obtained an interdict in the Sheriff Court in Inverness preventing Cameron from any movement of wool, produce, etc. over his property of Glenquoich or parts of Knoydart, or 'introducing himself in any manner of way unlawfully with such goods or commodities on the private ways, roads or paths of the complainer'.[147] In spite of this squabble, the tack of Inverguseran that Alexander Cameron had held for at least fifteen years was renewed in February 1818 for twelve years. Glengarry purchased the Scothouse estate at Whitsunday 1819. He then proposed to Cameron that he give up Aultfearn (part of Inverguseran) and take Torrery (part of the Scothouse land) instead. Cameron agreed

and began to pay rent for Torrery. In spite of this, Glengarry took the opportunity of his absence at markets to let Torrery to someone else. Cameron did not get into Torrery until 1821. Then Glengarry began a process to remove him from Inverguseran;[148] it is not clear why he did this. Thomas Gillespie had advanced £1,050 to Cameron on 12 November 1821, but there is nothing else to suggest any financial difficulties.[149] Cameron resisted the attempt to remove him and argued that the terms of his tack granted in 1818 entitled him to remain.

In 1822, while these legal proceedings were in progress, Glengarry led a physical attack on Torrery. He was rebuffed but returned the next day with more men led by himself and his factor, Major MacDonald. Some of Cameron's stock was scattered and driven away. It was said that 'these proceedings among many similar acts of injury and oppression form the subject of an action . . . against Glengarry'.[150] Alexander Cameron died about this time. His six children continued his disputes in the courts with Glengarry,[151] but their effort to hold on to Inverguseran must have failed. The lands he had held were advertised in the *Inverness Courier* of 6 July 1825 as being to let, with entry at Whitsunday 1826. On 2 January 1826, Robert MacKay, writer in Fort William, said to his colleague Donald Mackintosh in Edinburgh: 'We will now require to be thinking about the delivery of Inverguseran's Stock to Glengarry and the Arbiter to be appointed by us – I suppose there can be no objection in delivering the Flock to Glengarry according to the practice of the Country, or to the incoming tenant if a respectable man.'[152]

The Next Tenants in Kinlochnevis

While Alexander Dhu McDonald and Archibald Dhu MacDonell were in difficulties in Kinlochnevis, Hugh MacDonald, tacksman of Crowline in the north-west of Knoydart, was also in trouble. Glengarry had his possessions sequestrated in 1816 and removed him in 1818 because he owed almost two years' rent for Crowline and another farm, Ardnaslichnich.[153] [154] [155] Hugh MacDonald had other debts. He had taken the farm of Cross in South Morar, and owed £200 for stock to Donald MacLellan, the former tenant.[156] In spite of that record, Glengarry let Kinlochnevis to Hugh MacDonald and his eldest son Hugh. History repeated itself, and in May 1823 he obtained yet another decreet of irritancy and removal against them for failure to pay their rent for Kinlochnevis.[157]

Glengarry was evidently having trouble finding tenants who were financially sound to take his farms in Knoydart. His quarrelsome and irrational disposition no doubt aggravated the situation. Nevertheless, it has to be acknowledged that there was an underlying problem.

11

The Farm Concerns in the North

Early in the nineteenth century Alexander Macdonald of Glencoe and Adam Macdonald of Achtriachtan became entangled in complicated arrangements in a vast area straddling the boundary between Ross-shire and Inverness-shire. The land belonged mainly to Mackenzie of Fairburn, Fraser of Eskadale, Fraser of Lovat, and Mackenzie of Strathgarve. Several members of the MacDonald family from Innerigan in Glencoe moved into the area and became sub-tenants of both Alexander Macdonald of Glencoe and Adam Macdonald of Achtriachtan

As related in Chapters 1 and 6, Adam Macdonald inherited Achtriachtan at the end of the eighteenth century encumbered with large debts incurred by his father. Fraser-Mackintosh said:

> Through his close connection with Alexander Macdonald of Glencoe he doubtless imbibed the idea of making his fortune by sheep-farming, and lost frightfully by becoming tenant or liable for sub-tenants of large farms in Strathconon, belonging to General, afterwards Sir, Alexander Mackenzie of Fairburn.[1]

The interests of both Macdonalds, of Glencoe and Achtriachtan, in these northern farms began in 1803 when Alexander Mackenzie of Fairburn let several long tacks in the Forest of Monar and in Strathconon. In this chapter these lands will be described, and explanations given of how Alexander and Adam Macdonald became involved in them. This will be followed by accounts of the troubles which developed.

Monar and the Grazings of Monar (Ross-shire)

This is a large inland area in Ross-shire in which Alexander Macdonald of Glencoe became committed from 1803. The farm of Monar itself lies in the upper reaches of Glenstrathfarrar, on the Ross-shire side of the boundary with Inverness-shire. The middle and lower parts of Glenstrathfarrar are in Inverness-shire and the situation there will be explained shortly.

In the summer of 1792 the Forest of Monar was advertised in the *Caledonian Mercury*. It was 'supposed to be able to pasture 8,000 sheep'.[2] It is not known if this attracted an immediate response, but by the turn of the century it was leased by 'Murdoch MacLenan, and others'.

MacLenan was 'tenant of Auchintee of Lochcarron', but the others have not been identified. In late 1802 and early 1803, Mackenzie of Fairburn concluded a new tack for twenty-one years of 'the Lands and Grazings commonly called the Forest of Monar with the Houses, Gardens [etc.]'. The tenants named were Murdo MacLenan, the Rev. Dr Alexander Downie, minister of Lochalsh, John MacDonald, shepherd at Achnashien and later at Muilzie in Glenstrathfarrar, John Mitchell, tenant at Achnadaul and later at Donie in Lochaber, and Alexander Macdonald Esq. of Glencoe.[3] The Rev. Dr Downie did not sign the lease and was not mentioned in connection with later events. James Hogg, who was his guest in 1803, said: 'Besides the good stipend and glebe of Lochalsh, he hath a chaplaincy in a regiment, and extensive concerns in farming, both on the mainland and in the isles, and is a great improver in the breeds of both cattle and sheep.'[4] Alexander Macdonald of Glencoe and John Mitchell both signed the tack at Inverscaddle on 9 March 1803. It was suggested later that Alexander Macdonald was the cautioner for John Mitchell.[5] That was not, however, in the original document. Mitchell was one of the family at Achnadaul in Lochaber. John MacDonald was one of the Innerigan family from Glencoe. The reference to Achnashien was probably to the place of that name between Leanachan and Donie, in Lochaber.

There is uncertainty about the way in which the land was occupied. Mitchell was apparently at Monar in the early stages of the tack, but after a time he became bankrupt and returned to Donie in Lochaber in poor circumstances. John MacDonald was also at Monar, and there is a suggestion that this was a consequence of the bankruptcy of John Mitchell.[6] [7] In 1809 John MacDonald certainly had some responsibility for the payment of the rent.[8] Later, his brother Donald came from Glencoe to Monar and became known as 'Donald MacDonald of Monar'. A third individual who became involved in Monar was Alexander MacCallum of Culligran, a farm in lower Glenstrathfarrar, mentioned below.

Glenstrathfarrar (Inverness-shire)

Fraser-Mackintosh has described how the people of Glenstrathfarrar were removed at Whitsunday 1803 to make way for sheep.[9] Fraser of Lovat leased Glenstrathfarrar to Hugh Fraser of Eskadale and Robert Fraser of Aigas for seventeen years from 1804.[10] That lease consisted of the middle part of the glen. It lay below Monar, which belonged to Mackenzie of Fairburn, and above Culligran, which belonged to Hugh Fraser of Struy.[11] According to Fraser-Mackintosh, 'the after history of Glenstrathfarrar is of little interest, being a mere shifting and displacement of large sheep-tenants, Highland and Lowland, as they became bankrupt, or fell out of the race, or had their rents unduly raised'.[12] It is, however, of interest in the present context, because one of these tenants was Alexander

Macdonald of Glencoe. He leased the farm of Muilzie and placed upon it as his sub-tenants John MacDonald of the Innerigan family (mentioned above concerning Monar) and his brother Ronald.[13]

The lowest part of Glenstrathfarrar, consisting mainly of the farm of Culligran, belonged to Hugh Fraser of Struy. It was leased in 1801 to Alexander MacCallum for nineteen years at a rent of £180.[14] MacCallum had been overseer at Lubcroy, in Strath Oykell at the northern extremity of Ross-shire, to the celebrated Donald MacLeod of Geanies, who was sheriff depute for the counties of Ross and Cromarty for over fifty years.[15] [16] MacLeod's main claim to fame, or infamy, arises from the his role in quelling the 'Ross-shire Insurrection' of 1791–92.[17] He was a prominent sheep-farmer and presided over the Sheep Farm Association based in Inverness that existed in the latter years of the eighteenth century.[18] In due course MacCallum was drawn into the Monar complications, and into the Glenmarksie affair.

Glenmarksie (Ross-shire)

In 1806 William Mackenzie of Strathgarve in Ross-shire let the farm of Glenmarksie near the south end of Loch Luichart for nineteen years at a rent of £140 to Donald McBarnett from Achneich in Lochaber.[19] McBarnett was a nephew of William Mitchell of Achnadaul, whose wife was Margaret McBarnett.[20] Mitchell was on familiar terms with Alexander Macdonald of Glencoe and asked him to stand cautioner for his nephew. He agreed to do this. Mitchell offered Alexander Macdonald 'back security', which was accepted.[21]

McBarnett paid his rent for one year and then made no further payments. At the end of two years of non-payment, i.e. at the end of the third year of the tenancy, Alexander Macdonald was called upon as cautioner to make good the deficiency. William McKenzie would not release him from responsibility for the remainder of the lease, and Glenmarksie was to be troublesome to him, and in due course to his trustees, after his death.[22]

Strathconon (Ross-shire)

Strathconon belonged, like the Forest of Monar, to Mackenzie of Fairburn. In 1803 he let it out in six tacks, mostly for nineteen years, but in one case for twenty-one years. He gave four tacks on the higher ground in the upper reaches of Strathconon to his former tenants who were already there, or who held ground in the lower part of Strathconon which they had to leave. This made the lower ground available as two extensive tacks for new tenants. In each tack it was said that the ground was given 'for the purpose of covering the same with sheep, houses, biggings, yards'. In the four tacks going to the former tenants there was an additional provision.

> Whereas the said Brigadier General Alexander Mackenzie has introduced strangers as Sheep farmers into the country of Strathconon with which measure the country people do not seem to be well affected It is hereby specially provided and Declared that if . . . any of the tenants before named . . . or any person [employed by them] shall give the said strangers tenants or any of them any molestation or trouble . . . and that they or any of them shall be convicted of the same by any competent judge . . . these presents shall become void.

The lower of the two extensive new tacks consisted of three farms: Glacour, Annate and Easter Balnalt. That went to John Cameron and John MacLaren from Glenalmond for nineteen years at a rent of £150. Their cautioners were Alexander Cameron and Gilbert MacLaren, also in Glenalmond.

The other tack given to new tenants was an even larger one, for which Adam Macdonald of Achtriachtan was 'cautioner, security, and full debtor'. The tack consisted of 'the whole the Towns lands and Grazings of Inverchoran, Blarnabee, Balnacreig, that part of the land and grazings of Clashlumchulan lying to the south of the Water of Conon, Cranich, Knockdhu, Dalbreck, Wester Balnalt with the grazings of Luiblone and Corrienaslevich'. This was given to 'John MacCallum, Tacksman of Blaravan of Glenurchy and Duncan MacCallum his son' for twenty-one years from 1803 at a rent of £450 per annum.[23] James Hogg met MacCallum at Inveroran in 1803 and heard from him of the extensive farm which he had taken in Strathconon.[24]

The farms listed in this tack extended for several miles along the south side of the river flowing through Strathconon. It is curious that in the tack this is called the Water of Conon. That might seem the logical name for the river flowing through Strathconon, but it has apparently always been known as the River Meig in that part of its course.[25] The tack extended south across the watershed to the River Orrin, and west or south-west to touch the west end of Loch Monar and meet the Grazings of Monar in which Alexander Macdonald of Glencoe had a interest.[26] [27] [28]

Adam Macdonald's reason for becoming cautioner and accepting such a large potential liability is simply not known. As with the Forest of Monar, subsequent events are rather obscure. It was said many years later that the MacCallums were 'removed' by Adam Macdonald and replaced by tenants of his own choosing. That suggests that the MacCallums may have failed and left Adam Macdonald as cautioner with responsibility for the whole tack, in the same way as Alexander Macdonald of Glencoe had become responsible for Glenmarksie. In the process of multiplepoinding in the Court of Session when Adam Macdonald's affairs were being wound up, the tack was described as 'Achtriachtan's Farms in Ross-shire'.[29] They

do not appear to have been anything of the kind in the tack of 1803. In these later years Donald, John and Allan MacDonald of the Innerigan family were described at various times as tenants of Dalbreck.

The MacDonalds of Innerigan

Members of the MacDonald family from Innerigan in Glencoe had a key role in all these northern tacks. In the sheep-farming days the head of the family was Angus MacDonald of Innerigan, who died in 1814 or 1815 in his eighties. His wife was Mary Rankin, who died in 1829 at the age of eighty-eight. They had thirteen children.[30] [31] Five sons survived to adulthood. The first four, John (1766–1834), Allan (1771–1830), Ronald (1775–1811) and Donald (1780–?), became tenants or sub-tenants in one or several of the northern farms in which Alexander Macdonald of Glencoe and Adam Macdonald of Achtriachtan had interests. It is difficult to identify their precise positions at any given time. Allan seems to have been the last to arrive in Ross-shire, but by 1819 he was with Donald at Dalbreck.[32] Their mother, Mary Rankin, was also at Dalbreck with these two sons at the time of their removal in 1821.[33]

The youngest son, Archibald, was born in 1790 and followed a totally different career. On the recommendation of Alexander MacDonald of Dalilea, he was recruited by the Earl of Selkirk and went to North America. In due course he became a chief factor of the Hudson's Bay Company. His life has been well documented.[34]

Angus MacDonald and Mary Rankin had six daughters who reached adult life. Recent research has shown that one of these daughters, probably their second daughter Margaret who married a MacDonald from Abertarff, was the grandmother of Angus McDonald born at Craig in Wester Ross in 1816. Angus, like his great-uncle Archibald, went into the service of the Hudson's Bay Company in North America.[35] [36] Mary, the second youngest daughter, married a Duncan MacDonald in Beauly.[37] She was therefore fairly close to her brothers in Ross-shire. At one time, her brother Donald visited Beauly regularly to supply meal, barley and potatoes to householders there and probably resided with Mary and her husband when doing so.[38] [39] [40] Ann, the youngest daughter, married Alexander MacArthur in Glenroy in 1807 but died in 1810.[41] As shown in Chapter 9, MacArthur bought the New Inn in Fort William established by Donald McDonald, tacksman of Tulloch.

A gravestone on Eilean Munde remains as a reminder of this once extensive family in Glencoe.[42]

Chequered History of Monar

Alexander Macdonald of Glencoe had apparently tried to deal with problems at Monar during the last few years of his life. It is not clear how

these arose, but John Mitchell's bankruptcy may have been a factor. In 1813 Alexander Macdonald wrote to Alexander MacCallum of Culligran about 'keeping your half of Monar', which shows that MacCallum had already been drawn into the Monar affairs. He also authorised John and Donald MacDonald of the Innerigan family to take delivery of stock there, which shows that they were also involved.[43] An entry in the accounts of the Glencoe trustees suggests that Murdo MacLenan and Alexander Macdonald of Glencoe had become cautioners for John MacDonald as the tenant of Monar.[44] This is the only reference to MacLenan at this later stage of the tack. Alexander Macdonald of Glencoe may not have had authority to make these arrangements without reference to the other tenants, but that would not have troubled him.

That situation probably continued until Alexander Macdonald of Glencoe died in December 1814. It was unfortunate for the Glencoe trustees that in 1815, just after his death, General Mackenzie advertised the Estate of Monar for sale. It was purchased by Hugh Fraser of Eskadale on a disposition dated 18 August 1815. However, a dispute between Mackenzie and Fraser prevented the latter from taking entry until 15 January 1819.[45] During the intervening period of three and a half years, neither the outgoing nor the incoming proprietor seems to have concerned himself with the management of the estate. The minutes of the Glencoe trustees show that initially they were at a loss to understand Alexander Macdonald's commitments in all these distant properties. In particular, they were unsure of 'Glenco's liability for rents'. It may be assumed that Alexander Macdonald had followed his usual practice of committing little or nothing to paper. The trustees requested help in finding the answers from Campbell Mackintosh, writer in Inverness, who was much closer to the scene. At their third meeting in January 1817, they considered 'Mr. Campbell Mackintosh's exertions and explications thereon'. He continued to be of considerable help to the trustees because of his contacts in the area and the local information available to him.[46]

The trustees concluded that they had inherited a responsibility for Monar and went on to try to manage it. They allowed Alexander MacCallum to continue to occupy part of Monar on a year-to-year basis.[47] Donald MacDonald of the Innerigan family still held land there as well as becoming tacksman of Dalbreck in Strathconon.[48] The position of his brother John is less clear. His physical presence in Monar seems to have lapsed and he may also have been at Dalbreck, but later events show that he still had stock at Monar.

Around that time Donald MacDonald and Alexander MacCallum were on good terms and traded jointly. Quite substantial quantities of wool are mentioned. On one occasion MacCallum bought 282 stones of wool at Urquhart, and on another Donald MacDonald bought ten stones

of Cheviot wool. After these were sold, the profits were shared equally. By 1816 their joint trading ceased and a dispute about where they stood in relation to each other reached the Sheriff Court in Inverness. MacCallum produced detailed accounts to show that there was an outstanding balance of £17 7*s.* 8H*d.* due to him. Donald MacDonald countered with figures to show that he owed nothing to MacCallum. One minor matter was a charge by MacCallum against Donald MacDonald for 'the wintering of a horse belonging to you on the farm of Bunchrew including charge for sending the horse from Culigran'.[49] This showed that MacCallum had become the tacksman of Bunchrew shortly before his lease of Culligran was due to end. He remained at Bunchrew for the rest of his life and was succeeded there by his son Colin about 1825.[50] For some time before 1820 Alexander MacCallum was also tenant of lands and grazings in Kintail on the Chisholm estate.[51] There is a general impression that he was doing well in these years.

When Hugh Fraser of Eskadale obtained entry to Monar, he set about removing Alexander MacCallum and Donald MacDonald. Both prepared their positions for a fight. The terms of the lease that General Mackenzie had given in 1803 opened the way to a variety of defensive arguments. Eskadale and his legal advisers no doubt found these intensely irritating. They were driven to describing MacCallum as 'one of the most litigious and quibbling individuals in the Country'. John MacDonald wrote from Muilzie suggesting that MacCallum should give up his share of Monar to him 'as I think I have a better right to get it being one of the Principal Tacksmen and as that part of the farm possessed by you being once my place of residence and probably under necessity of going back to the said place next year for the benefits of my meliorations'. He promised that he would let MacCallum's sheep graze at two shillings each;[52] [53] that cheerful outlook suggests that he had not yet realised what the trustees had in mind.

The Glencoe trustees had consented, for their interest, to the actions for removal that Eskadale was taking. Their main hope must have been to get rid of any concern with the place. On 13 November 1821, Robert Scott said in a letter to Ewen Macdonald in India that Muilzie had been given up without loss and that Monar was the only remaining farm concern in the north.[54] Mr Fraser of Aigas, one of the original lessees of Glenstrathfarrar in 1804, paid £73 3*s.* 8*d.* to the Glencoe trustees on 10 April 1822 'on account of the Muilzie stock'.[55] That suggests that he had taken that farm back. There had been a hope from correspondence with Campbell Mackintosh that the trustees would have got clear of Monar at Whitsunday 1821, but this had not happened. The lease had two years to run, and all that Robert Scott could say was that he had reason to think 'it may be so managed as to prevent much, if any, loss'.[56]

That was a fairly optimistic prediction. Perhaps Robert Scott was assuming that the trustees would recover all the money that the Innerigan MacDonalds owed to Alexander Macdonald of Glencoe. In May 1820 various summonses were drawn against them. Angus, who was now dead, had left an unpaid debt of £182 10*s.* 0*d.*, and payment of this was demanded from his three surviving sons in Scotland, John, Donald and Allan. Allan got a separate summons for a debt of £255 recorded in an obligatory letter from him to Alexander Macdonald of Glencoe. Apparently the money had been paid on Allan's behalf to a John MacIntyre. John and Donald also got separate summonses for unidentified sums that may have been arrears of rent. While these actions were pending, the trustees were alarmed to hear reports that the MacDonalds were selling off stock at Monar. Presumably they were concerned that this would reduce the prospect that they could recover the debts by sequestration. Duncan Cameron wrote in confidence to Campbell Mackintosh, who reassured him that sufficient effects remained at Monar to pay the outstanding rent.[57]

The actions were disposed of briskly in the Court of Session in November and December 1820. No defence seems to have been offered, except perhaps by Donald.[58] At the time of these actions he was being made bankrupt, as will be explained shortly in connection with Dalbreck.

In 1821 a messenger called Stewart was sent from Dingwall to poind John MacDonald's stock at Monar. A vivid account of the difficulty of his task has survived because the charge for his services was disputed by the Glencoe trustees. John Cumming, a writer in Dingwall who had instructed the poinding, paid Stewart his charges in 1821, but in 1827 he had still not been reimbursed by the Trustees. He objected to a proposal that the account be submitted to the Auditor of the Court of Session, who

> cannot be a competent judge of the trouble the Messenger had been at in the proceedings against MacDonald, in respect the Auditor does not know the state of that part of the country the Messenger had to travel to MacDonald's residence, it being quite rugged and almost impassable, and the great fatigue occasioned by watching and driving the poinded Cattle and Sheep – I must therefore object to the proposition of submitting the account to the Auditor and . . . beg leave to suggest the Sheriff Substitute of this district as a fit person to tax the account, who from his local knowledge [. . .].[59]

Later in 1821, the Glencoe trustees had to pursue a couple of purchasers who had failed to pay for what they bought 'at the sale of effects poinded on behalf of the Pursuers belonging to John MacDonald, Tacksman of Monar'.[60]

It is not possible to tell if the trustees recovered all that they thought they

were due. It seems unlikely that they did. The Glencoe trustees had a John MacDonald incarcerated for debt in the tolbooth in Inverness in July 1822.[61] That would probably have been John MacDonald of the Innerigan family.

Troubles at Glenmarksie

When Alexander Macdonald of Glencoe was required in 1809 to make good the failure of Donald McBarnett to pay his rent for Glenmarksie, he expressed indignation that the non-payment had been allowed to go on for so long, so that he had to find a large sum at short notice.[62] However, he had no choice but to pay arrears of £295 15*s.* 10½*d.* on 11 June 1809. William Mitchell at Achnadaul had accepted a bill to Alexander Macdonald for £295 16*s.* 0*d.* dated 14 March 1809, thus honouring his commitment to give 'back security'. However, Mitchell owed substantial arrears of rent to the Duke of Gordon, as will be explained in Chapter 12, and the bill for 'back security' remained unpaid.[63] Alexander Macdonald hoped that William Mackenzie, then serving as a surgeon with the 5th British Militia at Woodbridge in Suffolk, would relieve him of responsibility for the remaining sixteen years of the lease. Mackenzie would not agree to this.[64] As Alexander Macdonald was committed to paying the rent for these years, he had to find a way of dealing with Glenmarksie.

With William Mitchell's concurrence, McBarnett was made bankrupt and his stock was rouped.[65] Alexander Macdonald purchased almost the entire stock at Glenmarksie for £340 16*s.* 9*d.* In 1810 he sub-let the farm to John and Ronald MacDonald, at Monar and Muilzie, and Alexander MacCallum, at Culligran, at a diminished rent of £120. The MacDonalds were responsible for £76 13*s.* 4*d.* of this and MacCallum for £43 6*s.* 8*d.* The stock that Alexander Macdonald had bought remained on the farm. After a time the MacDonalds sub-let their share to Donald and Angus MacMillan, who subsequently fell into arrears. However the MacDonalds remained responsible to Alexander Macdonald of Glencoe for their share of the rent of £120. He in turn was still responsible to Mackenzie of Strathgarve for the full rent of £140.[66] He never recovered the full value of the stock from the tenants.

The rent was not the only payment that McBarnett failed to make. He had a housekeeper, Margaret Ross, who took him to court in 1810 to try to recover unpaid wages. She had been due a total of £12 for three years from June 1806 to June 1809 and had received only £3 1*s.* 0*d.* By the time the case was heard in Dingwall Sheriff Court she was living in Inverness, but McBarnett was still in Ross-shire, somewhere in the parish of Contin.[67] Her loss was no doubt as serious for her in relative terms as that suffered by Alexander Macdonald of Glencoe.

After the death of Alexander Macdonald of Glencoe, his trustees received various payments of rent from MacCallum and the MacDonalds.

MacCallum paid £43 12*s.* 6*d.* on 1 March 1816 'for share of rent of Glenmarksie', followed on 6 March by £53 10*s.* 0*d.* 'by cash from Donald MacDonald per Mr. Mackintosh to further account of rents of Glenmarksie'. On 14 February 1817 there was another payment 'by cash from Mr. Campbell Mackintosh on account of rents received by him from Mr. MacCallum to account of the rents of Glenmarksie – £40'.[68]

The Glencoe trustees were probably dissatisfied with the arrangements at Glenmarksie, and they made a couple of attempts to find new tenants. An advertisement in the *Inverness Journal* of 29 April 1817 offered 'the Farm and Grazings of Glenmarksie, part of the estate of Strathgarve, as occupied by Mr MacCallum and the MacMillans' to rent from the following Whitsunday. Applications were to be made to Campbell Mackintosh or to Roderick Reach. The latter was then a writer in Inverness, who later moved to London as the first representative there of the *Inverness Courier.*[69] The advertisement must have failed, because a similar one appeared in the *Inverness Journal* of 13 March 1818. This time, the offer was for seven years from Whitsunday with immediate entry. Again this must have failed. The trustees then attempted to persuade a reluctant Mackenzie of Strathgarve to take the place off their hands without waiting until the lease ran out in 1825. He had resisted this from the beginning, but Duncan Cameron eventually succeeded. He first met Mackenzie on 26 December 1820. This was followed on 26 January 1821 by 'attendance on him [Mackenzie] for upwards of two hours arranging finally about the remaining years of the lease, and meliorations'.[70] Even these brief words convey a sense of achievement. Robert Scott said in November 1821 that 'the Trustees were pretty fortunate in being able to transact with the proprietor last spring upon tolerably reasonable terms for the four years from Whitsunday last'.[71]

Meanwhile, the trustees were trying to recover as much as possible of the money that they believed was due to them from the people involved in Glenmarksie. Alexander Macdonald of Glencoe had not pressed William Mitchell at Achnadaul for the 'back security' that he had given for his nephew's rent. William Mitchell was now dead and had been succeeded by his son George, but the bill for £295 16*s.* 0*d.* was still unpaid. At their second meeting in January 1816 the trustees considered proceeding against George Mitchell, as successor to his father. It was minuted that they 'did not feel at liberty to depart from the claim', and accordingly an action was raised against George Mitchell in the Court of Session.[72] The trustees failed. As Robert Scott reported to Ewen Macdonald in November 1821, 'after all that could be done, the Court relieved Mitchell from his Father's obligation to your Father, but view the case as difficult, and at the same time as so hard on both parties, Mitchell was entitled to no expenses'.[73]

There was also the question of liability for the cost of the stock at Glenmarksie that Alexander Macdonald of Glencoe had purchased for £340 16*s.* 9*d.* and left on the farm. On 18 February 1810 Alexander MacCallum and John and Ronald MacDonald had signed an obligatory letter for this amount, undertaking to pay it at Martinmas first. They had failed to do so, and Alexander Macdonald had tried to recover it in 1813. It fell to the trustees to pursue this. They sought recovery from MacCallum and John MacDonald, but as Ronald had died in 1811 they sought his share from his three brothers still in Scotland, John, Donald and Allan, 'sometime residing at Innerigan'.[74] [75] [76] In 1821 Robert Scott reported to Ewen Macdonald, with evident satisfaction, that the court had made MacCallum individually liable for the stock on Glenmarksie.[77] MacCallum was a man of greater substance than the MacDonalds, which no doubt explains Robert Scott's satisfaction. Eventually he did pay a considerable part, possibly all, of the money for which he was made liable, although there was a problem over a dishonoured bill for £300 and the trustees had to obtain a diligence against him.[78] MacCallum may have been in temporary difficulties around this time. There was an Alexander MacCallum incarcerated for debt in the tolbooth in Inverness in January 1823. He petitioned for aliment, and the Baillie who considered his petition accepted 'that the prisoner has no funds within his reach for supporting him'.[79] The phrase 'within his reach' seems to imply that he was not totally without resources.

Losses at Strathconon

Little can be said about events in Strathconon until the chaotic state of Adam Macdonald's affairs led, as described in Chapter 6, to the appointment of interdictors and trustees in 1816. A case in Dingwall Sheriff Court shows that in 1810 John MacCallum was still at Dalbreck, being pursued for an unpaid debt.[80] The Achtriachtan trustees appointed Robert Logan, agent of the British Linen Company in Inverness, as their factor for what they described as 'Achtriachtan's Farms in Ross-shire'. This was quite distinct from the appointment of Thomas MacDonald in Fort William to deal with affairs in and around Glencoe. The two factors reported independently to the Achtriachtan trustees, who were not very good at consolidating the information they received. On 4 August 1818 they directed Mr Logan 'to make arrears effectual and charge some of the sub-tenants who had fallen greatly in arrears'.[81] It is possible that it was after the appointment of Mr Logan, or after he had been instructed to pursue the arrears, that the MacCallums disappeared from the scene. It seems doubtful whether Adam Macdonald would have acted so decisively

The Innerigan MacDonalds then occupied the farm of Dalbreck. Donald's name was most closely and consistently associated with it, but his brothers John and Allan were each there at times and were sometimes

described as tenants. Dalbreck extended southwards across the watershed to reach a southern boundary formed by the River Orrin. The southern part, in Glenorrin, was held by Kenneth and Alexander Matheson as sub-tenants of the MacDonalds.[82]

Inverchoran, higher up Strathconon, was said in 1821 to be to be held on a sub-lease from Adam Macdonald by a Duncan and Donald MacDonald, who were claiming arrears alleged to be due to them from 1818 by a sub-tenant of part of the tack.[83] That suggests that these MacDonalds had been there since at least 1818. Their identity has not, however, been established.

It is not surprising, in the light of their difficulties elsewhere, that by 1820 the MacDonalds were having problems at Dalbreck. On 7 August 1820 a petition was presented on behalf of Donald MacDonald to invoke the legal provisions for merchantile sequestration. He was described as a grain and victual dealer at Monar. The petition was submitted with the consent of Allan MacDonald residing at Dalbreck who was a creditor of Donald to the extent required by law, being owed the sum of £137 1*s.* 0*d.* Creditors were invited to meet at the house of Kenneth Mackenzie, inn-keeper at Dingwall, on 16 August 1820 to name an interim factor.[84] Features of this procedure have been explained in Chapter 7 when describing the bankruptcy of Donald MacDonald at Drimintorran. It may seem strange that at Dalbreck Donald's brother Allan invoked the bankruptcy, but there were probably advantages for Donald in doing so.

In the *Inverness Courier* of Thursday 28 December 1820, there was an advertisement for the sale of 'the whole stock of Sheep, Black Cattle, and Horses belonging to the Sequestrated estate of Donald McDonald upon the Farms of Monar and Dalbreck'. Applications for particulars were to be made to Alexander M'Kenzie at Kinahaird, trustee, or John Cameron, writer, Dingwall. The next step was for Alexander MacKenzie as trustee on Donald MacDonald's sequestrated estate to have him removed from Dalbreck. A summons of removing was presented in the early months of 1821. The removal was sought with the consent of Adam Macdonald Esq. of Achtriachtan and his trustees. The people to be removed were Donald and Allan MacDonald and Mrs Mary MacDonald (their mother), all described as residing at Dalbreck, and Kenneth Matheson, a sub-tenant at Glenorrin.[85]

On 30 March 1821 there was an advertisement in the *Inverness Journal* offering Dalbreck 'at present occupied by the creditors of Donald MacDonald' to let for three years from Whitsunday next. The earlier attempt to sell the stock must have failed, because it was said that 'if considered an object, the present stock of Sheep and Black Cattle will be made over to the tenant either by private bargain or at valuation'. Again Alexander MacKenzie at Kinahaird and John Cameron, writer in

Dingwall, were handling the matter. MacKenzie's task continued for some time. In February 1824 he attempted in the Court of Session to recover money due to Donald MacDonald from the sequestrated estate of Patrick Walker, late wool merchant in Stirling.[86]

Before concluding this account of events in Strathconon, another Glencoe connection with this part of Ross-shire can be mentioned. On 2 February 1819, John Gillanders Esq. of Highfield drew a bill upon 'Mr. Donald MacDonald residing at Dalbreck in Stratconon and Mr. Donald Macdonald at Coricheran [Corrychurochan] in Lochaber' for £13 7*s.* 0*d.* This was accepted by them for payment in six months. The sum was not paid, and Gillanders initiated an action in Dingwall Sheriff Court in 1820. By November 1821, when Gillanders obtained a Warrant of Sale to recover the debt, the second Donald was described as 'late at Coricheran now at Scatwell'. Scatwell is in Strathconon, downriver from Dalbreck. Christopher McRae was the unfortunate messenger who had to go to Scatwell and poind two cows: a black long-horned one worth £5, and a black short-horned one worth £4 10*s.* 0*d.* In spite of various difficulties he removed them to Dingwall, where no buyer could be found. Gillanders ended in possession of the two cows when he would no doubt have preferred cash.[87]

Donald McDonald at Corrychurochan can be identified fairly confidently as a drover to whom Robert Downie of Appin sold 343 wedder and ewe lambs.[88] He was probably a son of John Ban McDonald, originally at Achtriachtan, whose removal by the Glencoe trustees from Knoydart and Glendessary to Corrychurochan was described in Chapter 10. He had three sons, one of whom was called Donald, and he was removed from Corrychurochan in 1820.

The MacDonalds of Innerigan seem to have left Glencoe to follow the distant sheep-farming interests of their two 'chiefs'. It is impossible to tell if they were motivated solely by commercial judgement or if they had been influenced by a sense of loyalty. If loyalty did play any part, they were badly led by those in whom they placed their trust. Very little is known about what happened to them after they were overtaken by financial disaster. Ronald had died in 1811 before the major difficulties emerged. He was not married. Allan died in Strathconon in 1830, also unmarried. John had married in 1807 and lived until 1834. Donald was unmarried at the time of these farming interests, but in 1830 he married a Miss Fraser in Inverness.[89] The date of his death is not known.

12

Fersit, Glenure and the Rankins

This chapter is about three concerns in the nineteenth century. A common factor is the involvement of Rankins from Glencoe in all of them. The first is the interest of Alexander Macdonald of Glencoe in Fersit. The second is his tenancy of Glenure and Barnamuck. The third is an action by the Glencoe trustees to recover an alleged debt from Angus and Duncan Rankin. In all of these it fell to Alexander Macdonald's trustees to deal with difficult situations after his death.

In Chapter 2 it was shown that Alexander Macdonald of Glencoe held the lease of Fersit in the late eighteenth century and was removed by the Duke of Gordon in 1798. He made several attempts in the early nineteenth century to regain the tenancy. These all failed, but he became a cautioner for others, including several Rankins, who obtained the tenancy of Fersit. He then created a very confused situation that had to be disentangled in the Court of Session after his death.

In Chapter 5 it was shown that Alexander Macdonald of Glencoe obtained the lease of Glenure and Barnamuck in 1788 and installed two brothers, John and Duncan Rankin, as his sub-tenants. They were still there when he died in 1814. The Glencoe trustees had to extricate themselves from Glenure and Barnamuck while the complexities at Fersit were being unravelled. The same John Rankin was involved in both places.

Finally, the Glencoe trustees embarked on litigation against Angus Rankin and his brother Duncan about an unpaid bill that purported to show that they owed £1,050 to Alexander Macdonald of Glencoe. This provides useful information about the way in which they and Alexander Macdonald of Glencoe conducted their business. This litigation became linked with the later stages of the Fersit case because Coll Macdonald WS acted for the Rankins in both.

Fersit in the Nineteenth Century

Alexander Macdonald's seven-year lease of Fersit began in 1778 and must have been renewed several times. His final lease expired in 1797. He did not, however, vacate Fersit then, and in 1798 the Duke of Gordon obtained a decreet of removal against him in the Court of Session.[1]

The tenants who replaced Alexander Macdonald of Glencoe did not

obtain entry until Whitsunday 1798. They were Captain Alexander MacDonell of Moy, Alexander McVean and James McCaul. The first two were related by marriage. Captain Alexander MacDonell at Moy was a younger brother of John MacDonell of Aberarder who had been the tenant of Torgulbin and was later at Killiechonate.[2] Alexander McVean was a son-in-law of John MacDonell, having married his daughter Sarah.[3] McVean and McCaul had been associated in the farm of Loch Treig, next to Fersit. McCaul had this farm for some years prior to 1797 as a sub-tenant of the MacDonells of Keppoch.[4] McVean was certainly at Loch Treig in 1793 when he had a dispute with Ranald MacDonell, late of Clianaig, over an alleged failure to pay for some sheep that his son Archibald had got from him.[5] [6] Ranald and Archibald had moved to Strathglass.[7] [8]

The new tenants of Fersit undertook to pay a rent of £300 per annum for a seven-year lease. To compensate them for their late entry, William Tod, the Duke's factor for Lochaber, agreed that they should have an extra year at the end. Their lease therefore continued until Whitsunday 1805.[9]

As Whitsunday 1805 approached, the Duke failed to make a timely move to invite offers for a further lease. Instead the tenants at Fersit were asked by letter of 25 March 1805 if they would agree to continue for a further year at an additional £100. Captain MacDonell and Alexander McVean agreed to this, but explained that it would not be easy for them. They said:

> Altho it is rather inconvenient for us to make any change for a year it must be done owing to one of the men . . . having got a place in Perthshire and we are obliged to take his stock as they will be valued which is absolutely necessary in Sheep-Farms, and we hope that His Grace the Duke of Gordon will cause the incoming Tenant to do us the same justice if we are so unlucky as to loose [sic] the Farm which we hope will not be the case.[10]

James McCaul had moved to Auchmore near Killin,[11] and so his name did not appear on the reply. The need to buy his stock might explain a debt that was to trouble McVean and later be a threat to the Glencoe trustees.

With the lease extended for a year, the Duke's factors invited offers for the farms to be let from Whitsunday 1806. Fersitmor and Fersitriach were listed separately as Lots 12 and 13.[12] On 3 September 1805 Captain MacDonell and Alexander McVean offered £900 to continue as joint tenants of both. That was three times the rent that had been sufficient in 1797. Alexander Macdonald of Glencoe attempted to get Fersit back with an offer of 1,000 guineas (£1,050) for Lots 12 and 13. but he wanted a lease for fifteen years. The factor noted, perhaps on the Duke's instructions, that if the present tenants would agree to pay £1,000 they

were to be preferred to Macdonald of Glencoe. Captain MacDonell and Alexander McVean did agree. They confirmed this by a letter of 29 October 1805 to William Tod at Fochabers accepting a formal offer of a lease for seven years at £1,000 per annum. Their letter of acceptance was sent by the Badenoch post, but just to make sure Alexander McVean sent another letter confirming their agreement by the Fort William post on 31 October.[13]

Alexander McVean and Captain MacDonell were soon in difficulties, and at the end of 1807 their rent arrears stood at £1,000. Rev J. Anderson of Kingussie, who was now the Duke's factor, noted: 'The undertaking [is] beyond their means. Will pay part soon.'[14] Towards the end of 1808 they were still well behind with their rent. On 12 November, Anderson wrote to Captain MacDonell. He began tactfully by complaining about McVean not being prompt with his rent, but then chided the Captain for the same fault. He put it to him that perhaps the Duke would 'rather call Strangers from the Tweed and give them his lands . . . than submit to enter into new arrangements with those on whose punctuality of payment he has so little cause to depend'. The captain apparently pleaded that he had sustained losses, and the winter of 1807/08 had certainly been a severe one. The Duke was prepared to modify his rents and to allow him more time to pay his arrears, if he found proper security for them. However, he expected that if the Captain's funds were inadequate he would say so and confine himself to his farms at Moy and Kylross, without the additional burden of Fersit.[15] Captain MacDonell secured an undertaking from Alexander Robertson of Struan, signed at Rannoch Barracks in February 1809, that he would guarantee his share of the rent.[16] (It may be significant that the Captain's wife, Jane (Juliet), came from Rannoch.[17]) McVean did not seem to have any such option open to him. By 1809 he owed £303 15*s.* 0*d.* in rent.[18]

Captain MacDonell and Alexander McVean were not the only tenants of the Duke who were in trouble then. William Mitchell of Achnadaul was £358 0*s.* 5*d.* in arrears in 1809. He wrote to the factor explaining his financial situation and resigning his farm. As at Fersit, the Duke proved to be a compassionate landlord. After discussion he offered Mitchell 'a handsome abatement' and persuaded him to give it a trial for a year.[19]

What happened next at Fersit was that the partnership between Captain MacDonell and Alexander McVean was dissolved. Fersit was then split into its two components. Alexander McVean did not have enough money to proceed on his own. He was joined by Donald Rankin at Kinlochbeg and John Rankin at Glenure in making an offer on 27 February 1809 of £500 per annum for a four-year lease of Fersitriach, to which Torgulbin was now added. Their offer was submitted in a letter written at Keppoch, where Alexander Macdonald of Glencoe then lived. It was accompanied by

an undertaking from him to be their cautioner and guarantee their rents.[20] Captain MacDonell kept Fersitmor, to which Inverlair was added. Thomas Gillespie of Ardochy, John MacNab at Gallovie and William Mitchell of Gordonhall were asked to fix the rents due by Alexander McVean and Captain MacDonell. They allocated £460 of the rent to Alexander McVean and his new partners, and £290 to Captain MacDonell, making a total of £750.[21] As the total rent had previously been £1,000, it may be surmised that, as at Achnadaul, the Duke was allowing 'a handsome abatement'.

In litigation over Fersit (see below), Donald Rankin at Kinlochbeg was identified as the brother of Angus Rankin at Dalness who was involved with him at Achnacon in Glencoe. John Rankin was the sub-tenant, along with his brother Duncan, of Alexander Macdonald of Glencoe at Glenure and Barnamuck. He was almost certainly related to Angus and Donald Rankin and was to some extent protected by Angus.

So eleven years after his departure from Fersit, Alexander Macdonald was at last getting a locus there again. He may not have got the tenancy in his own name, but he was soon to show that his role was not to be a passive one.

Before turning to that it is of interest that John MacNab at Gallovie, one of those who had fixed the rents at Fersit, had moved to Gallovie from Ballachulish in the early 1790s. In 1792 Allan Macdonald of Gallovie concluded a steelbow contract with Peter MacNab, tenant in Ballachulish, and his sons John and Robert, giving them occupation of Gallovie, Kinloch and Inverruddon for twelve years from 1793.[22] This move is corroborated in a letter of 9 May 1794 from Æneas Macdonald of the Achtriachtan family, presumably in his capacity as an agent supplying wool to an English buyer. He said that Peter McNab at Ballachulish 'must have a settlement with me . . . for his wool . . . Peter McNab goes to Badenoch to his possession there & wishes all his accts. in this Country to be as clear as possible'.[23] This is another example of a farming family moving a considerable distance within the Highlands during the eighteenth century.

The machinations of Alexander Macdonald of Glencoe at Fersit after he became involved as a cautioner created an extraordinarily complicated situation. The most useful source of information about events from 1809 onwards is a submission that was prepared some years later by Coll Macdonald WS. He was acting for the Rankins, who were his clients in an action of multiplepoinding in the Court of Session. A copy has survived with the papers of the Glencoe trustees.[24] According to his account, it was intended in 1809 that, because of McVean's shortage of capital, Donald Rankin should purchase a half share in the stock. However, other people wanted to be involved, and in the spring of 1809 a meeting was held at Kinlochleven to decide what was to happen. This was attended by Alexander McVean, Alexander Macdonald of Glencoe, and Angus,

Donald and John Rankin. Angus Rankin's position seemed to be that of a senior member of the family who had to be consulted about everything involving his relatives. In due course it was he who actually advanced to McVean the price of the sheep taken over by the other Rankins.

Alexander Macdonald wanted to muscle in, and it was agreed that he would purchase one third of McVean's half of the stock, i.e. one sixth of the total. Angus and Donald Rankin wanted to include John Rankin in the arrangements. They stipulated that, despite his tenancy at Glenure and Barnamuck, a considerable distance away, John Rankin should live on the farm at Fersit. He had no money to advance for stock, but 'in order to stimulate his industry he was admitted to partake of the profit and loss to the extent of one third of the half share which belonged to Donald Rankin' – again one sixth of the total. McVean was to manage the whole concern, and to be remunerated for this. It was also agreed that John Rankin was to be remunerated for superintending the herding of the sheep, but he was not to interfere with McVean.[25] The contemporary paperwork was scrappy in the extreme, and in setting all of this out Coll Macdonald presumably depended mainly upon what his clients, the Rankins, told him. His account may not be accurate in every detail, but it is probably good enough to give a picture of the disaster waiting to happen.

The lease set up in 1809 was a short one, and by the summer of 1813 offers were being received for a further lease to commence at Whitsunday 1814. Alexander Macdonald of Glencoe submitted an offer of £500 for Fersitriach and Torgulbin, that is to say for the land then held by McVean and the Rankins for which he was cautioner. He added a further offer of £360 for Fersitmor and Inverlair held by Captain Alexander MacDonell at Moy. This is quite astonishing, given the state of the market and of his own finances. The factor noted briefly: 'Offer declined – by order.'[26] Perhaps the Duke had had enough of Alexander Macdonald of Glencoe.

Fersit remained split. Captain Alexander MacDonell included an offer of £300 for Fersit (presumably Fersitmor) and Inverlair along with his offer for his other lands of Moy and Kylross. This was accepted. It is interesting that the factor's documents now call him Lieutenant A. MacDonell.[27] Why had the courtesy of 'captain' been withdrawn? The lease of Fersitriach and Torgulbin, previously held by McVean and the Rankins with Alexander Macdonald of Glencoe as cautioner, went to the MacDonells at Inch.

The MacDonells of Inch

The MacDonells of Inch were a significant family. The head for many years was Angus, an illegitimate son of Alexander MacDonell of Keppoch. He had been with his father at Culloden and was by now a very old man. The correspondence about Fersit was undertaken by his eldest surviving

son, Lieutenant Colonel Archibald MacDonell, who explained that these lands would be occupied by him along with his brothers Donald, John and Coll.[28] Archibald died soon after the lease was arranged, and John and Coll were the most active members of the family in relation to these particular farms.

William Tod's favourable comment on the MacDonells of Inch in 1772 was noted in Chapter 2. They achieved a powerful position in the early nineteenth century, partly on their own account and partly by acting for their Keppoch relatives.

In November 1807 John and Coll, two of Angus's sons, along with Allan MacDonald of Lochans in Moidart, probably the husband of their sister Alexandrina, became joint tacksmen of Glenfinnan and other lands. They undertook to pay £2,304 14*s.* 1*d.* to Mr John Cumming and Company, the former Tacksmen, for 'value received' – presumably the stock and other assets. Coll MacDonell handed over £700 at Annat on 10 December 1807. Their bill for the remainder was protested at the instance of John Cumming in Fort William in April 1808.[29] This suggests difficulty in producing all of the price when due. At this time their father, Angus, had fallen behind with the rent. He owed £165 on a rent of £365 at the end of 1807. The factor noted however that he 'has sufficient stock and will pay – the temporary deficiency occasioned by the bad times'.[30] Indeed the family had other assets, including several houses in the High Street of Fort William.[31] [32] The factor's assessment of the family's financial position was sound because they clearly continued to prosper. Their standing in the farming world was high. Locheil singled out Coll MacDonell along with Thomas Gillespie as the most reliable valuators in the Highlands.[33]

The adjoining farms to the east of Inch, i.e. Clianaig, Monessie, Achnacoichan, Inverlair and the Grazings of Loch Treig, were Gordon lands that had remained with the MacDonells of Keppoch after they lost Kilmanivaig and Brackletter in 1769 (see Chapter 2). Inverlair was later taken away from them and added to Fersitmor. When Major Ranald MacDonell of Keppoch died in 1785, his son Alexander was a minor. The confused situation about the management of his affairs was described in Chapter 4. This confusion meant that the farms on the Duke's land which remained with the Keppoch family were badly managed. With the coming of sheep-farming their value increased, and in 1799 the Duke's factor thought that, if properly managed, they could produce a higher rent. He said they were not well managed, and indeed he felt that young Keppoch was getting little benefit from what he saw as the Duke's indulgence in continuing the farms at a rent less than they were worth.[34]

The family circumstances changed after Major Alexander MacDonell of Keppoch died in 1808 and was succeeded by his brother, Lieutenant Richard MacDonell of the 92nd Regiment of Foot. He quickly ousted his

stepfather, Alexander Campbell, whose position had become dominant, and appointed the MacDonells of Inch as his managers.[35] John and Coll then attended to his affairs and thus became responsible for the management of these farms. The stock on Loch Treig and Clianaig alone was valued at £1,129 6*s.* 0*d.* when judicially ascertained in August 1810.[36]

When the MacDonells of Inch were about to take over at Fersitriach and Torgulbin, the stock on these two farms was valued at £3,504 5*s.* 9*d.* According to the custom of the country the MacDonells agreed to pay for it in two moieties, the first at Martinmas 1814 and the second at Martinmas 1815. However, Coll Macdonald recorded that Alexander McVean and Alexander Macdonald of Glencoe prevailed upon the MacDonells not to wait until the terms for payment but to advance £1,220 to McVean and £1,302 5*s.* 11*d.* to Alexander Macdonald. This obviously left only £981 19*s.* 19*d.* for the Rankins, who had continued to hold a half share of the stock.

It is tempting to try to understand why Alexander Macdonald and Alexander McVean behaved in this way. The dispute between the Glencoe trustees and Angus and Duncan Rankin over a bill for £1,050 (see below) shows that they owed Alexander Macdonald a substantial sum. John Rankin apparently owed money to Alexander McVean. Perhaps this was a crude attempt to recover some of these moneys by inducing the MacDonells to pay them excessive shares of the purchase price of the stock of Fersit.

After the Death of Alexander Macdonald of Glencoe

These events took place a very short time before Alexander Macdonald's death in December 1814. After that the MacDonells brought a process of multiplepoinding in the Court of Session. The defenders were the Glencoe trustees, Alexander McVean and Angus and Donald Rankin, who 'were also partners of Alexander McVean in the said farm'. The purpose of the action was to establish the amounts due by the MacDonells as incoming tenants to each partner and to adjust the very odd payments made to Alexander McVean and Alexander Macdonald. Most of the following information is drawn from the submission by Coll Macdonald WS mentioned above.[37]

The process was suspended when the partners agreed to let their agents try to settle matters amicably. Their agents were Duncan Cameron for the Glencoe trustees, Coll Macdonald for the Rankins and Duncan Stewart for McVean. They 'agreed that the best way of explicating the truth, in a matter where accounts were kept so irregularly, was to examine the parties in their presence'. The three Rankins and Alexander McVean were accordingly summoned to Edinburgh, where 'each underwent two separate examinations'.

These examinations in Edinburgh appear to have been inconclusive. Coll Macdonald explained: 'In the course of this investigation however it occurred as the preferable plan to devolve the submission to the late Dean of Faculty Ross as oversman, and this being done he accepted it.' Perhaps there was a need for an independent outsider, because as time passed awkward cross-currents were arising.

Coll Macdonald now had a personal stake in the outcome. As cautioner for Angus and Duncan Rankin, he had become liable for the amount they had to pay to the Glencoe trustees following the action over the disputed bill of 1809 for £1,050 (see below). In return, Angus Rankin 'granted him the conveyance to the sum in medio' in the Fersit case. So Coll Macdonald was hoping to recoup his outlay arising from that other dispute from the proceeds of the Fersit one.

It also became clear that Alexander Macdonald of Glencoe had not paid to Alexander McVean the original price of the stock allocated to him in 1809. He had made only a few minor payments.

Duncan Cameron sought, on behalf of the Glencoe trustees, to widen the Fersit issue by invoking a matter concerning sheep at two farms in Glenetive. It was alleged that after the Fersit tack had been taken by the MacDonells, John Rankin 'had no place of residence'. This is not consistent with the fact that he was still a sub-tenant at Glenure and Barnamuck, and indeed he returned there later. Be that as it may, 'His friend Angus Rankin . . . obtained from General Campbell of Monzie a lease of certain grazings . . . called Invercharnan and Invereolan, in favour of himself and John Rankin, each possessing or entitled to possess, half the concern.' Angus was tacksman of Dalness in Glenetive. Invercharnan and Invereolan were the next two farms lower down that glen. The story was that Angus and John had equal shares in the stock on these two farms, and John was to reside there and manage it. After a couple of years John gave up and sold his half of the stock to Angus for £600. At this point a brother of John called Duncan Ogg appeared on the scene. He claimed that he had been 'an obligant in the tack for the payment of the rent' and had an interest in John's share of the sheep which Angus had bought. Perhaps he was the Duncan Rankin involved with John at Barnamuck, although he was never called Duncan Ogg there. He brought an action against Angus, and persuaded the court that half of John's sheep had been his. Duncan promptly sold them to a Lieutenant McPhee, and Angus had to pay McPhee £340 for sheep he thought he had already bought.

The Glencoe trustees also had an interest in this stock. In November 1820, Duncan Cameron attended 'on a friend of John McVean, Invercharnan, regarding the late Glenco's obligation to him as cautioner for Angus and Duncan Rankin for the sheepstock of Invercharnan and his

claim for the balance remaining still due'.[38] The Rankins had apparently bought the sheep from John McVean and failed to pay him in full. This left the Glencoe trustees liable for the balance because of a commitment Alexander Macdonald of Glencoe must have made shortly before his death.

General Campbell of Monzie was not being paid the rent due for Invercharnan and Invereolan. In 1820 he presented a petition in Inveraray Sheriff Court against Angus, John and Duncan Rankin 'Conjointly and Severally' to recover 'the sum of £370 being the rent of the said lands for the year ending Whitsunday last 1820' and for the same amount in advance due for the next year. A landlord had to move quickly if he wanted to seize stock to cover a debt due before the tenant had time to make it disappear. In the first week of August 1820, Donald MacArthur, sheriff officer from Inveraray, seized 'Six cows, five calves, a two year old heifer, a Stirk, a Horse, all the crops of Oats, Barley and Potatoes, all the household furniture and farming utensils, 12 Ewe sheep, six lambs upon Invereolan and Invercharnan pertaining and belonging to Angus Rankin Tenant in Dalness, Duncan Rankin at Invercharnan and John Rankin in Gualchullin.'[39] Gualchullin was lower down Glenetive, near the head of Loch Etive.

What had happened to the sheep worth hundreds of pounds that had apparently been on those farms in the recent past? The General had been too late to catch the Rankins off guard, and the Glencoe trustees were left with a commitment to pay to John McVean the balance of the price of sheep that had vanished. It may not be surprising that Duncan Cameron wanted to try to bring all this into the Fersit litigation.

Alexander McVean then added yet another complication. He was pursuing a private debt due to him by John Rankin that had nothing to do with Fersit. It was accepted that the three Rankins were partners and entitled jointly to a sum of money yet to be determined, known as the 'sum in medio'. McVean knew that John Rankin had no money. He therefore wanted the Rankins bound jointly for the sum that John owed to him and tried to attach, by arrestment, the 'fund in medio'. At this point the Dean decided that if he was to continue as arbiter he wanted a minute of consent from all the parties. McVean declined to give his consent and the Dean declined to continue.

The court then resumed consideration of the process of multiplepoinding. By this time, around 1824, Angus Rankin was dead. Coll Macdonald made a proposal for settlement. The stock at Fersit had been valued at £3,504 5*s.* 9*d.* in 1814. Alexander McVean and Donald Rankin were each due a third of this (i.e. £1,168 2*s.* 7*d.*). Alexander Macdonald and John Rankin were each due a sixth (i.e. £584 1*s.* 3½*d.*). On this basis, the MacDonells had paid £635 19*s.* 8½*d.* too much to Alexander Macdonald and £85 11*s.* 11*d.* too much to Alexander McVean. The proposal

was that these excesses should be paid to the Rankins along with the 'sum in medio' of £1,027 8*s.* 11*d.* The arithmetic seems almost correct.

That left several unresolved matters. A balance of £325 2*s.* 10*d.* was still due from Alexander Macdonald to Alexander McVean for the purchase of one sixth of the sheep in 1809. The Glencoe trustees had known of this figure as early as January 1817 because it was mentioned in the minutes of their meeting in that month. The trustees also had a commitment to John McVean at Invercharnan for the balance of the price of the sheep that the Rankins had bought. Finally, there was the private debt due from John Rankin to Alexander McVean. Coll Macdonald suggested that these should not be brought into the general calculation, which seems very sensible.[40]

Another issue had been resolved in the Sheriff Court in Fort William in 1817. Alexander McVean, now tacksman of Drumfuir, Donald Rankin, still shown as residing at Achnacon, and the Glencoe trustees sought to recover £46 19*s.* 0*d.* plus interest from Captain Alexander MacDonald at Moy for 'Cattle Grazing etc. at the High Grass of Fersit' from 1809 until 1814. It was claimed that forty-one head of cattle were grazed in 1809 and twenty in 1810.[41] It is interesting that such precise figures could be produced. The probability was that the Captain had continued to use grazings to which he had access under the earlier lease. It is likely that the prime movers would have been the Glencoe trustees trying to lay their hands on any money to which they might have a claim. Incidentally, in March 1824 Captain MacDonell made an offer for a further lease of Fersit and Inverlair, and the MacDonells at Inch were offering again for Fersitriach and Torgulbin.[42] They were apparently surviving the rigours of the recession.

The Glencoe trustees had also to contend with a threatened liability for a debt of £999 (including interest) that Alexander McVean owed to James McCaul. Alexander Macdonald of Glencoe had given his security for this. By 1805 James McCaul had gone from Fersit to Perthshire, leaving his former partners, Alexander McVean and Captain MacDonell, obliged to purchase his share of the stock. This probably accounts for McVean's debt to him. The trustees were well aware of this liability. Alexander MacDonald of Dalilea, the agent of the Leith Bank who lived in Callander, wrote to Duncan Cameron about it in November 1815, and it was discussed at the meeting of the trustees in January 1816. The trustees were seriously concerned, but

> from what they had been informed as to the friendly interference of Mr. McDonald of Dalilea, in procuring delay they are equally certain that from his knowledge of the cautionary nature of Glenco's interest in this document that Mr. McDonald will . . . endeavour to recover from Mr. McVean, the real debtor, without operating on Glenco's security ad interim.

In June 1816 Duncan Cameron received a copy of a summons 'at the instance of Dalness [Coll Macdonald WS] to Ewen Macdonald [of Glencoe] and Alexander McVean on McCaull's Bill'. MacDonald of Dalilea may have been able to help, because nothing more was said about this at the next meeting of the trustees in January 1817. On the other hand, some Fersit business must have dragged on for a long time. Ewen Macdonald's account with James Arnott WS for 1831 and 1832 shows a small charge on 30 June 1831 'To account of Fersit Business since last January'. There was another on 30 June 1832 'To account for proceedings for recovery of interest on Fersit Stock'.[43] That was many years after Alexander Macdonald's first bid for the tenancy of Fersit in 1776. Why did the place have such a fascination for him?

Glenure and Barnamuck in the Nineteenth Century

After Alexander Macdonald died in December 1814 the Glencoe trustees had to continue to collect rent from John and Duncan Rankin, the subtenants he had installed there some twenty-five years earlier. They also had to continue to pay rent to Duncan Campbell of Barcaldine. They managed to collect money intermittently from the Rankins but had to pay Duncan Campbell regularly. Their difficulties were explained at their third meeting in January 1817. The first item considered was 'money due by Rankins in Barnamuck, besides Steelbow Stock of £1,000 to which the right has been constituted under the advice of Counsel'.[44] Clearly the trustees were seriously concerned.

On 3 February 1817 the Rankins paid £60 towards rent due, but this could not have been sufficient to meet the worries of the trustees. On 20 April that year they incurred legal costs of £15 16*s.* 1*d.* 'To sequestration agt. the Rankins of Barnamuck and Achnacone as charged agt. them'.[45] The position at Achnacon has been described in Chapter 6.

On 25 April 1817, the *Inverness Journal* carried an advertisement for the sale of 'A Beautiful Romantic Property in Argyllshire'. This was part of the estate of Glenure, consisting of the farms of Glenure, Barnamuck, Eleric and Derrylochan, the land leased to Alexander Macdonald of Glencoe and still held by his trustees. It was said to be an unusually beautiful property with a 'good Mansion House upon Glenure'. It was purchased by Dr Andrew Mackenzie Grieve, Inspector of Hospitals in North Britain, Edinburgh.[46] Dr Mackenzie Grieve seems to have been interested in Highland estates. Some years earlier he had lent £5,000 to John MacLeod in Skye on the security of parts of the Barony of Dunvegan.[47]

Patrick MacDougall of Dunollie remarked to his son John in a letter of 4 July 1817 that 'Barcaldine has sold half of his landed property'.[48] He linked this to the poor state of the Highland economy at that time. There may, however, have been other considerations. The Napoleonic war was

over, Duncan Campbell's military commitments would be reduced and he had married Elizabeth, daughter of James Dennistoun of Dennistoun in 1815. He embarked, probably about this time, on additions and expansions to Barcaldine House.[49]

By 1817 the lease that the trustees had to honour was about to expire.[50] Their accounts show that in February 1817 some rent for Glenure had been remitted by the trustees to 'Mrs. Elizabeth Campbell per her London agent'. On 3 December 1817 a further payment of rent was made to Dr Mackenzie Grieve 'as now Proprietor of Glenure and Barnamuck'. On 28 October 1818 there was an item 'Making up the State of Rents of Glenure and Barnamuck due to Barcaldine'. This suggests that the trustees had to clear off arrears. There were no further transactions.[51] The trustees had finally managed to rid themselves of this responsibility. It was to take a little longer to be finished with the Rankins.

In 1818 the trustees had to go to the Sheriff Court in Inveraray with a petition that told a very sorry tale. They explained about the lease of Glenure and Barnamuck and how John and Duncan Rankin had been there as sub-tenants of Alexander Macdonald of Glencoe. The Rankins had lost possession of the land and grazings at Whitsunday 1817, but they 'were allowed by the petitioners to retain possession of the houses, with the grazing of two Cows each for a year, at a rent payable by them to the petitioners'. That seems remarkably considerate. When the time came for the Rankins to vacate the houses another problem arose.

> By the terms of their agreement and the established custom of the Estate of Glenuir, not only were the said John and Duncan Rankin oblidged to leave the doors, windows, and fixed partitions in their dwelling houses and offices, but were also bound to leave these houses in a proper state of repair or otherwise make up the deficiency.

They had removed the doors, windows, and partitions. The trustees wanted an order from the Court requiring them to replace these. They also wanted £6 5*s.* 0*d.* from Duncan and £8 7*s.* 6*d.* from John.[52] Presumably this was to pay for putting the houses into good enough repair to be handed back to the landlord. Whether the trustees ever got satisfaction from the Rankins is not known. It seems unlikely.

The Rankin Dispute

The Glencoe trustees initiated several actions to secure money that they believed to have been due to Alexander Macdonald. Their attempt to recover money from Angus Rankin and his brother Duncan was particularly interesting. It illustrates how some Highland sheep-farmers went about their business.

One item found in Alexander Macdonald's repositories was a bill for

£1,050 due to him by Angus Rankin, tacksman of Dalness, and his brother, Duncan Rankin, tacksman of Caolasnacoan. The bill was dated at Glencoe on 6 April 1809 and was payable at Martinmas that year. Although it had been drawn by Alexander Macdonald, he had neglected to put his name on it. Nevertheless, Angus and Duncan Rankin put their names to it as acceptors. Angus's signature was holograph, but Duncan may have been unable to write. This was enough for the trustees to argue that the Rankins had accepted the liability and to raise an action in the Court of Session for £1,050. Coll Macdonald WS acted for the Rankins.[53]

The Rankins offered an explanation for the total of £1,050 that appeared on the bill. In the first few years of the nineteenth century they had 'acquired from Glencoe a parcel of sheep . . . at a price of about £500 or £600 sterling'. They accepted that they owed Glencoe that imprecise amount. Alexander Macdonald had then complicated the transaction. Two brothers from Glencoe, John and Donald Macdonald, moved at this time to Achranich in Morvern,[54] a farm that John MacDonald of Borrodale bought in 1803.[55] The Rankins had bought some sheep from them and therefore owed them money, but these Macdonalds owed money to Alexander Macdonald of Glencoe. He asked the Rankins 'to take the burden'. This apparently meant that, instead of the Rankins paying their debt to John and Donald Macdonald, they undertook to owe to Glencoe the money that the two MacDonalds were due to pay him. This would cancel the Rankin's own debt to John and Donald Macdonald. According to the Rankins, the £500 or £600 that they already owed to Glencoe along with what the Macdonalds owed to him amounted to £1,050, the amount of the bill now in dispute.[56] There is a delightful vagueness about these transactions. It seems improbable that the amount the Rankins owed to the Macdonalds was exactly the amount that the latter owed to Glencoe. Nothing was, however, said about any adjustment being necessary.

The Rankins produced two bills drawn by Angus Rankin and accepted by Alexander Macdonald of Glencoe. One was for £525 dated 8 March 1804 and the other was for £500 dated 27 April 1811. They explained that the first bill had been for an unspecified transaction. The second bill was a replacement for the first one when it was about to exceed its time limit. It did not include the odd £25, or take account of interest due on the first bill. The Rankins said they just kept both bills in the expectation that all would be resolved at some time. The substance of this point was that about £500 due by Alexander Macdonald to the Rankins should be set against the £1,050 that they were alleged to owe to him. However, it seems odd that this was not taken into account when the bill for £1,050 was drawn in 1809.[57]

The next argument deployed for the Rankins was that about 1812 they

had paid to Alexander Macdonald of Glencoe £400 in cash. According to his usual practice he had not given them a receipt. Evidence had therefore to be presented to try to persuade the court that this money had been paid. The story was that Angus Rankin had been at the Killin market, where he had obtained 'upwards of £400, chiefly from John Stewart, Achtrichtan'. Again the amount of money is not stated precisely. This may have been payment due for stock at Achtriachtan where, as shown in Chapter 6, John Stewart had replaced Angus Rankin as tacksman. Duncan Campbell of Glenkaitland in Glenetive was also at the Killin market and asked Angus Rankin to lend him £40 out of the money he had just received. Angus declined, saying that 'Glenco had made a peremptory demand on him for money'. Later Angus Rankin mentioned to Malcolm Robertson and John Robertson, who both lived in Glencoe, that he was on his way to pay the money to Alexander Macdonald of Glencoe. He subsequently told them that he had done so. Malcolm Robertson had a claim against Angus Rankin that he had hoped would be paid out of the money.

Angus Rankin was accompanied by one of his brothers when he went to Keppoch to give Alexander Macdonald the £400 in cash. On the following day, Alexander McVean came to Keppoch. As mentioned concerning Fersit, Alexander Macdonald of Glencoe still owed McVean money for the share of the stock that he had let him have in 1809. McVean expected 'that Glenco would allow some part of this money to be applied to the concern of Fersit; but when he applied for it Glenco told him, in a laughing manner, that he having got the money in his pocket, McVean must look elsewhere for funds to answer the demand on Fersit'.

Some months later Peter Campbell of Glenkaitland and Angus Rankin were at Keppoch. Glencoe said, in Campbell's presence, 'Angus, when are you to pay me the balance of your Bill?' Rankin replied, 'Good God, did I not pay you £400 last season?' Glencoe then said, 'You are very right.'

It was explained on behalf of the Rankins that Alexander Macdonald of Glencoe

> scarcely ever in the course of his life gave a receipt for payments to him and certainly did not keep any regular books of his extensive concerns. He trusted entirely to a most retentive memory for a record of his various concerns and those who dealt with him generally trusted to his honour and integrity which were perfectly spotless.

The lack of precision in many critical figures cited by the Rankins suggests that they were not particularly diligent book-keepers themselves.

A few months before his death, Alexander Macdonald employed Donald MacDonald from Monar (described as 'one of his dependents') for two days, 'to make out states of his accounts with various persons with whom he had dealings'. Donald was one of the Innerigan family whose affairs were

explained in Chapter 11. The court granted an order for the production of Glencoe's documents. Donald MacDonald's statement of Glencoe's account with the Rankins was then produced in court. It showed a claim of £433 against them, but Donald MacDonald was not brought before the court to speak to his conclusion.

The court had decided at an early stage that the bill for £1,050 was valid in spite of the absence of Alexander Macdonald's signature. They were prepared to accept the validity of the bill for £500 drawn by Angus Rankin in 1811 and accepted by Alexander Macdonald. The bill for £525 drawn by Angus Rankin in 1804 was no longer valid. They were quite unsympathetic about the £25 dropped from the later bill, and about interest. These matters should have received attention in 1811. The story about the £400 in cash made no impression at all. So, the Rankins owed £1,050 to Alexander Macdonald and he owed £500 to them.[58] The Rankins were therefore due to pay £550 to the Glencoe trustees; interest would probably have been allowed from 1809. In 1819, Coll Macdonald WS agreed to join them as cautioner in a bond to the Glencoe trustees for a debt of £831 8*s.* 11*d.* Presumably this was the £550 plus interest at around 5 per cent. He paid this sum to the Glencoe trustees in February 1820, anticipating that he would recoup his outlay from the money that the Rankins would recover from the Fersit affair.[59] Perhaps he did.

The Glencoe Macdonalds in Morvern

That really concluded the issue between the Glencoe trustees and the Rankins. However, the move by John and Donald Macdonald from Glencoe to Achranich cited in the court proceedings explains the presence of Glencoe Macdonalds in Morvern (see Introduction). In later years, John MacDonald of Borrodale referred to them in his correspondence. In 1806 he was interested in the possibility of buying the farm of Keill in Morvern. He said that he would 'send an express for John MacDonald one of my Tenants to come over to me here and he will be able to give me a tolerable correct idea as he knows all these farms and understands their true value much better than any of the Old Residenters in Morvern'. John may have disappointed him, because he said in March, 'consequently I have sent him back with instructions to him and his brother to examine the premises privately and minutely'.[60] Nine years later, in 1815, he told Mr James Park Harrison of the Lorn Furnace, Bunaw [Bonawe], that 'John MacDonald, one of my tenants at Achranich, will show the woods to any person you may send to look at it'.[61] In 1827 MacDonald of Borrodale wrote to Peter Campbell, writer in Inveraray, about the removal of tenants. These included Donald Macdonald at Achranich, 'one of the former tenants allowed to remain with two cows grass'.[62]

Although John and Donald Macdonald apparently went to Achranich

on their own initiative, there is a hint that Alexander Macdonald of Glencoe may have had some financial interest in Achranich. In 1821 the Glencoe trustees received £21 2*s.* 1*d.* from MacDonald of Borrodale, which was described as 'Balance of Dividend due Glenco from the Estate of Achranich'.[63] That is a useful reminder that the complex dealings of Alexander Macdonald of Glencoe were even more extensive than revealed in this study.

13

The Glencoe Trustees, 1814–1831

The actions taken by the Glencoe trustees about individual leases held by Alexander Macdonald have been described in the relevant chapters. This chapter is concerned with the generality of their task, and with matters not related to particular leases.

When Alexander Macdonald of Glencoe died on 14 December 1814 his affairs were in a chaotic state. His debts were very large in relation to his assets, and he had commitments to long leases at high rents. His only surviving daughter Jane was still unmarried, and his youngest son Ronald had not completed his education.

The trustees nominated by Alexander Macdonald in 1804 were named in Chapter 6. Only a few took an active part in the work of the trust after his death. Mary Cameron, his wife, had already died, Ewen Cameron of Fassifern was quite old, Major John Cameron was killed in 1815 and Peter Cameron was away at sea for much of the time. Duncan Cameron WS became the dominant figure who shouldered most of the work, with such help as he could get from those trustees who were available and willing. A shortage of trustees readily available in Edinburgh led to two additional nominations in 1822. John Henderson, a linen draper in Edinburgh who had lent the trust a total of £2,000 in 1817 (see below) was appointed on 27 May. Robert Scott, the first factor who had subsequently moved from Fort William to Edinburgh and become a partner of Duncan Cameron WS, was appointed on 3 June.[1 2 3]

Early in 1815, Duncan Cameron and Angus Cameron of Kinlochleven went to Keppoch, Alexander Macdonald's home for at least a decade, 'to open Glenco's Repositories'. Their travelling expenses were carefully recorded, and Duncan Cameron charged £37 17*s.* 0*d.* 'To my trouble in opening the repositories and attending to Trust affairs – 12 days'.[4]

It is unlikely that Duncan Cameron was pleased with what he found at Keppoch. Alexander Macdonald's poor record keeping was a source of difficulty to the trustees and to Duncan Cameron in particular. However, Alexander Macdonald did have the sense to keep some pieces of paper in which various people acknowledged that they owed him sums of money. These were the basis of claims raised by the trustees against the individuals concerned.[5]

The trustees held three meetings, in May 1815, January 1816 and January 1817. They met at Perth, 'for convenience', but no more than three trustees were present at any meeting. Following the first meeting, Robert Scott, then a writer in Fort William, was appointed as the trustees' factor. Colin Macdonald, second surviving son of Alexander Macdonald who was home on furlough, was present at the second and third meetings.[6] From the beginning of 1816 the trustee accounts refer to Colin as 'Captain Macdonald', but there is no evidence that he attained this rank. His final promotion was to the rank of Lieutenant on 22 April 1809.[7]

Alexander Macdonald had based his trust disposition and settlement on the assumption that he would be worth £10,000 at his death.[8] The trustees soon noted that 'instead of a reversion of £10,000 the debts would nearly extinguish the whole capital'. Before long they discovered that the situation was even worse. In 1816 they agreed that the heir, Ewen Macdonald in India, was to be 'informed fully of the circumstances of the estate'.[9]

By the time of the third meeting in January 1817, Robert Scott had become a partner in Duncan Cameron's firm in Edinburgh and another factor had to be appointed. Colin Macdonald proposed Donald MacDonald, tacksman of Drimintorran. He was appointed later in 1817.[10] The relationship of the Drimintorran family to the Macdonalds of Glencoe was considered in Chapter 7.

Alexander Macdonald's Finances

The trustees had to settle Alexander Macdonald's personal accounts. On 22 December 1815 they paid £951 17*s.* 8½*d.* 'To balance due Mr. Cameron [Duncan Cameron WS] on his accounts with Glenco preceding his death plus interest'. On 20 January 1816, £20 was paid to Dr Taylor, presumably in Fort William, followed by £44 on 8 February 1817. More personal accounts were settled in 1817. On 5 February, accounts 'due by Mr. MacDonald' for £1 8*s.* 6*d.* to his tailor, £3 15*s.* 0*d.* to his shoemaker and 18 shillings to his hatter, all in Edinburgh, were paid. On 31 May an account for £30 4*s.* 10*d.* was paid to 'Mr. John Cameron, Merchant, Fort William, for furnishings to the deceased's funeral'.[11]

Alexander Macdonald's Borrowing and Lending

The trustees faced a difficult task in trying to unravel Alexander Macdonald's complex transactions and discover what money was owed to him and by him. Their accounts contain many references to transactions which cannot all be identified and explained, but they give a measure of the extent and complexity of his affairs. In broad terms, Alexander Macdonald had debts exceeding £8,000 at the time of his death in December 1814. This total includes the loans mentioned in Chapter 6, and the £1,000 lodged with him by John MacDonald of Drimintorran mentioned

in Chapter 7. It also includes £317 16*s.* 3*d.* owed to the Leith Bank and £515 0*s.* 5*d.* to the British Linen Company.[12]

There was an outstanding loan of £100 by 'Miss Stewart, Keppoch'.[13] She was an old family servant, and £100 must have been a very large amount of money to her. In 1821 Robert Scott asked Ewen for 'any wish you may have with regard to Miss Stewart upon Keppoch's being given up . . . she has become very frail, but anxious about everything that concerns you'.[14] She must have died shortly after. Her loan of £100 had been due for repayment at Martinmas 1819. In February 1824 it was repaid to her representatives with £4 9*s.* 9*d.* interest. They also received £47 10*s.* 0*d.* arrears of annuity owing to her 'under Glenco's letter of 8 June 1804'. Her representatives had then to pay £9 18*s.* 2*d.* – 'the expense of discharge and correspondence'.[15]

An extraordinary transaction with Peter Cameron appeared in the accounts for 1825, when the trustees were about to wind up their affairs. On 24 January 1810, when Alexander Macdonald had recently borrowed £3,000 and was about to borrow another £1,000, he lent £1,000 to his brother-in-law Peter Cameron. This was 'at the time of his obtaining a command', presumably of an East Indiaman. There is no indication of why this money was advanced. Alexander Macdonald clearly received neither interest on it as a loan nor a return on it as an investment. Nothing seems to have been known about this money by the trustees until September 1825. Their accounts then show the repayment of the original £1,000 plus interest meticulously calculated from the date of the advance in 1810. The interest amounted to £1,065 4*s.* 3*d.*[16]

The Trustees Obtain Additional Funds

From time to time the trustees had to meet requests from lenders for the repayment of loans and to pay off accumulated arrears of rents. This required further borrowing. In 1817 they obtained two loans each of £1,000 in quick succession from John Henderson, linen draper in Edinburgh. They also got £1,500 from Miss MacLeod of Bernera, who had already lent £1,000 to Alexander Macdonald.[17] [18] [19]

During 1818 the trustees received a total of £5,000 from India. They described this as 'a remittance of £5,000 from the said Ewen Macdonald Esq., now of Glencoe, to be applied towards the extinction of his father's debts'.[20] The money came in two instalments, £2,000 arriving in May, followed by £3,000 in October.[21] Both were transmitted by 'McIntosh and Co. of Calcutta'. Ewen Macdonald was then aged thirty and had been an assistant surgeon in the service of the East India Company for nine years. Although fortunes were made in India, it seems unlikely that as a relatively junior officer he could have accumulated so much wealth by 1818. It almost certainly did not come from his own resources. The probable source was

Lieutenant-General Sir John Macdonald of Glencoe. He spent a long life in the service of the East India Company in Bengal and died at Calcutta in 1824. His will included the following provision:

> And I do hereby release the said Ewen Macdonald . . . from the Repayment of all Sum or Sums of money due to me or which may be due to me at the time of my decease from him in respect of a certain Bond of his for securing the sum of five thousand Pounds sterling or thereabouts and direct the said Bond to be cancelled and then delivered up to the said Ewen Macdonald.[22]

Lieutenant-General Sir John Macdonald, KCB

John Macdonald was born around 1747 or 1748 and became a cadet of the East India Company in 1766. He was admitted to the service in 1767 and progressed steadily through the commissioned ranks in the Bengal Establishment until he became a Lieutenant-General in 1813. On 7 April 1815 he was appointed a Knight Commander of the newly enlarged Order of the Bath. He died in Calcutta in May 1824 at the age of seventy-six.[23] [24]

Sir John was granted arms at Edinburgh in 1818. He was then described as 'the only surviving son of James Macdonald Esquire sometime an Officer of Marines deceased, who according to the constant tradition of the family was descended from the ancient Family of Macdonald of Glencoe'.[25] This is the only information found about his parents. His gravestone in South Park Cemetery, Calcutta, says only 'Sacred to the memory of Lieut-General Sir John Macdonald KCB, who, after an honourable and faithful service of more than half a century, died on the 29 May 1824'.[26]

There was contact in India between Sir John and both Ewen and Colin. When Colin wrote from London to his grandfather Sir Ewen Cameron on 18 April 1817, he mentioned some Madeira wine that General Macdonald had given him as a present when he came home two years earlier. He thought it had been lost, but he managed to recover it in London and sent some to Sir Ewen.[27]

Besides releasing Ewen Macdonald from the bond for £5,000, Sir John left him 10,000 rupees. That would have been worth about another £1,250. Ewen was one of Sir John's executors, but it is frustrating that there is no hint of what their relationship might have been.[28] Sir John's position in the family remains obscure.

The Search for Funds Again

The arrival of £5,000 from India in 1818 probably accounts for a lull in the search for further loans, but within a couple of years more money was needed. During 1821 and 1822 the loans from the Rev. Alexander Rose in Inverness, Mr John Henderson in Edinburgh and the Drimintorran family

had to be repaid. These totalled £4,000. A loan of £1,500 was obtained from the trustees of Mr John Dod of Seatonhill and another of £3,000 from tutors of Chisholm (William Alexander Chisholm). A further £1,000 may also have been obtained from Sir Alexander Grant of Dalvey, Bart., but the records about this are confusing.[29]

In 1822 the trustee accounts contain a list of the bonds and loans on which interest had to be paid. The total was £9,400, on which annual interest of £462 10*s.* 0*d.* had to be paid[30].

Alexander Macdonald's Younger Children

It will be recalled from Chapter 6 that when Alexander Macdonald of Glencoe died in December 1814 he was survived by four of his children. Ewen, now the heir, and his brother Colin were both in the service of the East India Company. Colin returned to India in 1817 after more than two years on furlough and died at Cuttack on 29 April 1818.[31] He had fathered an illegitimate daughter, Louisa Jane, while at home on furlough. She had been left in the charge of a Mrs Henderson in Edinburgh when he returned to India. His brother Ewen immediately accepted responsibility for her maintenance and eventually, in 1823, secured for her an allowance from the orphan fund of the East India Company.[32] [33] [34] Louisa Jane is mentioned from time to time in the trustee accounts.[35]

Ronald Macdonald

The youngest of Alexander Macdonald's surviving sons, Ronald, was fourteen years old at the time of his father's death. He was a pupil of Mr Mesnard, a schoolmaster in Berwick, to whom the trustees paid fees and maintenance.[36] In February 1819 he joined the 18th (South Staffordshire) Regiment of Foot, a commission as an ensign having been purchased for him.[37] In 1821 Ronald wrote to his uncle, Duncan Cameron, about obtaining a commission as a lieutenant in India, but nothing came of this.[38]

On 13 November 1821, Robert Scott told Ewen Macdonald that Ronald had 'gone from Gibraltar to Corfu . . . and this station will I think be well suited to his constitution, which is not very strong, perhaps from his having of late got so much taller'. He added, 'the money for your Brother's Lieutenancy will at once be lodged upon notification from the Agents that there is an opportunity for purchasing for him'.[39] However, Ronald had to wait until January 1835 before he got his Lieutenancy, and he was then with the 89th Regiment of Foot.[40] His allowance from Ewen seems to have been settled, because on 18 May 1825 there was an entry reading 'Remitted to Mr. Ronald McDonald to account of his allowance by London bill – £27'. A further payment of £16 8*s.* 6*d.* was recorded on 8 August.[41]

Towards the end of 1825, Ronald was in Manchester.[42] In October 1826 he married Maria, the daughter of Dr Thomas, a retired naval surgeon, of Ballacosnahan, Patrick, Isle of Man. They had two children: a daughter, Ann Cosnahan, was born in July 1828 and a son, Alexander James John, was born in October 1829.[43] Later, Ronald was posted with his Regiment to Australia, and was sent to Norfolk Island to guard convicts – unlikely to be a coveted posting. He returned to Sydney towards the end of 1839, was promoted to captain, and retired from the army.[44] The *Inverness Journal* of 24 December 1841 reported that Ronald's wife Maria had died in Sydney on 20 April and that Ronald died there on 20 August in the same year.

Jane Cameron Macdonald

When her father died, Jane must have been approaching an age when it would be important to launch her in society. The trustees did their best for her. She lodged with a Miss Humphrey in Edinburgh from late 1815 or early 1816 until about the end of 1817. For most of that period the trustees paid £26 5*s.* 0*d.* a quarter to Miss Humphrey.[45]

The payments that the trustees made for Jane in the three years or so between her father's death and her marriage give some insight into the life that she could lead in Edinburgh. This was in spite of the very large debts that the trustees were discovering as they delved more deeply into her late father's affairs. In 1816 the trustees paid £1 4*s.* 6*d.* for 'freight and canal dues of Miss McDonald's pianoforte to Edinburgh'. Accounts totalling almost £170 were paid for her to various traders and shopkeepers in Edinburgh. The trustees also provided her with cash, usually in amounts of £5 from time to time. Exceptionally, she was given £10 in November and again in December 1817, perhaps because of her marriage towards the end of that year.[46]

In his trust disposition, her father had stipulated that his unmarried daughters were to receive an allowance of £80 per annum. Jane's expenditure was obviously running well above that level, but she apparently had expectations from another source. On 27 January 1816 Duncan Cameron wrote to 'Major Macpherson about Miss Macdonald's share of Miss Kennedy's legacy'. This seems to have been a reference to her cousin, Jane Kennedy, the daughter of Rev. John Kennedy, whose affairs were described in Chapter 5. Jane Kennedy was living at Duthil in 1806,[47] but presumably she died later. In October 1816 Miss Macdonald received her legacy, and Duncan Cameron made a copy for her 'of state of Settlement with Dalness'. Coll Macdonald WS of Dalness had been one of John Kennedy's trustees, along with Major Charles McPherson and Alexander Macdonald of Glencoe. Duncan Cameron also made up an account 'by her [Jane Macdonald] to the Trustees as representatives of

her father for sundries paid on her account'.[48] In his meticulous fashion he intended to recover some of the payments he had made for her out of the Glencoe trustees' non-existent funds.

Before the end of 1817 Jane Macdonald married Captain Coll MacDougall of the 42nd Regiment.[49] On 23 January 1818 an account 'due by Miss Macdonald, now Mrs MacDougall' was paid.[50] Marriage did not, however, bring to an end the trustees' concerns with Jane. Under the terms of her father's trust disposition she was entitled to a payment of £1,000 on marriage. This was to be paid from 'any surplus or rent or interest' from the £10,000 that he had optimistically expected to be worth at his death.[51] The trustees plainly did not have £1,000 to give to Jane. Instead Duncan Cameron prepared a heritable bond for £1,000.

Suddenly Jane's position changed with an entry on 13 April 1821 noting Captain MacDougall's death.[52] Duncan Cameron moved quickly, perhaps more as an uncle than a trustee. On 15 April he set about sending £50 to Jane, but first he had to write to Miss Rankin at Keppoch to get her address. Jane was in France, but he succeeded in reaching her and got confirmation in due course that she would 'draw the £50'.[53]

Jane must have returned fairly soon from France to Keppoch, because on 5 June Duncan Cameron received 'Post from Miss Rankin informing of birth of Mrs. MacDougall's daughter'.[54] However, she was not to reside at Keppoch for long. When Robert Scott wrote to Ewen Macdonald on 13 November 1821, he said:

> I heard lately from France of Mrs. MacDougall and her son and daughter. All are well . . . and of her having gone from Boulogne to St. Omer with intention to remain there during the winter'.[55]

In December 1825 Jane and her children were in Guernsey.[56] In his Disposition of 1837 her brother Ewen directed that Jane was to receive an annuity of £20 out of the revenues of Glencoe.[57]

The Trustees Wind up Their Affairs

On 2 December 1825 a summons of declarator was issued at the instigation of the trustees. An action followed in the Court of Session lasting from 1826 until 1831. The case advanced by the trustees was that they had 'executed all the practicable purposes of the Trust' and now wished to transfer the estate to Ewen Macdonald as the heir. Ewen was not opposed to this. Much of the time taken to conclude the matter must have been due to the need for communication to pass to and fro between the lawyers in Edinburgh and Ewen in India. Lord Meadowbank pronounced an interlocutor on 17 June 1828, and the ownership of the estate was then passed from the trustees to Ewen Macdonald.[58] He was, however, required to give a personal bond of relief to the trustees, freeing them from any encumbrance. This he did, the

bond being signed by him at Neemuch in Raghpootanah on 13 August 1829. This enabled the Court proceedings to be concluded in 1831, and the trust finally ended.[59]

Information submitted to the court and included in the bond signed by Ewen Macdonald showed that the debt he assumed with the estate was £6,500. This legacy from Alexander Macdonald's ill-judged sheep-farming was to cast a long shadow over his family and over the subsequent history of Glencoe.

Postscript

Ewen Macdonald as Owner of Glencoe

At the time of his father's death Ewen was assistant surgeon with the 24th Regiment of Native Infantry. He remained in that position until 9 August 1823 when he was promoted to the rank of surgeon. Thereafter he was with the 48th Regiment of Native Infantry for a short time, but by about 1825 he became surgeon to the 9th Regiment of Light Cavalry, with whom he remained for the rest of his career. He was present at the lengthy siege and the storming of Bharatpur in 1825–26.[1] [2]

In 1829 Ewen moved with his regiment from Cawnpore to Neemuch, where he remained for four years. It was in Neemuch that his daughter Ellen, who was to inherit the estate of Glencoe, was born in 1830. The register of baptisms within the Chaplaincy Station and District of Neemuch in the Archdeaconry and Diocese of Calcutta shows that on 4 August 1832 Ellen Caroline Macpherson Macdonald, daughter of Ewen Macdonald Esq., surgeon, and a native woman was baptised at Neemuch by the Rev. Charles Parker. The record said that she had been born at Neemuch on 5 January 1830,[3] but her date of birth was always given subsequently as 5 July 1830.

Shortly after his daughter was baptised, Ewen moved with his regiment to Meerut and then to Karnaul. He returned to the United Kingdom on furlough in the spring or early summer of 1836, apparently in poor health, but he did not retire formally from the service of the East India Company until 1 May 1838.[4] [5] On 19 June 1836 he wrote from London to Andrew Belford, his solicitor in Inverness. 'It was my intention to have long ere now proceeded to the Highlands but something always comes in the way to prevent it. However, I have now taken my passage for Inverness on the *Duchess of Sutherland* which departs on Saturday 2nd Prox.' One of Ewen's problems had been that he could not manage to get his baggage collected in time for the *Duchess*'s previous trip.[6] The *Duchess of Sutherland* was the first steamship to ply between London and Inverness and had made her first voyage in March 1836.[7] Ewen did well to secure a passage on 2 July. The *Inverness Courier* of 6 July reported: 'There were ninety passengers on board that splendid vessel the 'Duchess of Sutherland' on her last trip to Inverness – a proof that the summer tourists have begun to set in Northwards.' Those who could not get a passage had to come north by the Aberdeen steamer. One of Ewen's

fellow passengers was the Duchess-Countess of Sutherland herself, making her way north for the season. She was landed at Dunrobin Castle at 3 o'clock on Tuesday morning, presumably a special call for her convenience. The shore was said to be lined with the populace of Sutherland to welcome her. Ewen did not receive such special treatment, but when he reached Inverness he would have found a well-established service of steamboats on the Caledonian Canal to take him in relative comfort towards Fort William and beyond.[8]

There were several indications, soon to be confirmed, that Ewen was not in good health. By the end of 1836 he was in Cheltenham with his uncle, Sir Duncan Cameron. Both were said to be trying to recuperate, Sir Duncan presumably from his customary gout, from which he suffered all his life.[9] A letter of 7 January 1837 from James Macgregor, solicitor in Fort William, was addressed to Ewen Macdonald 'Care of Captain Peter Cameron' at a coffee house in London and was redirected to him at the post office in Cheltenham.[10]

On 28 February 1837 Robert Downie, who had been a merchant in Calcutta before buying the former Stewart estate of Appin,[11] wrote from London to Ewen on estate business. He recalled that they had met in Calcutta and said that Mrs McColl from Lismore had told him: 'what I can scarcely believe, that you are ailing. I trust not seriously so, for I saw you at a distance at Oban when you appeared to be in high health, at least as far as countenance went.'[12] Perhaps Ewen had been on his way to or from Cheltenham when Robert Downie saw him at Oban. Steam navigation was by then well established on the west coast and would have been the preferred means of travel.[13]

Ewen must have returned from Cheltenham to Scotland fairly early in 1837, because Joseph Mitchell records that he dined with Sir Duncan Cameron at Fassifern on 30 April 1837 and that Dr Macdonald, his nephew, was present. He described him as the proprietor of the small estate of Glencoe. Another guest was a Colonel Ross, also of the East India Service. Mitchell said that both had served some thirty years in India and 'were suffering in health from their long residence in that enervating climate'. The Colonel regaled the company with a graphic account of his fight with a tiger, when he had thrust his right arm down the throat of the animal, and dispatched him with his dagger in his left hand. After that it may not be surprising that Mitchell fails to record any contribution that Ewen Macdonald might have made to the conversation.[14]

Ewen's health must have deteriorated soon after Mitchell met him. In the summer of 1837 he was in Edinburgh receiving medical treatment. He wrote from there to Sir Duncan Cameron on 1 August 1837.[15] He had apparently written an earlier letter telling him about 'the unfortunate state

of my health'. By 1 August he said that he was spitting up less blood but was troubled by a cough. He felt weak and debilitated and said, 'my constitution is breaking up'. He had seen Dr Davidson, 'who examined my chest most minutely and is to call this evening with Dr. Craigie when they are both to examine my Chest with a sort of instrument the name of which I do not recollect'. The instrument was probably the stethoscope, which was then of fairly recent origin and was the subject of active interest in leading medical circles in Edinburgh during the 1830s.[16] Ewen was fretting to know whether or not his medical advisers would say he was fit to leave Edinburgh. One of his concerns was that he had already missed the general election held in July 1837, following the death of King William IV on 20 June. He lamented his 'absence from the election. This attack is most unfortunate, particularly at the present moment.' He was not to have another opportunity to exercise his right to vote – the next election did not take place until 1841, the year after his death. The election that he missed because of his illness would have been a bonus for him. It was the last occasion on which parliament was dissolved following the death of the sovereign.[17]

In October 1837, perhaps before leaving Edinburgh, Ewen arranged to purchase some silver from Marshall and Sons at 87 George Street. He had it engraved, which brought the total cost to £134 14*s.* 6*d.*, but he returned to them some old silver for which he was allowed £68 9*s.* 3*d.*, leaving £66 5*s.* 3*d.* to be paid.[18]

A year later Ewen was in better health, and Mitchell noted that he was among a group of the leading chiefs in the north who attended the funeral of The Chisholm in September 1838. The funeral left from Wilson's Hotel in Inverness and made its way to the burial place at Erchless Castle, some eighteen miles away. After what was described as a sumptuous dinner, the mourners returned to Inverness.[19] If Ewen stayed the whole course, it would have been quite a strenuous day for a man who had said a year earlier that his constitution was breaking up.

Ewen Macdonald as proprietor of Glencoe and Sir Duncan Cameron as proprietor of Callart on the north side of Loch Leven, which his father Ewen Cameron had purchased about 1796,[20] were joint owners of Eilean Munde, the burial island in the loch. Around this time they had it planted with trees which, according to a contemporary account, 'will give it an interesting and romantic appearance, and will add greatly to the beauty of the surrounding scenery'.[21] This probably coincided with the building of the present Callart House for Sir Duncan Cameron between 1835 and 1837.[22]

On 19 August 1840 Ewen Macdonald died at Callart House, not quite two years after attending the grand funeral of The Chisholm. There was a similar but perhaps more modest gathering at his burial on Eilean Munde.[23]

In a codicil added to his will on the day before his death, Ewen said that Drs Davidson and Craigie, the two physicians who had attended him in Edinburgh, were each to receive three dozen bottles of Madeira.[24] The prolonged and fluctuating course of his illness and the account he gave of it when he was in Edinburgh in the summer of 1837[25] suggest that he may well have been suffering from advanced pulmonary tuberculosis when he returned from India.

The value of Ewen's personal estate in Scotland was £702 17*s.* 5*d.*, including moveable possessions worth £374 7*s.* 6*d.* The sum of £94 5*s.* 5*d.* was due to him as a pension from the East India Company and as a portion of an annuity under the Bengal Medical Fund.[26] These sums were, however, insignificant when set against the debt of £6,500 that he had inherited along with Glencoe. Most of that was still outstanding when he died in 1840.

After Ewen Macdonald's Death

In his disposition and settlement of 31 August 1837, Ewen directed that three trustees, Sir Duncan Cameron, Captain Peter Cameron, and Mr Andrew Belford were to hold Glencoe in trust for his daughter Ellen Caroline Macpherson Macdonald. She was then aged seven and living at Fassifern, but was about to be placed in a boarding school in Inverness.[27]

Ewen apparently did not decide that Ellen should inherit Glencoe until shortly before making his disposition in 1837. There is a much amended draft bond of annuity in the Fraser-Mackintosh Collection in the National Archives of Scotland which indicates how Ewen's intentions developed.[28] The draft was prepared initially in 1833, and in the first instance it was intended to provide his niece Louisa Jane Macdonald, the illegitimate daughter of his late brother Colin, with an annuity of £50 from the estate of Glencoe after his death. Ellen, for whom no provision was proposed at that stage of the draft, would then have been about three years old. Subsequently, at an unknown date, the draft was amended in pencil to include provision for Ellen, so that both girls were to have annuities of £50.

Further amendments were inserted in 1836, probably by Andrew Belford. These introduced provisions such as the nomination of trustees, which would make the draft into a more satisfactory formal document. There are also some notes in pencil at the end of the draft which appear to be jottings to remind Andrew Belford of his client's wishes. Bearing in mind that Ewen reached Inverness in July 1836, it is probable that he wrote the original text in India in 1833, added the pencil amendments to include Ellen while still there, brought it home in that state and handed it to Andrew Belford when he reached Inverness, with instructions to develop it further.

The most significant of the jottings reads 'if [??] succeeds, he must pay

off all the debts out of rents'. The obvious male successor would have been Ewen's only surviving brother Ronald, who was then in Australia. The draft as contemplated at that stage would have ensured that if Ronald obtained possession of Glencoe, Ellen and Louisa Jane would be assured of support out of the revenues of the estate. The jottings also show that Ewen was concerned about how to impose on his successor a requirement that the debts on the estate should be paid off. Bearing in mind the extent of the debt and the modest rental of the estate, this was an onerous requirement, but he persisted with it.

Within about a year Ewen had bypassed Ronald, and on 31 August 1837 he signed a disposition which made Ellen his heir under the provisions of an entail.[29] It is impossible to say why he chose to do this. Perhaps he hoped that by leaving it in trust to his young daughter he could impose conditions on the trustees which would fulfil his wish that the debt should be paid off. He tried to ensure this by stipulating that his trustees were 'to pay off all my debts . . . and when the said debts are fully paid . . . to complete the title deeds in the person of the existing heir of Entail'.[30] That is to say, the trustees could not transfer the estate to Ellen until all of the debts had been paid off.

The Tasks Facing Ewen's Trustees

The trustees were in a difficult position in two respects. First, they had to satisfy themselves that Ewen was entitled to leave the estate of Glencoe to his illegitimate daughter and thereafter to her heirs. In a memorial for the opinion of counsel in November 1840, they said, 'the Memorialists understand that doubts are entertained by some of the parties interested under the destination contained in Mr. Alexander Macdonald's Marriage Contract and Trust Disposition' about how far it was competent for his son Ewen to change that destination. Alexander Macdonald had directed that the estate be left to 'heirs male of my body . . . whom failing heirs female of my body . . . whom failing to my nearest lawful heirs'. The introduction by Ewen of his natural daughter changed the whole course of the destination laid down by his father. Counsel was, however, of the opinion that Ewen was not bound by the content of his father's deeds.[31]

Second, the requirement that all the debts should be paid before the estate could be handed over to Ellen must have seemed almost impracticable. Ewen had inherited a debt of £6,500 in 1828. When he died in 1840, it had been reduced by about £1,000. This modest reduction had been achieved during a period when Ewen was drawing a salary and latterly a pension from the East India Company, and he may not have needed to draw a substantial personal income from the estate. After his death a different situation arose. The rental of the estate was about £505 in 1841,[32] but defaults by tenants and other causes would ensure that the

full rental was never collected. Interest at 5 per cent on the outstanding debt of over £5,000 would amount to about £260 or £270 per annum. Ewen directed that Ellen was to receive an allowance not exceeding £100 and his sister Jane was to receive £20 per annum.[33] There would also be expenditure on maintenance of the estate, and the usual public burdens. There simply could not be significant sums of money available to apply to the reduction of the debt.

Ewen completed his trust disposition and settlement in the summer of 1837, close to the time when he was so ill that he almost despaired of his life. His decisions may be more easily understood if the urgency which must have arisen from his illness is taken into account.

When Ewen died Sir Duncan Cameron and Captain Peter were no longer young men. Sir Duncan was sixty-five and there had been references in the past to ill-health and to visits to England and France for health reasons.[34] He was already limiting his professional work and had persuaded Mr Belford to move from his office in Edinburgh and set himself up in Inverness to look after his Highland clients.[35] When Mr Downie of Appin wrote to him in February 1841, he immediately passed his letter to Mr Belford, 'who is not only my agent, but one of Glenco's Trustees and who attends to the business parts of the Trust'.[36] Captain Peter Cameron was a couple of years younger than his brother, but he died in 1843,[37] leaving only the elderly Sir Duncan and Mr Belford.

Marriage of Ellen Macdonald

Towards the end of the 1840s the trustees became involved in arrangements for Ellen to marry Archibald Burns jnr. of 'Gilston and the Isle of Ceylon'. The reference to Gilston is quite obscure; there is no evidence of any landed interest or inherited wealth. Archibald jnr. was the eldest son of Archibald Burns, manager of the Central Bank of Scotland.[38] This was a small joint stock bank founded in Perth in 1834, which became part of the Bank of Scotland in 1868.[39] Archibald Burns snr. had moved from a position with the Commercial Bank of Scotland in Dumbarton to manage the Central Bank when it was established in Perth. He took an active part in the movement which led to the establishment of the Free Church of Scotland in 1843 and was an office bearer in the Free St. Leonard's congregation in Perth.[40] It is not known what young Archibald was doing in Ceylon (Sri Lanka), but two of his younger brothers pursued careers in banks overseas – Benjamin in the Colonial Bank of New Zealand and Frederick in the Bank of Bengal.[41] Although his father was to give financial help to Archibald when he needed it, he had a large family and had to consider the interests of his other children. This was certainly not a marriage that would bring a promise of an end to the financial problems of the estate of Glencoe.

The prospect of marriage raised questions about Ellen's position in the light of Ewen's stipulation that the lands and estate of Glencoe could not be handed over to her until all of the debts had been paid. The debt had been about £5,500 when Ewen died in 1840. In 1848 it stood at £5,360,[42] and by 1852 it was £5,500.[43] No real progress was being made in reducing the debt.

In 1848 the trustees requested the opinion of counsel on a number of questions. For example, could Ellen make provision for her husband, in the event of her death, out of the revenues of the Glencoe estate? Counsel was of the opinion that 'Miss Macdonald . . . may . . . validly exercise any power contained in the entail and legally competent to her, including provision for husband and children'. However, 'the validity must await the survivance of the period when the Trust purposes being fulfilled she will have the right to the possession of the estate under Tailzie'.[44] That is to say, the debts had to be repaid.

There was also the question of whether she might be able to call upon the trustees to hand the estate over to her. Counsel recognised some difficulty in determining the construction to be placed upon various provisions, but thought that 'she cannot call upon the Trustees to make over the lands to her, and the Trustees are not at liberty to do so'. He added that the trustees had no power to sell any part of the land for payment of debt.[45] It would seem that in his enthusiasm to ensure that the debts would be paid off, Ewen had succeeded in denying any room for manoeuvre to his trustees or his daughter.

Counsel's opinion was dated 4 November 1848. Around that time discussions were in progress between Archibald Burns snr. and Sir Duncan Cameron's son-in-law, Alexander Campbell of Monzie and Inverawe, about the affairs of the Glencoe estate in the light of the proposed marriage. The latter may have been involved simply because Monzie was conveniently close to Perth, but his wife, Christina Cameron, was on very close terms with Ellen Macdonald and was later described as her 'life long companion'.[46] [47] Archibald Burns jnr. was presumably going about his business in Ceylon while these discussions were taking place.

Campbell of Monzie reported to Sir Duncan Cameron in a letter of 20 November 1848, indicating that he had become involved because of 'the attempt to sell', which had of course been ruled out by counsel's opinion. He wrote: 'Mr. Burns is now of the opinion that the Trust had better remain as it is but he is quite willing that the extra £100 should be allowed to accumulate in the young couple's name to pay off the debt.' This was presumably a reference to the annual allowance of £100 that Ellen was entitled to receive from the revenues of Glencoe. Campbell said that he 'had been advising the sending out of one of the daughters, which has been well received and he [Mr Burns] is now all for an early start to

Ceylon'. That was a reference to the need to make suitable arrangements to get Ellen to Ceylon. Having taken the matter thus far, he did not wish to be involved further and suggested, 'the better plan is to leave all the matters (as far as I am concerned) to Burns and Belford'. He gave the excuse of 'too many cooks etc.'[48]

Sir Duncan Cameron and Mr Belford gave their consent to the marriage and applied themselves to the preparation of a marriage contract.[49] Meantime, Ellen was living in Perth, presumably with her future husband's parents. She wrote from there to Mr Belford on 22 January 1849 acknowledging the safe receipt on 25 December of plate and crystal which her father had directed that she was to receive on her marriage. She went on to say that she supposed Mr Belford had heard from Mrs Burns of the good opportunity which had presented itself: a Mrs. Bell who was going to India 'has kindly offered to take charge of me'. This would avoid the need for one of the Burns daughters to make a return trip. Passages were taken for 29 March by steamer to Alexandria, whence they would proceed by the overland route to Suez and there join the P&O steamer *Bentinck*, which would take them to Ceylon on its way to Calcutta. Ellen added: 'I was in Edinburgh during the early part of last week getting my things.'[50]

The marriage took place on 1 May 1849 at Point de Galle in Ceylon, the port of call used by the P&O steamers on their recently established Suez–Calcutta service.[51] In accordance with the conditions laid down by Ewen Macdonald, Archibald Burns took the surname Macdonald.[52] From then on he was known as Archibald Burns Macdonald of Glencoe.

Early Married Life of the Burns Macdonalds

The activities of the Burns Macdonalds during the 1850s cannot be traced completely. There were celebrations in Glencoe at the end of October 1850 when news was received of the birth of an heir.[53] Information on the birth certificates of later children shows that Ellen had had a son who did not survive, and there is no other record of him. The Burns Macdonalds were back in Scotland in the early part of 1851. On 10 February 1851, Archibald Burns Macdonald wrote from Perth to Mr Belford in Inverness, regretting that he had not 'had an opportunity of making your acquaintance and having a talk on Glencoe matters' but saying this was now out of his power as he was to leave on 17 February on his way back to Ceylon. However, he said that he would 'be glad to know from you the amount of the debt now on Glencoe, and if you could give me a notion of what annual reduction may reasonably be looked for from the property'.[54] This reveals that after almost two years of marriage Archibald Burns Macdonald was surprisingly ignorant about the financial affairs of the Glencoe estate. His father ought to have been in a position to enlighten him but perhaps had not done so. It would be interesting to know what Mr Belford said in reply. Would this

have revealed to Archibald Burns Macdonald for the first time the true state of the finances of the Glencoe estate?

Ellen did not go back to Ceylon with her husband. At the census on 31 March 1851 she was with his parents and eight of his brothers and sisters at their home in St. John's Street, Perth, where the head office of the Central Bank of Scotland was situated. There was no infant with her. The probability is that, having lost her first child, her husband brought her home and there is nothing to indicate that she ever returned to Ceylon.

The New Trustees

With the prospective duration of the trust stretching far into the future, it must have become desirable for Sir Duncan, now in his mid-seventies, and Mr Belford, his colleague of many years' standing, to find a way of disengaging themselves. This was achieved in 1852. The first step was the assumption of additional trustees. These were Archibald Burns, manager of the Central Bank of Scotland in Perth (father-in-law of Ellen), Archibald Burns Macdonald of Glencoe, presently in Ceylon (husband of Ellen), John McLeish and Robert Peter, agents of the Central Bank of Scotland in Crieff and Aberfeldy respectively. The next step was for Sir Duncan and Mr Belford to retire as trustees. The new trustees relieved them of

> all risk and trouble in regard to a debt of £5,000 constituted by Bond and Disposition in Security over the Trust Estate granted in favour of the Trustees and Executors of Thomas Gilzean of Bunachton dated 19 and 23 January 1844 and another debt of £500 constituted by a Bill and Promissory Note in favour of Mr. Hugh Kennedy, Merchant in Fort William dated 7 April 1850.[55]

This concluded Sir Duncan's second period of association with the estate of Glencoe as a trustee. He lived, however, until January 1863 and apparently continued to take an interest in the affairs of his nephew's daughter and her family. This interest was maintained after his death by his daughter Christina and her husband Alexander Campbell of Monzie and Inverawe.

The Burns Macdonalds Settle in Scotland

Archibald Burns Macdonald must have rejoined his wife in Scotland before the end of 1852, as their next child, Ellen, was born in Perth in the summer of 1853. Then the Burns Macdonalds seem to have settled for a few years at Fanans on the Inverawe estate. Three sons were born there. The eldest, Archibald Maxwell, who was to be the heir, was born on 1 January 1855.[56] The next, Ewen, was born on 6 December 1855. Thereafter there was a gap until Duncan was born on 21 September 1859. There is some confusion because the register records the name of this male child born in 1859 as

Archibald Burns Macdonald. However, successive censuses are consistent in calling him Duncan.

At the census on 8 April 1861, the Burns Macdonalds were living at Glen Nevis House, a property that Sir Duncan Cameron had purchased in 1851.[57] Archibald Burns Macdonald was described as 'landed proprietor and factor', the only indication of any occupation that he may have pursued in these years. Shortly after this the family settled in Perth, where Archibald held the office of Distributor of Stamps and Collector of Taxes for many years.[58] The rest of the family were born there – Stuart in 1862, Caroline in 1863 and Allan in 1868.

The Burns Macdonalds were faced with an apparent impasse over the Glencoe estate. It had been left to Ellen, but it must have seemed improbable that she would ever gain possession of it. The debt to be extinguished remained stubbornly above £5,000. In 1863 they put into effect a scheme to overcome the difficulty.

Ellen Burns Macdonald Gains Possession of Glencoe

In May 1863 Archibald Burns Macdonald effected a policy on the life of his wife Ellen for £6,000 with the Life Association of Scotland. This was a Class A policy, entitling the assured 'to participate in the surplus funds of the Association'. The premium was £161 10*s.* 0*d.* a year. He also borrowed £5,700 from the Life Association of Scotland at 5 per cent interest. The security given to the Life Association for this loan took three forms. First, the life policy for £6,000 was assigned to the Life Association. Second, Ellen agreed to assign the estate of Glencoe to the Life Association, but the security that this offered was limited by the terms of her father's entail. In effect, all she could assign was the rental of the estate during her lifetime. The Life Association could not take possession of the estate because that had to pass intact to the next heir of entail after Ellen's death. Third, Archibald undertook to pay an annual sum of £120 to be used for the payment of any penalties incurred by the late payment of the premium on the life policy or the interest on the loan. The Life Association was, however, to be accountable to him for any of this money not used for the purpose specified.[59]

The next steps were for Ellen to present her father's disposition and deed of entail, and for his trustees to renounce the estate in her favour.[60] The debt stood at exactly £5,000 and the trustees felt able to hand over the estate on the grounds that the debt was to be discharged immediately.[61] Archibald had secured a loan large enough to enable him to do this and still have £700 left in his hands.

The principal advantage of this arrangement must have been the certainty that when Ellen died sufficient capital would be available from the proceeds of the insurance policy to pay off the debt on the estate. It

could then pass free of debt to her heir. A serious disadvantage, however, was that the annual outlays to be met were virtually doubled. Archibald Burns Macdonald had to pay £285 interest on his loan of £5,700, which was broadly equivalent to the amount of interest previously paid on the debt of about £5,000 or so on the estate. However, he had also to pay a premium of £161 10*s.* 0*d.* on the life assurance policy, and £120 to be held to meet any penalty payments. These payments came to a grand total of £566 10*s.* 0*d.* each year. By 1863 the rental of the estate of Glencoe after payment of public burdens was £595 5*s.* 5*d.*[62] Bad debts, empty property and the costs of factoring and maintenance would all have to be met out of that sum. It is most unlikely that the yield from the estate would meet Archibald's commitments.

It is not clear why the Burns Macdonalds waited until 1863 to make these arrangements. There are two possible reasons. The first is that it was about this time that Archibald Burns Macdonald obtained secure employment as Distributor of Stamps and Collector of Taxes in Perth. The second is that they may have felt inhibited while Sir Duncan Cameron was alive. He died in January 1863.[63]

It is difficult to judge whether or not it was prudent of the Burns Macdonalds to accept the additional commitments involved in these arrangements. Relatively little is known of any other assets and income they may have had, but Archibald Burns snr. must have made a substantial contribution to their finances at some stage. When he died in January 1887 leaving an estate valued at £22,144 13*s.* 2*d.*, he left £50 to his son Archibald for mournings and declared that

> on payment thereof he shall have no further claim upon my estate, having already received from me much more than he could look for and claim from my estate, and in consideration of which he has granted me a discharge in my favour of all claims on my estate.[64]

In 1863 the intention of the Burns Macdonalds may well have been to ensure that in due course the estate of Glencoe would be freed from debt and retained in the family. Perhaps they could have succeeded, albeit with difficulty, if they had been able to let the 1863 arrangement stand without resorting to further borrowing. But that was not to be, as will be explained shortly.

When Ellen took possession of the estate in 1863 she was able to make provision for her husband, which she had undertaken to do in their marriage contract of 1849. So on 3 May 1864 she made a deed of provision and disposition granting her husband a restrictable annuity of £400 during his lifetime if he should survive her, to be paid from the lands and estate of Glencoe.[65] The restriction derived from legislation which permitted the owner of an entailed estate to make provision for a spouse but limited the

annuity to half the rental. The deed was eventually entered in the Register of Sasines in February 1877, just before the estate was disentailed, so that the annuity was secured as an obligation to be met by the next heir when he succeeded. This could have diverted a substantial part of the future rent of Glencoe from Ellen's successor to her husband who was to outlive her by thirty years, but other events were to intervene.

The Burns Macdonalds in Perth

During the 1860s the Burns Macdonalds set up home in Bridgend, by then a fashionable suburb of Perth lying on the east of the Tay, in the parish of Kinnoull. Initially they lived in Mansfield Place. However, in 1864 Archibald Burns senior took possession from Lt. Col. David Scott Dodgson of the southern part of land in Bridgend known as Comely Bank, with the dwelling house and buildings thereon.[66] The house was then called Earnoch Bank. It is believed that it had been built shortly after 1823.[67] Later it was known simply as Earnoch. Leslie's Directory for Perthshire shows that by 1866 Archibald Burns Macdonald was the occupant of the house, and on 4 November 1867 he took possession of the land and house from his father.[68] A few days later, on 8 November 1867, he borrowed £1,000 on the security of the land and house from William Ross of Main Street, Bridgend, Perth.[69] This loan remained in existence until 1884.

Further Loans on the Security of Glencoe

In 1874 the Burns Macdonalds went to the Life Association of Scotland for a further loan of £1,000. This was said to have been required 'to repay the expense of rebuilding the mansion house of the said lands and estate of Glencoe . . . also furnishing the said mansion house'.[70] [71] The arrangement was similar to that of 1863. Archibald Burns Macdonald received the loan at 5 per cent and took out a further life assurance policy on his wife for £1,100 at a premium of £43 14*s.* 6*d.* The policy was assigned to the Life Association, and Ellen agreed that the estate of Glencoe should also be assigned as security, as limited by the entail. The additional sum that the Life Association required to be paid to cover any penalties was £160. The gross rental of Glencoe had risen to £750 11*s.* 6*d.*[72] However, the total annual payments to the Life Association had also increased and clearly could not be sustained indefinitely.

New legislation offered the prospect of disentailing the estate and getting rid of the constraints that Ewen Macdonald had placed upon his daughter and her successors. Ellen's eldest son and heir, Archibald Maxwell Burns Macdonald, who attained the age of twenty-one in January 1876, gave his consent.[73] However, before disentailment took place Ellen and her husband took steps to ensure that certain family provisions were firmly secured on

the estate. The deed of 3 May 1864 which provided Archibald Burns Macdonald with an annuity of up to £400 was entered in the Register of Sasines on 3 May 1877. Another bond making provision for the children of the family other than the heir was drawn up on 30 April 1877 and also entered on 3 May.[74] The text of this bond explained that a petition for disentailment had been lodged on 1 December 1876, and it was necessary that the provisions of this bond should be 'previously secured over the lands'. The family, which was then complete, consisted of Ellen, who had married Andrew Hunter Ballingal WS of Perth on 6 September 1876, Archibald Maxwell (the heir who was not to benefit from this deed), Ewen, Duncan, Stuart, Caroline and Allan. The sum of £3,000 was to be provided for the children other than the heir, and this was to be paid one year after Ellen's death, with a penalty and interest at 5 per cent if not paid. Trustees, headed by David Lumsden of Fincastle in Perthshire, were appointed to administer this money. There was a statutory provision restricting the sum paid if it was found that it would exceed three years' free rental at the time of payment.

Disentailment and Further Loans

The instrument of disentail was recorded on 16 May 1877.[75] Ellen was then free to deal with the estate as her own unrestricted property, and in particular to allow the value of the estate and not just the rent to be used as security for loans. The loan arrangement with the Life Association of Scotland was renegotiated. The sum of £5,700 had been borrowed in 1863, and £1,000 in 1874. In 1877 a further £1,300 was borrowed, making a total of £8,000 at 5 per cent. The borrowing was undertaken as usual by Archibald, but Ellen was now able to offer the estate with right of sale as security for all of these loans. The Life Association's interest was to take precedence over the provision of Archibald's annuity. With this security, the Life Association was able to relinquish the assignment of the two life policies, and control over these was returned to Archibald Burns Macdonald.[76] He was no longer required to make annual payments to cover any penalties he might have incurred. The total annual outlay was now reduced to the interest charge of £400 per annum on the loan of £8,000. For once, this was well below the rental of the estate, but the certainty under the previous arrangements that the capital debt would eventually be extinguished on Ellen's death had been lost. It becomes difficult to see how the debt could be discharged without exercising the right which Ellen now had to sell some or all of the estate.

In 1884 more money was borrowed, but the source was changed from the Life Association of Scotland to the trustees of the late James MacCulloch of Trees near Barrhead in Renfrewshire. The sum borrowed was £11,000 at 5 per cent. The MacCulloch trustees were given power of sale

over the estate of Glencoe, but the bond for £3,000 for the Burns Macdonald children was excepted, and their entitlement remained intact.[77] The loan of £8,000 from the Life Association of Scotland was repaid and a discharge recorded.[78] The net effect was that an additional £3,000 had been borrowed, the debt was now £11,000, and the annual interest was increased to £550.

No reason is mentioned for this additional borrowing, but a couple of months later Archibald Burns Macdonald repaid to William Ross in Perth the £1,000 which he had borrowed on the security of his house and land there in 1867.[79] In practical terms he had switched the security for £1,000 from his house and land in Perth to his wife's estate of Glencoe.

The Trust Disposition and Assignation of 1885

In June 1885 Ellen Burns Macdonald and her husband Archibald signed a trust disposition and assignation which gives insight into some of the problems facing the family. Although it was not open to Ellen to entail the estate and restrict her successors as her father had done, she was able nevertheless to impose severe constraints on those who had to administer her affairs after her death. Her primary objective seems to have been to ensure that the estate did not fall into the hands of her eldest son and heir, Archibald Maxwell Burns Macdonald, but she also had in mind a desire to maintain at least a small part of the estate in the hands of the family. This latter wish was to be defeated within a few years of her death by the power of sale held by the MacCulloch trustees.

The first provision of the trust disposition of 1885 was that her husband was to relinquish any claim on the estate of Glencoe. He gave up his right to the annuity of £400 and to any moveable property at Glencoe which might have been his.

The next provision concerned their eldest son. Archibald Burns Macdonald undertook that while his wife and their eldest son both survived he would

> pay to or for behoof of the said Archibald Maxwell Burns Macdonald . . . such sums as he, Archibald Burns Macdonald shall . . . deem to be advisable and necessary for the aliment and support of the said Archibald Maxwell Burns Macdonald and towards getting him employment including any necessary outfit it being at the same time binding on the said Archibald Maxwell Burns Macdonald always to use his best endeavours to earn his livelihood for himself.

Ellen appointed her son-in-law Andrew Hunter Ballingal WS of Perth as trustee to carry out the complicated provisions that she laid down for the administration of her estate of Glencoe after her death. He was to hold it in trust, maximise the revenue and pay all necessary expenses, including

its upkeep. Having done all that, he was 'to pay the free income . . . along with the free income from any investments which might be made to Archibald Maxwell Burns Macdonald until his death and in the event of his leaving no issue or his issue dying without any of them becoming entitled to the said lands and estate in absolute property then to Ewen Burns Macdonald', and so on through the remaining sons and the two daughters. The effect of these elaborate provisions was to ensure that possession of the estate would miss a generation and be inherited, if at all, by one of Ellen's grandchildren.

The trustee was authorised to sell any part of the lands or use them as security for borrowing money, except the mansion house with the fishings and other amenities. The member of the family entitled to receive the free income was also entitled to occupy the mansion house, providing that arrangements were made to reimburse the trustee income lost by not being able to let it out on the market.[80]

Glencoe after the Death of Ellen Burns Macdonald

Ellen Caroline Macpherson Burns Macdonald died at Perth on 3 March 1887. The *Oban Times* of 12 March said that she and her husband had never been able to take up their abode at Glencoe, but

> her Highland connections enabled her to keep in touch with her ancestral home . . . She continued withal to make herself an abiding influence in the Glen . . . apart from her private benefactions she busied herself with the erection of the memorial which now marks the scene of the massacre.

She was buried on Eilean Munde. Her sons Duncan and Allan were present at her funeral.

The terms of the bond of provision of 1877 for the children other than the heir required that payment should be made a year after Ellen's death.[81] The two survivors of the trustees appointed in 1877 (David Lumsden of Fincastle had died) took action in the Court of Session against Andrew Ballingal as trustee, appointed under the disposition of 1885, to have it found and declared that the sum due under the bond of 1877 was now payable. The court pronounced to this effect on 5 June 1888. The amount due was limited by statute to three years' free rent of the estate, and this came to £2,023 16*s.* 9*d.*

The capital sum of £2,023 16*s.* 9*d.* had to be borrowed. The lender was Miss Anne Inglis Dove, sole trustee of the late David Dove, Solicitor in the Supreme Court. The trustees under the bond of 1877 assigned to her the security which they had over the estate of Glencoe. The trustee of the Glencoe estate had to pay yet more interest. Another £100 or so was added to the £550 due to the MacCulloch trustees. This must have brought the

total uncomfortably close to the rental of the estate and may have been the last straw that was to break the camel's back.

The details of the assignation to Miss Dove reveal that the Burns Macdonald's elder daughter Ellen was in Perth with her husband, and Ewen, the second son, was at Adavale in Queensland, Australia. Duncan was a captain in the Royal Artillery, Stuart was at St. Louis, USA, Caroline was in Perth and Allan was in Glasgow. Both of the witnesses to Stuart's signature were officials of the St. Louis and San Francisco Railroad Company.[82]

Shortly after his mother's death the eldest son, Archibald Maxwell Burns Macdonald, came home from Ceylon. At the time of the 1891 census he was living with his elder sister and her husband, Andrew Ballingal, at 4 Mansfield Place in Bridgend, Perth. This was only a short distance from Earnoch where his father was living with his second wife. His younger sister, Caroline, was also in the Ballingal household. It was reported that Archibald Maxwell Macdonald had 'returned in feeble health' and never recovered fully. He died on 9 June 1894 at Taifletts, adjacent to Mansfield Place, which by then had become the home of his brother-in-law and elder sister. Like his mother, he was buried on Eilean Munde.[83]

The notice of Archibald's death appeared in the *Oban Times* of 16 June 1894. The same issue carried the advertisement for the sale of the estate of Glencoe 'under the Power of Sale granted in Bond for £11,000'. In other words, the sale was instigated by the MacCulloch trustees. The upset price was £18,000, later reduced to £13,330. The rental, after deduction of various public burdens, was £720. The estate extended to about 6,300 acres.[84]

The best offer was £15,900 made on behalf of the Honourable Sir Donald Alexander Smith, KCMG, of Montreal in Quebec, and this was accepted. Two rights were reserved to the Macdonald family of Glencoe: the use of the burial island, and access to the site of the memorial to the victims of the Massacre that Ellen Burns Macdonald had erected. Entry was to be given at Martinmas 1894, but some dispute about what was included in the sale delayed final settlement into 1895.[85]

The purchase price of £15,900 was more than enough to enable the MacCulloch trustees to recover their loan of £11,000 and repay to Miss Dove the £2,023 16*s.* 9*d.* that she had advanced to fund the settlement with the family in 1888.[86] It may be surmised that a sum of almost £2,900 less expenses would be left over for the Glencoe trustee, Andrew Hunter Ballingal. This seems to be confirmed by the fact that in 1898 he and two legal colleagues whom he had assumed as fellow trustees were able to lend £2,480 to Charles Lyon Cameron of Williamston in the County of Perth.[87] The land of Glencoe may have been lost, but a small fund remained for the modest benefit of the family in accordance with the complicated provisions of Ellen Burns Macdonald's settlement of 1885.

Archibald Burns Macdonald's Later Life

One of the effects of that 1885 settlement had been to distance Archibald Burns Macdonald from the affairs of Glencoe, and after the death of his wife Ellen all transactions were carried out by his son-in-law, Andrew Hunter Ballingal, as trustee.[88]

On 10 July 1888, the year following Ellen's death, Archibald Burns Macdonald married Jane Hope Walker, the daughter of William James Walker of Shandon in Dunbartonshire. She was the same age as Ellen, the elder daughter of the Burns Macdonalds, and was presumably the individual of that name who had signed the register as one of the witnesses at Ellen's marriage to Andrew Hunter Ballingal in 1876.

Archibald Burns Macdonald continued to live at Earnoch in Perth, was widowed again in 1912, and died there on 30 September 1917 at the age of eighty-eight. His personal estate was valued at £12,519 8*s.* 11*d.* At the time of his death he was in receipt of a retired allowance from the Inland Revenue, and he was still able to draw some income from the trustees of the marriage contract set up when he married Jane Hope Walker in 1888.[89] The obituary in the *Perthshire Courier* did not mention his connection with Glencoe, and the *Oban Times* carried only a formal intimation of his death.[90] [91] He was buried in Greyfriars Cemetery in Perth, and the words 'late of Glencoe' appear in very small letters below his name on the gravestone.

Conclusion

Archibald Maxwell Macdonald is remembered in a memorial plaque placed by his sister Ellen on the rear wall of St. Mary's Church in Glencoe. This reads: 'He became Laird of Glencoe upon the death of his mother who had been owner for 47 years. He was the last laird of his race . . . [the estate] passed out of the family on his death.' That delicate wording suggests that it was his death which had somehow caused the estate to pass out of the family, rather than the debts that they had placed upon it. To be fair to the Burns Macdonald family, they began their tenure substantially handicapped by the debt still outstanding from the ill-judged activities of Alexander Macdonald in the first heyday of Highland sheep-farming. The loss of the estate might be regarded as the final posthumous episode in his sheep-farming career, but that was not the whole story. In the last dozen years or so of Ellen's life, the Burns Macdonald family had used the estate as security for yet more debt and left no margin to meet the contingency that arose after Ellen's death when the younger children required payment of the sums due to them by the deed of 1877 in their favour.

Appendix A

Financial Transactions

Payments between individuals were often made in cash, but many transactions depended upon bills of exchange, often called simply bills. They served much the same purpose as cheques today, but they did not depend upon a bank account and differed in other important respects.[1]

A bill of exchange was a written order addressed by one person to another, requiring the latter to pay a sum of money either to the party giving the order, or to a third party, at a given place, and usually within a given time. The person requiring the money to be paid was called the drawer of the bill. The person ordered to pay was called the drawee before he subscribed, and the acceptor afterwards. By subscribing or accepting he acknowledged his commitment to pay. Thus far, the procedure is almost the reverse of that for cheques, in that the person to whom payment is due prepares the bill and the person who has to pay then subscribes to it.

The drawer of a bill, who held a piece of paper entitling him to payment of a specified sum, could indorse that bill to another person. That person was then called the indorsee, and the acceptor must then make payment to him instead of to the drawer when it became due. An endorsee could indorse the bill to yet another person, who in turn became the one entitled to receive the payment, and so on. Haldane has described how 'Many of these bills remained for long unpaid before they could be cashed, and meanwhile passed from hand to hand, fulfilling to some extent the function of the banknotes of later years.'[2] Among the droving fraternity, a bill could be indorsed many times. One of the complications of this system was that if the bill was not paid when it was eventually presented for payment, the drawer and all the others who had indorsed it were potentially liable to the unfortunate individual who was the final holder of the bill. Therefore, action in the courts to recover payment could be against a very long list of individuals. If a bill was not paid when due it was incumbent upon the holder to 'protest' it to preserve his recourse upon the drawer and the indorsers. This was a procedure undertaken by a notary and recorded in a prescribed way. Therefore, a record of bills being protested against an individual usually suggests financial difficulties.

Bills could be discounted by banks and often were. For most of the period of this study, that involved taking or sending them to Inverness.

The Bank of Scotland established an office there in 1775 and appointed as their agent a local merchant, Baillie Shaw. He conducted the business of the bank from the house which had belonged to Provost John Hossack at the time of the 1745 rising.[3] The British Linen Company, which developed into banking, established branches by 1785, and appointed William Inglis as their agent in Inverness.[4] He was also a merchant, a baillie, and for a time provost of the burgh.[5] The names of these two agents occur quite frequently in references to financial transactions. 'Baillie Shaw, Banker' or 'Baillie Inglis, Banker' are often mentioned rather than the names of the banks themselves.

The continuity of banking was dependent upon maintaining agents, and there were sometimes gaps between appointments. Thomas Gillespie described such a situation in a letter of 23 February 1801 to Alexander Macdonell, writer in Inverness.

> I mentioned in my last that I would have occasion for a little money will you please say if the Bank of Scotland is likely to open at Inverness and under whose management. I have lately been informed that Campbell and Thomson & Co. Bankers, Stirling, are to give up business, with whom I have had a credit for a number of years if this proves to be the case I would like to open a credit at Inverness in case the business gets into the hands of a man of liberality and honesty. I had an offer from the Agent of the Leith Bank at Callander but think that Inverness might suit me rather better if I can get security that will please. It must be a great loss to this County and Ross-shire the death of that truly worthy man Baillie Shaw.[6]

William Inglis also died in 1801,[7] so the two banks in Inverness lost their agents at the same time. The agent of the Leith Bank at Callander was of course Alexander MacDonald of Dalilea. He obtained business in the Highlands for that bank, no doubt helped by his own connections. In 1808 the Leith Bank did have an agent, Hugh Cobban, in Inverness, but it is not known when he was appointed.[8]

The next agent of the Bank of Scotland in Inverness was John Fraser, who was involved in various businesses which included tanning, shoe-making, brewing and wool manufacturing in Inverness. He was also a partner in the Caledonian Coach Company and took to distilling at Ferintosh. In 1815 the bank petitioned for the sequestration of his assets in an attempt to recover over £11,000 that he owed to them.[9] Gillespie was probably being prudent when he wanted to know who would get the agency before he placed his business. However, Fraser left one solid achievement. In 1804 he acquired Baillie Shaw's property and replaced the old house with the building that served the Bank of Scotland for a time and then became the offices of *The Inverness Courier*.[10]

When the time came for the banks to open branches in Fort William, it had become fairly common for lawyers to be appointed as agents. Thomas MacDonald was agent for the National Bank there, James Macgregor for the British Linen Company, and Robert Flyter for the Bank of Scotland.[11] [12] [13]

In the late eighteenth century, the banks lent money to some of their sheep-farming customers. The usual mechanism was by means of a bond of credit. This allowed the customer to borrow up to a specified maximum at the current rate of interest. Angus Macdonald of Achtriachtan had one for £300 from the short-lived Ayr Bank (Douglas, Heron & Company), operated from the bank's Edinburgh branch (see Chapter 1). Donald McDonald at Tulloch had one from the Bank of Scotland for £300 (see Chapter 4). Captain John MacPherson of Ballachroan and Lieutenant Evan Macpherson at Cullachy each had one for £500 from the British Linen Company (see Chapters 2 and 5). These three bonds were operated from the Inverness offices of the banks. Merchants may have been another source of credit. The firm of McDonald and Elder, merchants in Sleat, seem to have given Archibald Dhu MacDonell in Knoydart access to funds and credit when he needed it (see Chapter 10).

Much larger sums of money were, however, advanced by private arrangements between the lenders and the borrowers. People like Alexander Macdonald of Glencoe and Angus Macdonald of Achtriachtan were able to borrow large sums from relatives and individuals in the Highlands who may have been acquaintances. Perhaps it was relatively easy for them to borrow because of their social standing and because they each had some land that could be used as security. However, it was shown in Chapter 5 that it was also possible for someone like Lieutenant Evan Macpherson at Cullachy, who was not a landowner but belonged to a well-known family, to borrow on a lesser but nevertheless imprudent scale.

Appendix B

Names of People and Places

The archival material used was found to contain a variety of anglicised versions of names of people and places, and there was a lack of consistency in their use. Much of this variety and inconsistency has been retained in the text, but in some instances it seemed necessary to impose a degree of standardisation. The following explanations may be helpful.

Names of People

Macdonald and Macdonell are anglicised versions of the same name and were regarded as interchangeable. With both names, Macd—, MacD—, McD—, M'D—, or even Mcd— may be found.

Many of the people concerned were unable to write. It may be assumed that the spelling of their names was chosen by whoever wrote the documents in which they appear. There was, however, a lack of consistency. For example, Alexander Dhu McDonald, who could not write, is mentioned thirty-seven times in the course of a precognition in 1783. On nineteen occasions his name is given as McDonald, on seventeen as Macdonald and on one as MacDonald. His father, Archibald, is mentioned eleven times, seven as McDonald, three as Macdonald and once as MacDonald,[1] but in a tack of 1786 he is both McDonald and MacDonell.[2]

Among those who could write, inconsistency was found between different members of the same family, and occasionally in the same individual over time. The Glencoe and Achtriachtan families both demonstrate this.

John of Glencoe who died in 1787 used McDonald.[3] [4] In 1776 and 1777, while he was still alive, his son Alexander also used McDonald.[5] [6] However, in all correspondence after the death of his father, Alexander invariably used Macdonald, as did his sons Ewen and Colin.

In the Achtriachtan family, Angus who died in 1798 always signed himself McDonell.[7] [8] [9] At a roup at Kinlochmoidart in May 1782, he and Alexander Macdonald, then Younger of Glencoe, bought two lots of cattle jointly. Achtriachtan is listed as McDonell and young Glencoe as McDonald.[10] Angus's second son Æneas used McDonald.[11] [12] His other sons also appear to have used Macdonald rather than Macdonell, but

their preferred spellings have not been established clearly. In his paper of 1899 on the Achtriachtan family, Fraser-Mackintosh adopted the expedient of standardising all their names to Macdonald.[13]

Other examples could be cited. The old tacksman family of Tulloch is usually known as MacDonell, yet Angus of that family signed himself MacDonald in a tack of 1786.[14] Alexander Macdonell, writer in Inverness, always signed his name in that way, but was more often than not addressed in a variety of spellings of both Macdonell and Macdonald.

It would be courteous to have regard to the way in which individuals chose to spell their names, when that could be ascertained, but strict adherence to individual preferences would be likely to cause confusion. For example, it has become common practice to call the Achtriachtan family Macdonald rather than Macdonell, and the Keppoch family and their cadets (including the Tulloch family) are customarily known as MacDonell, although most of them seem to have used Macdonald from time to time. The practice adopted has been to use the forms preferred by the individuals when these are known, unless that would cause confusion. Hence the Achtriachtan family is known as Macdonald and the Keppoch people and their cadets are MacDonell.

Christian names present fewer difficulties, but Alexander sometimes appears as Alastair, or the familiar Sandy, or the less familiar Sanders.

Names of Places

Place names are usually anglicised forms of names derived from Gaelic, and to a lesser extent from other languages. The spelling of those place names still in current use is now well standardised. However, in the period of this study, and for some time thereafter, standardisation was far short of its current level. In addition, many names fell into disuse as small holdings of land were absorbed into bigger neighbouring ones, and they were not affected by later standardisation.

Commonly, one syllable may have a different vowel or combination of vowels. Glencoe frequently appears as Glenco. Tulloch is sometimes Tullich. The latter seems to have been the preferred version used by Campbell Mackintosh, writer in Inverness. Carnoch is sometimes Carnach. Names beginning with Kin or Kinloch often begin Ken or Kenloch

More than one syllable and some consonants may be affected, as in Balachulish, Balliehuish or Ballacholish for Ballachulish.

One name of unusual difficulty is Innerigan, in Glencoe. The Gaelic spelling is Inbhirfiodhan. Of the early map makers, in the late sixteenth century Pont rendered this as Innerigan.[15] In the mid-eighteenth century General Roy used Inverigan, but he used Innerelan for Inbhir-fhaolain in Glen Etive. The choice between 'Inver' and 'Inner' as a rendering of 'Inbhir' seems to have been an arbitrary one.

Of nineteenth-century map makers, George Langland's map of 1801 and John Thomson's of 1832 both used Inverigan. An Ordnance Survey map published in 1897 uses the Gaelic Inbhirfiodhan. Later maps tend not to show Innerigan at all.

An estate plan, undated but apparently from the early years of the nineteenth century, also uses Inverigan. The same plan also gives Auchtreearton for Achtriachtan and a few other spellings that seem strange to the modern eye.[16]

In the archives studied, Innerigan appears on eight occasions, and Inneregan, Inneriggan, Inverigan, Inverrigan and Invereegan each appear once. There is yet another spelling, Inverighan, on two gravestones on Eilean Munde.[17]

Inverrigan has been used in three well-known modern accounts of the Massacre of Glencoe.[18] Its frequent appearance in print may give an impression of authenticity. However, Donald J. Macdonald used Innerigan in the index, Inverigan in the text and Inverrigan on the map of Glencoe.[19] In this book Innerigan has been used throughout, mainly because of its relatively frequent use in the papers of the Glencoe trustees. It is also pleasing to follow the precedent set by Pont.

Kilmonivaig presents a simpler problem. It usually appears as Kilmanivaig in the archives studied. The modern spelling, Kilmonivaig, came into use later. Rev. Thomas Ross used Kilmanivaig in the account of the parish that he wrote in 1795 for the *Old Statistical Account*.[20] Rev. John M'Intyre used Kilmonivaig in his account written in 1842 for the *New Statistical Account*, but Kilmanivaig appears on the map of Inverness-shire published with it.[21] A Philip's atlas published in 1858 still used Kilmanivaig.[22] It seems reasonable to follow Rev. Thomas Ross and use Kilmanivaig throughout.

Other place names in the text may be accompanied by a more modern version in brackets or in adjacent text, if that seems necessary.

Appendix C

Principal Characters

Anderson, Rev. John. 1759–1839. Ordained to Kingussie 1782. Appointed commissioner (factor) to the Duke of Gordon 1806. Translated to Bellie (Fochabers) 1809.

Bean, George. Writer in Inverness. In practice from late 1770s. His daughter Ann was the wife of Alexander Macdonell, writer in Inverness.

Burns, Archibald. See Macdonald, Archibald Burns.

Cameron, Alexander. Tacksman of Inverguseran in Knoydart from *c.*1794. Took tack of Glendessary after the death of Alexander Macdonald of Glencoe in December 1814. Died *c.*1823.

Cameron, Allan. A MacMartin-Cameron from Glengloy who became tacksman of Meoble on Lovat's land in South Morar, probably in the early 1780s. A thrusting and ambitious sheep-farmer. Obtained a share in Kinlochmorar from about 1805.

Cameron, Major Angus of Kinlochleven. Born *c.*1760. A Glencoe trustee. Living in St. Andrews in 1818 and in Isle of Man in 1821. Helen, wife of Angus Rankin, tacksman of Dalness, may have been a sister.

Cameron, Lieutenant Donald, Braeroy. A MacMartin-Cameron, nephew of Donald mor og Cameron, tacksman of Annat in Braeroy, and cousin of Donald McDonald, drover at Tulloch. Commissioned 1778; discharged 1783. Alleged to have led a gang of thieves who stole livestock from the Highlands and sold it in the Lowlands. Fled overseas when a justiciary warrant was issued for his arrest.

Cameron, Sir Duncan WS of Fassifern. 1775–1863. A Glencoe trustee; carried the main burden of managing the estate. Second son of Sir Ewen Cameron of Fassifern and Louisa Campbell, daughter of Duncan Campbell of Barcaldine and Glenure. Elder brother died 1815. Succeeded in 1828. Writer to the Signet in Edinburgh, but much of his business was in the Highlands.

Cameron, Sir Ewen of Fassifern. 1740–1828. A Glencoe trustee, but not active because of his age. Son of John Cameron of Fassifern and Jean, daughter of James Campbell of Achalader. Succeeded his father in 1785. Created baronet in 1817. His sister Catherine was the wife of John McDonald of Glencoe. Daughters married to Alexander Macdonald of Glencoe, Roderick McNeil of Barra and Duncan Macpherson of Cluny. Succeeded by his son Duncan.

Cameron, Mary. Daughter of Ewen Cameron of Fassifern and niece of Alexander Campbell of Barcaldine. Married Alexander Macdonald of Glencoe in October 1786. Died 1804.

Campbell, Alexander of Barcaldine. 1745–1800. Succeeded his father Duncan in 1784. Sister Louisa (or Lucy) was the wife of Ewen Cameron of Fassifern and hence the mother of Mary Cameron who married Alexander Macdonald of Glencoe. Owner of Glenure, Barnamuck, etc. which he leased to Alexander Macdonald of Glencoe from about 1788.

Campbell, Alexander. 1746–1824. Musician and writer. Married Sarah Cargill, widow of Ranald MacDonell of Keppoch who died in 1785. Lived at Clianaig and managed the farms of his stepson Alexander of Keppoch, absent on military service. Ousted after Alexander of Keppoch died in 1808. Withdrew to Edinburgh. He and Sarah Cargill separated by 1810.

Clark, Thomas. *c.*1764–1851. Tenant of Drumnadrochit; became tenant of Auchteraw near Fort Augustus in 1806 when Lovat removed the representatives of the deceased Rev. John Kennedy. Married Ann, daughter of John MacKay, messenger and constable at Fort Augustus.

Downie, Robert of Appin. A former merchant in Calcutta; purchased the Appin estate in 1817. This included superiority over Glencoe. Bought Leacantuim and parts of Achtriachtan from Adam Macdonald of Achtriachtan in 1817 and 1818. Was MP for Stirling Burghs.

Flyter, Robert. 1769–1836. Writer in Fort William from *c.*1799. Sheriff substitute there for over twenty-five years. Appointed baron baillie in Lochaber by the Duke of Gordon in 1806.

Gillespie, John. Brother of Thomas and associated with him in sheep-farming. Joint tacksman with Thomas at Greenfield in early 1790s. At Scothouse in Knoydart from 1795–1804, then went to Kilmuree in Strathaird on the Isle of Skye. His only son Thomas was heir to his brother Thomas.

Gillespie, Thomas. Died 1824. Son of Thomas, a pioneering sheep-farmer on the Annandale estate. Came to Glenquoich on Glengarry's land about 1782 and established a large sheep-farm. Joined by his brother John and later by James Greig. Acquired extensive sheep-farming interests in the Highlands. Acknowledged authority on sheep-farming.

Greig, James. *c.*1789–1837. Sheep-farmer associated with Thomas Gillespie. Possibly related to Gillespie's wife, Christian Greig (nephew?). Best known as tacksman of Tulloch, Brae Lochaber.

Kennedy, Rev. John. Died *c.*1801/02. Graduated King's College, Aberdeen, March 1770. Schoolmaster at Ruthven, parish of Kingussie, 1767–76. Went to East Florida, returned early 1780s. Married (1) Elizabeth, daughter of Evan McPherson, tacksman of Laggan (died in or before

1782); (2) Christina, daughter of Hugh Macpherson of Ovie (died 1786); (3) Jean, daughter of John Macdonald of Glencoe. Became a sheep-farmer from about 1785. Best known as tacksman of Auchteraw near Fort Augustus.

McBarnett, Donald. Son of a tenant in Achneich in Brae Lochaber and a nephew of William Mitchell tenant of Achnadaul. In 1806 took a nineteen-year lease of Glenmarksie in Ross-shire, with Alexander Macdonald of Glencoe as his cautioner. Failed to pay rent. Removed and made bankrupt at end of third year. Alexander Macdonald of Glencoe became responsible for the remaining years of the lease.

MacCallum, Alexander. Died *c.*1825. Overseer at Lubcroy, in Strath Oykell to Sheriff Donald MacLeod of Geanies. In 1801 took a nineteen-year lease of Culligran in Glenstrathfarrar from Hugh Fraser of Struy. Was involved with Alexander Macdonald of Glencoe and the Glencoe trustees in Monar and Glenmarksie. Became tacksman of Bunchrew on the Forbes of Culloden estate *c.*1816. Succeeded there by his son Colin.

McCaul, James. Died at Killin 1825. Sub-tenant (steelbow) of MacDonell of Keppoch on the farm of Loch Treig prior to 1797. Then tenant of Fersit in partnership with Captain Alexander MacDonell of Moy and Alexander McVean. Withdrew from that and settled at Auchmore near Killin.

Macdonald, Adam of Achtriachtan. Succeeded his father *c.*1799; estate encumbered with heavy debts. Trustees appointed on two occasions. Became cautioner in 1803 for tenants of Mackenzie of Fairburn in Strathconon who failed. The last of his property was eventually sold in 1821. He married Helen, eldest daughter of Ewen Cameron of Glennevis.

Macdonald, Alexander (16th) of Glencoe. Mid-1750s–1814. Son of John McDonald of Glencoe and Catherine Cameron, daughter of John Cameron of Fassifern. Married cousin Mary, daughter of Ewen Cameron of Fassifern in 1786. Succeeded his father in 1787. Became tenant of Fersit in 1778 and obtained other leases from 1787 onwards. Expanded his sheep-farming interests greatly between 1802 and 1807 by acquiring long and expensive leases in Inverness-shire and in Ross-shire. Borrowed heavily, and left large debts when he died in December 1814.

McDonald, Alexander Dhu. Died 1822. Son of Archibald McDonell alias McAlister, sub-tenant at Tulloch in Brae Lochaber. Indications of possible connection with Glencoe. Became tenant of part of Achintore in 1796; removed around 1804. Became tacksman of Rifern in south Morar in 1801; removed in 1815. Became joint tenant (with Archibald Dhu MacDonell at Kyles Knoydart) of farms at the head of Lochnevis previously held by Alexander Macdonald of Glencoe. Removed from

there in 1819. Died suddenly in Fort William in June 1822 during the course of prolonged litigation against Glengarry.

MacDonald, Allan (Innerigan). 1771–1830. A son of Angus MacDonald of Innerigan and Mary Rankin. By 1819 he was at Dalbreck in Strathconon in association with his brothers Donald and John. Was removed from there in 1821. Unmarried.

Macdonald, Angus of Achtriachtan. (Always used the spelling McDonell – see Appendix B.) Succeeded to Achtriachtan *c.*1750. Died December 1798. Leased several farms on the Duke of Gordon's land in Lochaber from at least 1769. Faced increasing debts in the 1790s. Married Anne, a daughter of John Campbell of Ballieveolan. Succeeded by his third son, Adam.

MacDonald, Angus (Drimintorran). Born *c.*1799. Son of John MacDonald of the Drimintorran family and Elizabeth, daughter of Angus Macdonald of Achtriachtan. Studied law in Edinburgh – apprenticed to Cameron, Scott and Arnott. Began an action in the Court of Session around 1825 claiming that he had been denied the inheritance to which he was entitled.

McDonald, Angus, at Fersit and Auchteraw. Probably a Glencoe man who was a sub-tenant of Alexander Macdonald of Glencoe at Fersit. Occupied Auchteraw as sub-tenant of Rev. John Kennedy from 1798. Alexander Macdonald of Glencoe was cautioner.

McDonald, Archibald (Innerigan). 1790–1853. Youngest son of Angus MacDonald of Innerigan and Mary Rankin. On the recommendation of Alexander MacDonald of Dalilea, he was recruited by the Earl of Selkirk and went to North America. Became a chief factor of the Hudson's Bay Company.

Macdonald, Archibald Burns. 1829–1917. Eldest son of Archibald Burns, banker. Married Ellen, daughter of Ewen Macdonald of Glencoe in 1849, and assumed surname Macdonald. Distributor of Stamps, etc. in Perth for many years. Married again after death of Ellen in 1887.

Macdonald, Coll WS of Dalness. 1756–1837. Son of James Macdonald, a younger brother of Alexander Macdonald of Dalness. Admitted WS 1788. Succeeded his cousin John who died in 1775, but did not obtain possession of Dalness until 1794. Based in Edinburgh but had extensive practice among Highlanders.

McDonald, Dr Donald of Cranachan. A son of John of Cranachan, perhaps the eldest. Practised medicine in Fort Augustus, but was tenant of various farms in Brae Lochaber, Knoydart and near Fort Augustus.

MacDonald, Donald (senior) of Drimintorran. 1730–1802. Claimed to be of the Glencoe family, possibly an illegitimate ofshoot. May have farmed at Kingairloch before settling at Drimintorran. Married twice and had two families. First wife Flora was related to the Macdonalds of

Glencoe. Second wife, also Flora, was a sister of Duncan Campbell, factor on the Ardgour estate.

MacDonald, Donald (junior) of Drimintorran. 1791–1849. Eldest of the second family of Donald MacDonald snr. Took over Drimintorran in 1809. Appointed factor to the Glencoe trustees in 1817. Became tenant of several other farms and prospered before failing in 1837. Substantial assets were sold and he went to New Zealand. Married twice. First wife died without issue. Second wife was Ann, daughter of Adam Cummings of Pallinsburn, Northumberland.

MacDonald, Donald (Innerigan). 1780–?. A son of Angus MacDonald of Innerigan and Mary Rankin. Became a sub-tenant of both Alexander Macdonald of Glencoe and Adam Macdonald of Achtriachtan on their farms on the land of Mackenzie of Fairburn and Mackenzie of Strathgarve in Ross-shire. Became bankrupt in 1820; removed from Dalbreck in Strathconon 1821.

McDonald, Donald Dhu at Tulloch. *c.*1748–1808. A son of Archibald McDonell alias McAlister, sub-tenant at Tulloch in Brae Lochaber. First noted 1773. Originally a drover. Used by Æneas Mackintosh of Mackintosh in developing and managing his Lochaber estate. In 1795 he became tacksman of Tulloch and was a substantial sheep-farmer there and on the Duke of Gordon's land. Owned the New Inn in Fort William.

Macdonald, Ellen Caroline Macpherson. 1830–1887. Illegitimate daughter of Ewen of Glencoe. Born in India. Inherited Glencoe in 1840, but encumbered by debts and held in trusteeship until 1863. Married Archibald Burns in Ceylon (Sri Lanka) in 1849. Died in Perth 1887.

Macdonald, Ewen of Glencoe. 1788–1840. Second son of Alexander Macdonald of Glencoe and Mary Cameron, daughter of Ewen Cameron of Fassifern. Served as a surgeon in the Bengal Establishment of the East India Company from 1809. Was heir when his father died in December 1814. Accepted the estate of Glencoe in 1828, encumbered by debts of £6,500. Returned to UK in 1836 and died in August 1840. He left his estate to his natural daughter, Ellen Caroline Macpherson Macdonald, who subsequently married Archibald Burns.

McDonald, John at Achachar (Acharr) and Gart. A lieutenant in the 105th Regiment, raised in 1761 and disbanded in 1763. Became a farmer and a dealer in cattle, and was at Achachar by August 1764. Married a daughter of James Small, one of the factors on the annexed estates. Obtained half of the farm of Gart, near Callander, in 1781. Failed in October 1783.

Macdonald, Lieutenant-General Sir John, KCB. *c.*1747–1824. Joined service of the East India Company in 1767. Progressed through the ranks of the Bengal Establishment; became a lieutenant-general in 1813.

Knight Commander of the Order of the Bath, 1815. Granted arms at Edinburgh in 1818. Described as 'the only surviving son of James Macdonald Esquire sometime an Officer of Marines deceased, who according to the constant tradition of the family was descended from the ancient Family of Macdonald of Glencoe'. Died in Calcutta.

MacDonald, John Ban. At one time a tenant in Achtriachtan. Managed the farms in Knoydart and Glendessary leased by Alexander Macdonald of Glencoe. Given the tenancy of Corrychurochan after Glencoe's death. Failed to pay his rent and removed from there in 1820. Had three sons, Angus, Donald and Allan.

MacDonald, John of Borrodale. *c.*1754–1830. Went to Prince Edward Island 1772. Merchant in Quebec by 1780. Returned to Scotland 1785. Developed trade and had shipping links with Liverpool. Factor for Clanranald, but also bought land on his own account. Progressive and widely respected.

Macdonald, John of Dalness. Died 1775. Son of Alexander Macdonald of Dalness. Merchant in Jamaica. Obtained possession of Dalness in 1764. Died without issue. Dalness then passed to his cousin Coll (see above).

MacDonald, John (Drimintorran). Died 1800. Son of Donald MacDonald senior of Drimintorran by his first wife and cousin german of Alexander (16th) of Glencoe. Married Elizabeth, daughter of Angus Macdonald of Achtriachtan. One son, Angus, born *c.*1799.

MacDonald, Thomas (Drimintorran). 1792–1856. Second son in the second family of Donald MacDonald senior of Drimintorran. Became a writer in Fort William and agent there for the National Bank of Scotland. Undertook legal work in and around Lochaber for the Glencoe trustees. Became tenant of Achnadaul, probably *c.*1840. A son, John, was manager of *The Times*.

MacDonell, Colonel Alexander Ranaldson. 1773–1828. Son of Duncan MacDonell of Glengarry and Marjory, daughter of Sir Ludovic Grant of Dalvey. Succeeded his father in 1788. A Glencoe trustee. Extensive lands included Glengarry and Knoydart. Overbearing, quarrelsome and resorted occasionally to violence. Had substantial debts by early nineteenth century, which increased to about £80,000 at his death. Married Rebecca, daughter of Sir William Forbes, banker, Edinburgh. Succeeded by his son Æneas Ranaldson.

MacDonell, Alexander of Keppoch (Young Keppoch). Succeeded, as a minor, his father Ranald 1785. Curators were appointed, but his mother Sarah Cargill and later his stepfather Alexander Campbell intervened in the management of his affairs. The lease of his Mackintosh lands (Keppoch, etc.) ceased in 1795, but he retained his Gordon lands (Clianaig, etc). Commissioned in 1794, and his lands

were managed by his stepfather. Died in 1808. Succeeded by his younger brother Richard.

Macdonell, Alexander of Milnfield, writer in Inverness. In practice for thirty-seven years. Died 1820. Acted professionally for Lovat, Glengarry and other Highland landowners. Married Ann, daughter of George Bean, writer in Inverness, in 1786. Brother Allan was for a time parish priest in Fort William.

MacDonell, Captain Alexander of Moy. A younger son of Ronald of Aberarder. Commissioned in the 82nd Regiment in 1778. In 1786 he became a joint tenant of Dalnaderg, with his brother-in-law Angus MacDonell of Tulloch, but lost that in 1795. Tenant for many years of Moy and Kylross on Gordon land. Also for a time involved in Fersit, and had Kilmanivaig, Brackletter and Highbridge. His military rank was lieutenant.

MacDonell, Angus of Inch. *c.*1724–c1815. Illegitimate son of Alexander of Keppoch who was killed at Culloden. Looked after the family's affairs for a time and became a successful sheep-farmer at Inch.

MacDonell, Captain Angus of Tulloch. 1759–1819. Son of Alexander of Tulloch. Succeeded sometime between 1783 and 1786. Served in 84th Regiment. Married Margaret, daughter of Ronald MacDonell of Aberarder. Did not succeed to lands of Tulloch in 1786 but became a joint tenant of Dalnaderg with his brother-in-law Alexander MacDonell of Moy. Lost that in 1795 and eventually went to Claggan in 1799. His military rank was ensign.

MacDonell, Archibald Dhu in Knoydart. Known to have been at Shenachy in Knoydart in 1792. By 1796 he was at Carnoch at the head of Lochnevis. Succeeded there by Alexander Macdonald of Glencoe in 1802. Went to Inverie. Tried and failed to get a share of Kinlochmorar. Obtained Sandaig, but moved to Kyles Knoydart in 1811. Joined Alexander Dhu McDonald in 1815 in taking over a share of the leases left vacant by the death of Alexander Macdonald of Glencoe. Still alive at Kyles Knoydart in 1824. Believed to have had sons Alexander, Archibald, Duncan, John and Donald, and a daughter Mary who married Roderick Macdonald jnr. at Lee in Knoydart.

MacDonell, Archibald alias McAlister. Born *c.*1723. Probably died early 1790s. First noted in 1773 as a sub-tenant of Alexander MacDonell of Tulloch in Brae Lochaber. Three sons known: Alexander Dhu, John and Donald. In 1783 his wife was Sarah McGlaserich alias Campbell. She was stepmother of Alexander Dhu and John, but may have been mother of Donald. Later affairs of his son Alexander Dhu suggest a possible link with Glencoe.

MacDonell, Coll (Inch). Born 1778. Still alive 1841. A younger son of Angus of Inch. An active and successful sheep-farmer.

MacDonell, John of Aberarder. 1748–1818. Eldest of three sons of Ronald of Aberarder. Noted at Torgulbin in 1782 and later at Killiechonate, with which he is most often associated. Married Catherine, daughter of Alexander MacDonell of Keppoch who died at Culloden.

MacDonell, John (Inch). 1767–1850. A younger son of Angus of Inch. An active and successful sheep-farmer.

MacDonell, Richard of Keppoch. Younger son of Ranald of Keppoch. Succeeded his brother Alexander in 1808. Lieutenant in 92nd Regiment of Foot. Removed his stepfather, Alexander Campbell, from the management of his affairs and put them in the hands of John and Coll MacDonell at Inch.

MacDonell, Ronald of Aberarder. Died *c.*1785. Was in the 1745 rising and escaped from Culloden. Removed from Aberarder 1770. Became tenant of Torgulbin and Moy. Had three sons: John, Alexander and Archibald. Along with John and sometimes Archibald, Ronald was the subject of several criminal charges in the early 1780s. Found guilty of deforcement of a messenger in 1783. Ronald's younger brother Alexander was tenant of Garvamore in upper Strathspey and kept the inn there.

MacDougall, Coll. Captain, 42nd Regiment. Married Jane, daughter of Alexander Macdonald of Glencoe, towards the end of 1817. Died 1821.

McGlaserich, Sarah alias Campbell. Born *c.*1723. Wife, probably second, of Archibald MacDonell alias McAlister, sub-tenant at Tulloch.

MacIntyre, Rev. Duncan. 1757–1830. Mother was Margaret, daughter of Angus Macdonald of Achtriachtan. MA King's College, Aberdeen 1779. Briefly schoolmaster in Kilmanivaig. Maryburgh Mission (Fort William) 1784. Minister of Laggan 1816.

MacKay, John, tacksman of Inchnacardoch. Died 1821. Head constable and messenger, County of Inverness, based at Fort Augustus. Held office from early 1780s. Son Robert became a writer in Fort William.

Mackenzie, Brigadier Alexander of Fairburn. Subsequently promoted to general and knighted. A landowner in Ross-shire. Succeeded his grandfather in 1789. Owner of Forest of Monar which was let to several tenants, including Alexander Macdonald of Glencoe in 1803. Also owned Strathconon, where Adam Macdonald of Achtriachtan became cautioner for John and Duncan MacCallum in 1803.

Mackenzie, William of Strathgarve. A landowner in Ross-shire. Let Glenmarksie to Donald McBarnett in 1806, with Alexander Macdonald of Glencoe as cautioner. Served as a surgeon with the 5th British Militia stationed in 1809 at Woodbridge, Suffolk.

Mackintosh, Æneas of Mackintosh. Landowner and chief of Clan Mackintosh. Succeeded his uncle (also Æneas) in 1770. Served in American

War of Independence with 71st (Frazer's Highlanders). Discharged and returned home in 1783. Over the next twenty years he transformed his Lochaber estate into large sheep-farms. Alexander Macdonald of Glencoe was a tenant from 1802 onwards. Knighted in 1812. Died 1820.

Mackintosh, Alexander. Merchant in Maryburgh. Born *c.*1756. Still alive in 1834. Related to Sir James Mackintosh (1765–1832) of the Kyllachy family. Established himself as a merchant in Maryburgh (Fort William). From 1786 until 1807 he was also a tenant of various farms on Æneas Mackintosh's Lochaber estate.

Mackintosh, Campbell. Writer and town clerk, Inverness, Collector of Taxes, Inverness-shire. *c.*1758–1835. In practice for fifty-one years. Acted for many landowners, including Æneas Mackintosh of Mackintosh, Glengarry and Lovat. Purchased the estate of Dalmigavie in Strathdearn in 1819. Son Donald, a writer in Edinburgh.

McLean, Neil. Land surveyor. Acted at times as a factor for Glengarry. Tried unsuccessfully to remove 'Bitag' (Alexander Dhu McDonald) from Sourlies in Knoydart in 1817. Member of original board of the Caledonian Bank, Inverness.

MacPherson, Alexander. Writer in Inverness. Son of Donald MacPherson, baron baillie on Duke of Gordon's Lordship of Lochaber. Qualified as a writer in Edinburgh. Was in practice in Maryburgh (Fort William) *c.*1777 but moved to Inverness *c.*1780. Alleged to have had improper relationship with the Kennedys of Glengarry and with Lieutenant Donald Cameron, Braeroy.

MacPherson, Captain/Major Charles of Gordonhall (Kingussie). Inspector General of Barracks for Scotland. Was a cautioner for Lieutenant Evan Macpherson (Ovie) and a trustee for Rev. John Kennedy, tacksman of Auchteraw.

MacPherson, Donald. Duke of Gordon's baron baillie in Lochaber. Active in the 1770s but came under suspicion of improper relationship with the Kennedys of Glengarry and with the Camerons of Braeroy. Lost his position and was replaced *c.*1780. Father of Alexander MacPherson, writer in Inverness.

Macpherson, Lieutenant Evan. *c.*1760–1823. A younger son of Hugh Macpherson of Ovie. Brother-in-law of John MacPherson of Ballachroan and of Rev. John Kennedy of Auchteraw. Went into partnership with Kennedy in 1787 as joint tenants of Auchteraw, Cullachy and other lands. Partnership ended in 1791 when Macpherson became tenant of Cullachy and other lands. Became bankrupt in 1794 and joined the Gordon Highlanders.

MacPherson, Captain John of Ballachroan. 1724–1799/1800. Second son of Alexander MacPherson of Phoness. His mother was of the Aberarder family in Lochaber. Married Ann, a daughter of Hugh Mac-

pherson of Ovie. Served in the 82nd Regiment and then became tacksman of Ballachroan. Known as the Black Officer. Tenant of Fersit 1773–1778. Brother-in-law of Lieutenant Evan Macpherson, and was one of his cautioners. Died with several companions in a snowstorm at Gaick at the turn of the year 1799/1800.

McVean, Alexander. Married Sarah, daughter of John MacDonell of Aberarder. At Loch Treig in the 1790s. Joint tenant from 1798 of the whole of Fersit with James McCaul and Captain Alexander MacDonell of Moy, brother of John of Aberarder. In 1809, became tenant of Fersitriach in partnership with Donald and John Rankin. Alexander Macdonald of Glencoe was their cautioner. This lease ended in 1814, with all parties in difficulties. Last noted as tacksman of Drumfuir in 1817. Subsequently in Canada.

Mitchell, David. Merchant in Doune. In the 1780s he extended his business into droving. He advanced money to several drovers, including John McDonald at Acharr in Lochaber and Gart near Callander, and Donald McDonald at Tulloch in Brae Lochaber, with whom he had some partnership arrangement.

Mitchell, John. Of the family at Achnadaul. In 1803 he became one of several joint tenants, including Alexander Macdonald of Glencoe, of the Forest of Monar in Ross-shire. He was soon in financial difficulties and returned to Lochaber.

Mitchell, William. Tacksman of Achnadaul. In Achnadaul in the 1790s; succeeded by his son George around 1812–15. Wife was Margaret McBarnett, related to the McBarnetts in Achneich. On good terms with Alexander Macdonald of Glencoe, who was cautioner for his nephew Donald McBarnett as tenant of Glenmarksie from 1806.

Oliver, Robert. Died 1823. A sheep-farmer originally at Langburnshiels south of Hawick. With his brother Thomas and Thomas Stavert, became tenant of Cullachy near Fort Augustus in 1795, after the failure of Lieutenant Evan Macpherson. Stavert died shortly after. In 1809 Thomas Oliver withdrew from Cullachy. Robert became the sole tenant, and also became tenant of Torlundy on the Duke of Gordon's land.

Oliver, Thomas. Brother of Robert, also originally at Langburnshiels. By 1809, when he renounced his interest in Cullachy, he was tenant of Shankend a couple of miles north of Langburnshiels. His main interest was probably in the Borders, while Robert looked after Cullachy.

Rankin, Angus. Tacksman of Dalness. Died *c.*1822. Probably the senior member of the Rankin family in Glencoe and thereabouts. He had numerous involvements and entanglements in sheep-farming, on his own and with others, including other members of the Rankin family. Wife Helen may have been a sister of Major Angus Cameron of Kinlochleven.

Ross, James. Usually described as Cashier to the Duke of Gordon. Functioned as the most senior official in the Duke's estate office at Fochabers. The local factors, such as William Tod, reported to him. Much correpondence to and from him in late 1770s and early 1780s.

Ross, Rev. Thomas. 1747–1822. Born in Ross-shire. Educated Marischal College, Aberdeen. DD, Aberdeen, 1818. Schoolmaster of Nigg. Ordained to parish of Kilmanivaig 1776. Active as a sheep-farmer. Married Lucy, a daughter of John Cameron of Fassifern and a sister of Ewen Cameron, hence related by marriage to Alexander Macdonald of Glencoe.

Scott, Robert. Writer in Fort William and Edinburgh. Factor to the Glencoe trustees from 1815 until he moved to Edinburgh about 1817. He continued to deal with the business of the Glencoe trustees as a partner of Duncan Cameron. Was co-opted as a trustee in 1822. Died a few years later before the business of the trustees was finally completed.

Stavert, Thomas. A sheep-farmer at Coliforthill, two miles south of Hawick. Joined with Robert and Thomas Oliver in taking the lease of Cullachy in 1795, but died shortly after.

Tod, William. Factor on the Duke of Gordon's estate who dealt with Badenoch and Lochaber, reporting to James Ross. Replaced by Rev. John Anderson in 1806.

Bibliography

Adam, F. *The Clans, Septs, and Regiments of the Scottish Highlands*. Eighth Edition. Edinburgh & London, 1970.

Adam, R.J. (Ed.) *Papers on Sutherland Estate Management 1802–1806*, Vols.1 & 2. Edinburgh, 1972.

Armstrong, R. *Powered Ships: The Beginnings.* London & Tonbridge, 1975.

Barron, J. *The Northern Highlands in the Nineteenth Century*, Vols.I, II & III. Inverness, 1903, 1907 & 1913.

Beattie, A.G. & Margaret H. *Pre-1855 Gravestone Inscriptions in Lochaber and Skye*. The Scottish Genealogy Society, 1990.

Beattie, A.G. & Margaret H. *Pre-1855 Gravestone Inscriptions in Inverness District West.* The Scottish Genealogy Society, 1993

Bell, R. & W. *Dictionary of the Law of Scotland.* Third Edition. Edinburgh, 1826.

Brash, J.A. *Scottish Electoral Politics 1832–1854*. Edinburgh, 1974.

Bremner, D. *The Industries of Scotland.* 1869 (Reprinted Newton Abbott, 1969).

Buchan, J. *The Massacre of Glencoe*. London, 1985 (First published 1933).

Bumsted, J.M. *The Peoples' Clearance*. Edinburgh & Winnipeg, 1982.

Campbell, R.H. *Scotland since 1707. The Rise of an Industrial Society.* Second Edition. Edinburgh, 1985.

Checkland, S.G. *Scottish Banking: A History, 1695–1973*. Glasgow & London, 1975.

Clerk, Rev. A. *Notes of Everything: Kilmallie Parish Minister's Diary of c.1864*. Kilmallie Parish Church, 1987.

Cole, Jean M. *Exile in the Wilderness: The Biography of Chief Factor Archibald McDonald, 1790–1853*. Toronto, 1979.

Cole, Jean M. (Ed.) *This Blessed Wilderness: Archibald McDonald's Letters from the Columbia, 1822–44*. Vancouver & Toronto, 2001.

Comrie, J.D. *History of Scottish Medicine*. London, 1932.

Craven, The Ven. J.B. (Ed.) *The Journals of Bishop Forbes.* Second Edition. London, 1923.

Crawford, D.G. *Roll of the Indian Medical Service 1650–1830*. London, 1930.

Dobson, D. *The Original Scots Colonists of North America*. Baltimore, 1989.

Dobson, D. *Scottish American Court Records 1773–1783*. Baltimore, 1991.

Dodgshon, R.A. 'The Economics of Sheep Farming in the Southern Uplands during the Age of Improvement', in *Economic History Review*, Vol.29, 1976, 552–54.

Dodwell, E. & Miles, J.S. *Alphabetical List of Officers of the Indian Army 1760–1834*. London, 1938.

Dodwell, H.H. (Ed.) *The Cambridge History of the British Empire. Vol.4: British India 1497–1858*. Cambridge, 1929.

Douglas, H. *Flora Macdonald, The Most Loyal Rebel*. London, 1994.

Duckworth, C.L.D. & Langmuir, G.E. *West Highland Steamers*. London, 1950.

East India Register and Directory.

Fairweather, Barbara. *A Short History of Ballachulish Slate Quarry*. North Lorn and Glencoe Folk Museum, undated.

Fairweather, Barbara & Cargill, D.C. *Pre-1855 Tombstone Inscriptions on Isla Munda, The Burial Island in Loch Leven*. The Scottish Genealogical Society, undated.

Fasti Ecclesiae Scoticanae, Vols.4, 6 & 7. Edinburgh, 1923, 1926 & 1928.

Fergusson, J. *Argyll and the Forty-Five*. London, 1951.

Forbes, F. & Anderson, W.J. 'Clergy Lists of Highland District. 1732–1828', in *The Innes Review*, Vol.XVII, 1966, 129.

Fraser, A.C. *The Book of Barcaldine*. London, 1936.

Fraser-Mackintosh, C. 'The MacDonells of Scotus', in *Transactions of the Gaelic Society of Inverness*, Vol.XVI, 1889–90, 79–97.

Fraser-Mackintosh, C. *Letters of Two Centuries*. Inverness, 1890.

Fraser-Mackintosh, C. 'The Camerons of Letterfinlay', in *Transactions of the Gaelic Society of Inverness*, Vol.XVII, 1890–91, 31–45.

Fraser-Mackintosh, C. *Antiquarian Notes*. Second Series. Inverness, 1897.

Fraser-Mackintosh, C. 'The Macdonalds of Achtriachtan', in *Transactions of the Gaelic Society of Inverness*, Vol.XXIII, 1898–99, 136–45.

Garnett, T. *Observations on a Tour through the Highlands and part of the Western Isles of Scotland*, Vol.1. London, 1811.

Gaskell, P. *Morvern Transformed*. Cambridge, 1968.

Gifford, J. *Highlands and Islands: The Buildings of Scotland.* London, 1992.

Grant, Anne. *Letters from the Mountains.* London, 1806. (Also Sixth Edition, edited by J.P. Grant, London, 1845.)

Grant, I.F. *Every-day Life on an Old Highland Farm.* London, 1924 (Revised 1981).

Grant, I.F. *Highland Folk Ways.* London, 1961.

Grant, I.F. *Along a Highland Road.* London, 1980.

Haldane, A.R.B. *The Drove Roads of Scotland.* Edinburgh, 1952.

Haldane, A.R.B. *New Ways Through the Glens.* London, Edinburgh, etc., 1962.

Heron, R. *Observations made in a Journey through the Western Counties of Scotland in the Autumn of MDCCXCII,* Vol.I. Perth, 1793.

A Highland Newspaper: The Inverness Courier 1817 – 1967. Inverness, 1969.

Hodson, V.C.P. *List of Officers in the Bengal Army 1758–1834.* 4 Volumes. London, 1927–47.

Hogg, J. *A Tour of the Highlands in 1803.* Edinburgh, 1888 (Reprint 1986).

Hopkins, P. *Glencoe and the End of the Highland War.* Edinburgh, 1986.

Hunter, J. *A Dance Called America.* Edinburgh, 1994.

Hunter, J. *Glencoe and the Indians.* Edinburgh & London, 1996.

Johnson, Christine. *Developments in the Roman Catholic Church in Scotland 1789–1829.* Edinburgh, 1983,

Larkin, A. (attributed to) *Sketch of a Tour in the Highlands of Scotland . . . September and October 1818.* London, 1819

Leyden, J. *Journal of a Tour in the Highlands and Western Islands of Scotland in 1800.* Edinburgh & London, 1903.

Lindsay, Jean. *The Canals of Scotland.* Newton Abbott, 1968.

Linklater, M. *Massacre: The Story of Glencoe.* London & Glasgow, 1982.

Macdonald, A. & A. *The Clan Donald,* Vol.III. Inverness, 1904.

MacDonald, A.G. 'Some Australian 19th Century Records', in *Clan Donald Magazine,* No.4, Edinburgh, 1968, 46–48.

MacDonald, A.G. 'The Last of the Glencoe Chiefs?', in *Clan Donald Magazine,* No.7, Edinburgh, 1977, 52–57.

MacDonald, D. 'Alexander Macdonald of Glencoe and the Forty-Five', in *Clan Donald Magazine,* No. 10, Edinburgh, 1984, 79–84.

Macdonald, D.J. *Slaughter under Trust.* London, 1965.

Macdonald, D.J. *Clan Donald.* Loanhead, 1978.

McDonald, D.R. 'Glencoe – The Passing of the Chiefs', in *Clan Donald Magazine*, No. 8, Edinburgh, 1979, 55–57.

Macdonald, I.S. 'Alexander Macdonald Esq of Glencoe: Insights into Early Highland Sheep-Farming', in *Review of Scottish Culture*, No. 10, 1996–7, 55–64.

Macdonald, I.S. 'Some Highland Lawyers and Their Clients', in *Review of Scottish Culture*, No. 12, 1999–2000, 85–92

Macdonald, I.S. 'The Dalness Macdonalds', in *Clan Donald Magazine*, No.14, Edinburgh, 2001, 55–59.

MacDonald, N.H. *The Clan Ranald of Knoydart & Glengarry*. Edinburgh, 1995.

Macdonald, S. *Back to Lochaber*. Edinburgh, 1994.

MacDonell, Ann & MacFarlane, R. *Cille Choirill, Brae Lochaber*. Spean Bridge, 1986.

MacDonell, Josephine M. *The MacDonells of Keppoch and Gargavach*. Glasgow, 1931.

MacDougall, Jean. *The Correspondence of Four MacDougall Chiefs 1715–1864*. London, 1984.

MacFarlane, R. 'The MacDonells of Aberarder', in *Clan Donald Magazine*, No. 12, Edinburgh, 1991, 147–57.

Macinnes, A.I., Harper, Marjory-Ann D. & Fryer, Linda G. (Eds.) *Scotland and the Americas, c.1650–c.1939: A Documentary Source Book*. Edinburgh, 2002.

Maciver, E. *Memoirs of a Highland Gentleman*. Edited by Rev. G. Henderson. Edinburgh, 1905.

Mackay, D.N. (Ed.) *Trial of James Stewart. (The Appin Murder)*. Second Edition. Edinburgh & Glasgow, 1931.

MacKay, I. 'Clanranald's Tacksmen', in *Transactions of the Gaelic Society of Inverness*, Vol.XLIV, 1964–66, 61–93.

Mackay, M.S. *Doune. Historical Notes.* Stirling, 1984.

Mackenzie, A. *The History of the Highland Clearances*. Second Edition. Glasgow, 1883. (Reprinted 1946).

Mackenzie, A. *History of the Clan Cameron.* Inverness, 1884.

Mackenzie, A. *History of the Clan Mackenzie.* Inverness, 1894.

Mackenzie, A. *History of the Frasers of Lovat.* Inverness, 1896.

Mackenzie, A. *History of the Macdonalds*. Inverness, 1881.

Mackenzie, Alison. 'The Gaick Memorial', in *Journal of the Clan Chattan Association*, Vol.9, No.6, 1994, 325–6.

Mackintosh, A.M. *The Mackintoshes and Clan Chattan.* Edinburgh, 1903.

Maclean, Loraine. *The Indomitable Colonel.* London, 1986.

Macmillan, S. *Bygone Lochaber.* Glasgow, 1971.

J.R.N. Macphail (Ed.) *Letters written by Mrs Grant of Laggan concerning Highland Affairs and Persons connected with the Stuart Cause in the Eighteenth Century.* Scottish History Society, 1896, Vol.26.

Macpherson, A. *Glimpses of Church and Social Life in the Highlands in Olden Times.* Edinburgh, 1893.

Macpherson, A. *Captain John Macpherson of Ballachroan and the Gaick Catastrophe of the Christmas of 1799 (O.S.) A Counterblast.* Kingussie, 1900.

Mitchell, Alison. *Pre-1855 Gravestone Inscriptions from Speyside.* Revised Edition. Scottish Genealogical Society, 1992.

Mitchell, J. *Reminiscences of my Life in the Highlands*, Vols. 1 & 2. 1884 (Reprint Newton Abbott 1971).

Mitchison, Rosalind. *The Old Poor Law in Scotland: The Experience of Poverty 1574–1845.* Edinburgh, 2000.

Munn, C.W. *The Scottish Provincial Banking Companies.* Edinburgh, 1981.

Murray, Sarah. *A Companion and Useful Guide to the Beauties of Scotland etc.*, Vol.1. London, 1799.

The New Statistical Account of Scotland, Vols.VII & XIV. Edinburgh & London, 1845.

Orde, Anne (Ed.) *Matthew and George Culley. Travel Journals and Letters, 1765–1798.* Oxford, 2002.

Osborne, B.D. *The Last of the Chiefs, Alastair Ranaldson Macdonell of Glengarry 1773–1828.* Glendaruel, 2001.

Pennant, T. *A Tour of Scotland.* Third Edition. Warrington, 1774.

Perth Burgh. *List of Buildings of Architectural or Historical Interest.* In Public Library, Perth.

Prebble, J. *The Highland Clearances.* London, 1963.

Prebble, J. *Glencoe.* London, 1966.

Register of the Society of Writers to Her Majesty's Signet. Edinburgh, 1983.

Richards, E. *The Leviathan of Wealth.* London, 1973.

Richards, E. *A History of the Highland Clearances.* London & Canberra, 1982.

Richards, E. *A History of the Highland Clearances*, Vol. 2. London, etc., 1985.

Richards, E. *The Highland Clearances.* Edinburgh, 2000.

Richardson, I. *Laggan – Past and Present.* Laggan Community Association, 1990.

Robertson J. *Kingussie and the Caman.* Kingussie, 1994.

Robertson, J.I. *The First Highlander, Major-General David Stewart of Garth CB. 1786–1829.* East Linton, 1999.

Rolt, L.T.C. *Thomas Telford.* London, 1958.

Roughead, W. *Glengarry's Way and other Studies.* Edinburgh, 1922.

Scarlett, Meta H. *In the Glens Where I Was Young.* Moy, Inverness-shire, 1988.

Sinclair, Sir John. (Drawn up . . . under the Direction of) *The General Report of the Agricultural State, and Political Circumstances, of Scotland,* Vol II. 1814, Appendix.

Sinclair, Sir John (Ed.) *The Statistical Account of Scotland 1791–1799,* Vols. XII, XVII & XX. Wakefield, 1983.

Sinton, T. *By Loch and River being Memories of Loch-Laggan and Upper Spey.* Inverness, 1910.

Smout, T.C. *A History of the Scottish People 1560–1830.* London, 1969.

Smith, Annette M. *Jacobite Estates of the Forty-Five.* Edinburgh, 1982.

Somers, R. *Letters from the Highlands.* 1848 (Reprint Inverness, 1977).

Stewart, D. *Sketches of the Character, Manners, and Present State of the Highlanders of Scotland.* Edinburgh, 1822. (Reprinted 1977.)

Stewart, J. *The Camerons.* Stirling, 1974.

Taylor, W. *The Military Roads in Scotland.* Revised Edition. Colonsay, 1996.

Telford, T. 'A Survey and Report of the Coasts and Central Highlands of Scotland', in *Reports on Highland Roads and Bridges, 1803.* Parliamentary Papers, 1803–1821, Vol.I, Section IV, 16.

The Pocket Lawyer: A Practical Digest of the Laws of Scotland. By a Member of the Faculty of Advocates. Edinburgh, 1830.

Timperley, Loretta R. *A Directory of Land Ownership in Scotland c.1770.* New Series 5. Scottish Record Society, 1976.

Vining, Elizabeth G. *Flora MacDonald: Her Life in the Highlands and America.* London, 1967.

Watson, J.A.S.'The Rise and Development of the Sheep Industry in the Highlands and North of Scotland', in *Transactions of the Highland and Agricultural Society of Scotland,* Fifth Series, Vol.44, 1932, 1–12.

Watson, W.J. *Place Names of Ross and Cromarty.* Reprint, Evanton, 1996.

Wight, A. *Present State of Husbandry in Scotland,* Vol.2. Edinburgh, 1778.

Wigston, Rev. K. 'Glencoe', in *Clan Donald Magazine*, No. 12, Edinburgh, 1991, 161–65.

Wilson, C.R. (Ed.) *Indian Monumental Inscriptions. Vol.1: Bengal.* Calcutta, 1896.

Wordsworth, Dorothy. *Recollections of a Tour made in Scotland A.D. 1803.* Edited by J.C. Shairp. Edinburgh, 1874.

Youngson, A.J. *Beyond the Highland Line.* London, 1974.

Glossary

Note: The definitions and meanings given here are intended to relate to the period of the book. Additional or alternative meanings that are not relevant to the text have generally been omitted.

Works consulted include the following

J. Barron, *The Northern Highlands in the Nineteenth Century, Vol.1*, 1903

Bell's Law Dictionary, 1826 edition

Chambers's Scots Dictionary, 1911

The Concise Scots Dictionary, 1985

G. Donaldson and R.S. Morpeth, *A Dictionary of Scottish History*, 1977

Christine Johnson, *Developments in the Roman Catholic Church in Scotland 1789–1829*, 1983

A.R.B. Haldane, *The Drove Roads of Scotland*, 1952

The Pocket Lawyer, 1830

The Scots Thesaurus, 1990

The Shorter Oxford Dictionary, 1983Annette M. Smith, *Jacobite Estates of the Forty-Five*, 1982

Ann E. Whetstone, *Scottish County Government in the Eighteenth and Nineteenth Centuries, 1981*

Acts of Sederunt: enactments by the Court of Session, intimating an opinion on matters of law, in the form of declaratory judgements to be pronounced in all similar cases in future.

advocate depute: A member of the Faculty of Advocates appointed by the Lord Advocate to prosecute under his direction.

advocation: a process by which the decision of an inferior court may be brought under the review of the Court of Session.

agent (bank): a manager, usually part time.

agent (law): a writer (solicitor).

aliment: maintenance or support claimed from another.

annexed estates: the estates of Jacobite supporters annexed following the rising of 1745. Initially estates were forfeited, but in 1752 they were annexed 'inalienably' to the Crown. The annexation was rescinded in

1784. (Exceptionally, Lovat got his lands back in 1774.) The annexed estates were administered by the Board of Commissioners and Trustees for the Annexed Estates.

annuity: the grant of an annual sum, chargeable upon the grantor and his heirs.

arrestment order: enables a creditor to attach (seize) the debt due, or moveables belonging to his debtor, in the hands of a third party.

assoilzie: to free a party from the conclusions of an action, or to find an accused person not guilty.

avizandum: to make avizandum is to remove a case from the public court to permit further consideration by the judge.

baron baillie: appointed by a baron to preside over his court and generally administer the barony. With the abolition of heritable jurisdictions, the functions diminished.

bigging: a building.

bill of exchange: a writing ordering a person to pay a sum of money at a given place, and within a given time, either to the person making the order or to a third person. (See Appendix A for further information.)

boll: a dry measure of weight or capacity varying according to commodity or locality. A boll of meal = approximately 63.5 kg.

bond: a written obligation to pay or to perform, as various as the circumstances for which it may be granted.

bond of caution: an undertaking by one party to become surety or cautioner for the obligations of another. This may be for purposes such as payment of rent, good behaviour or answering bail.

bond of credit: an arrangement entered into between banks and individual customers permitting the latter to borrow up to a specified maximum sum at the current rate of interest. Customers required to have cautioners. (See Appendix A for further information.)

caption: *see* letters of caption.

cattle: a collective name for the bovine genus, also formerly used for live animals reared as food, for their milk, skin, wool, etc. Sheep may therefore be included on occasion.

cautioner: one who stands surety for another; *see* bond of caution.

cessio bonorum: a legal process before a court that allowed a debtor to be released from prison if he surrendered all of his means and was innocent of fraud.

Circuit Court of Justiciary: the Justiciary Court had three circuits that were visited twice yearly. These were the South, West and North Circuits. The last-named sat in Perth, Aberdeen and Inverness.

Collector of Taxes: an official appointed by the Commissioners of Supply (predecessors of the County Councils) to collect taxes in the counties. His greatest concern was the land tax.

constable: a officer of the justices of the peace, entrusted with the execution of their warrants, decrees or orders. Has the duty of apprehending offenders against the peace, vagrants, etc., and bringing them before the next justice. Also to suppress riots and apprehend the rioters. Constables may also hold the office of messenger.

compearance: appearance before a court or other authority.

composition: an agreement for the settlement of a dispute, or a sum fixed by mutual agreement paid in settlement of a claim.

comprisement: an appraisal or valuation.

conveyance: a deed by which a right is either created, transferred or discharged.

Court of Session: the Supreme Civil Court of Scotland.

cousin german: first cousin.

curator (to a minor): a person appointed, possibly with others, in the absence of a father, to protect the interests of a minor, i.e. a male child between the ages of fourteen and twenty-one years or a female child between the ages of twelve and twenty-one years. Tutors have a similar role in relation to pupils, i.e. children under these ages. The same individuals were commonly appointed as tutors and curators to maintain continuity.

decreet: a judgement or decree of a court or judge.

decreet of declarator: a form of action by which some right of property or of servitude, or some inferior right or interest, is judicially declared.

decreet of preference: a decision in an action of multiplepoinding which settles the order of preference to be given to the various competing claims; *see also* multiplepoinding.

deed: a formal written instrument executed and authenticated according to certain technical forms, setting forth the terms of an agreement, contract, or obligation. Deeds may be registered in the books of a court; *see* Register of Deeds.

defender: the party against whom the conclusions of a process or action are directed. Latterly used only in civil cases.

deforcement: an act of contempt of the law, consisting of a violent opposition and hindrance to an officer of the law in the execution of his official duty.

depone: to testify, give evidence on oath or declare on oath.

desertion of the diet: to desert the diet in a criminal process is to abandon the proceedings. Desertion simpliciter will put a final stop to all further proceedings. Desertion pro loco et tempore does not bar the prosecutor from raising a new libel.

diligence: a warrant issued by a court for enforcing the attendance of witnesses or the production of documents. Also the process of law by which persons, lands or effects are attached (apprehended or seized) on execution, or in security for debt.

dispone: to dispose of, hand over.
disposition: a unilateral deed of alienation, by which a right to property is conveyed.
disposition and settlement is the name usually given to a deed by which a person provides for the general disposal of his property after his death.
Distributor of Stamps: an official who issued or sold government stamps. The paper or parchment on which deeds and bills and bonds were written had to bear a government stamp. (William Wordsworth held the office of Distributor of Stamps for the County of Westmoreland at a salary of £200 a year. He resigned on being appointed Poet Laureate.)
drover: a man who drives cattle or sheep.
dyke: a ditch, or a wall of stone; sometimes a boundary wall.
ejection: *see* letters of ejection.
entail: (or tailzie) the settlement of heritable property on a specific line of heirs, thus regulating the course of succession.
exculpation: *see* letters of exculpation.
fank: a fold or a pen.
fast day: usually a day in the week preceding the celebration of half-yearly communion, with a service of preparation for the sacrament. National fasts might occasionally be proposed or declared for special reasons. For example, in 1853 the Presbytery of Edinburgh proposed to Lord Palmerston, then Home Secretary, a national fast 'under the visitation of Asiatic Cholera'. Palmerston thought that cleaning the place up would be a better idea.
fee: generally used in the phrase 'fee and liferent', where fee refers to the succession and liferent to the use and enjoyment of the subject during life. Thus a widow might be given a liferent of property that is to pass to a son after her death.
feudal superior: one who has made the original grant of heritable property, under the condition that the grantee shall annually pay to him a certain sum of money, or perform certain services. The superior has the right to feu duties and other services stipulated in the grant.
fiscal: *see* procurator fiscal
forfeited estates: *see* annexed estates.
fugitation: or outlawry, a sentence pronounced by a court on a person accused of a crime who fails to appear. All of his moveable property is forfeited.
heritable bond: a bond for a sum of money to which is joined, for the creditor's further security, a conveyance of land or heritage, to be held by the creditor in security of the debt.
High Court of Justiciary: supreme criminal court of Scotland. Sits in Edinburgh, but *see also* Circuit Court of Justiciary.

hog: a young sheep, from weaning until shorn of its first fleece; a yearling.
horning: *see* letters of horning.
inhibition: *see* letters of inhibition.
invalid pensioner: a soldier no longer fit for service on account of age, infirmity or disability.
interdict: a court order prohibiting some action complained of as illegal or wrongful, until the question of right is tried in a proper court.
interdiction: the means by which the actions of a facile person are restrained either voluntarily or by a court.
interlocutor: a judgement or judicial order, pronounced in the course of suit, which may not finally determine the matter.
judge ordinary: a judge with a fixed and regular jurisdiction.
judicial valuation: a valuation determined in a court of law falling under the cognisance of a judge.
Justiciary Court: *see* High Court of Justiciary; Circuit Court of Justice.
letters of caption: a warrant for the arrest of a person for debt.
letters of ejection: letters authorising and commanding the sheriff to eject a tenant who has disobeyed a charge or decreet to remove.
letters of exculpation: a warrant granted to the pannel or defender in a criminal prosecution, for citing and compelling the attendance of witnesses in his defence.
letters of inhibition: a warrant prohibiting a debtor from burdening or alienating his heritage to the prejudice of his creditor(s).
letters of horning: a warrant in the name of the sovereign charging the persons named to act as ordered, e.g. to pay a debt, under the penalty of being put to the horn, i.e., proclaimed as an outlaw or bankrupt. (Derived from the trumpet formerly used in making the proclamation.)
libel: a formal statement of the grounds on which either a civil or criminal prosecution takes place.
liferent: *see* fee.
marches: the boundary line of a property.
Martinmas: Scottish quarter day, falling on 11 November. One of two legal term days. The effective date for purposes such as payments was taken as 28 November. *See* term; Whitsunday.
merchantile sequestration: *see* sequestration.
melioration: an improvement made by a tenant on a property rented. Also the allowance, in money, for such improvements due to the tenant on the termination of the lease.
merk: an old Scottish coin, of the value of thirteen shillings and fourpence Scots, i.e. two thirds of a pound Scots. *See also* Scots money.
messenger: also messenger-at-arms. The main function was the execution of all summonses and letters of diligence, both in civil and criminal matters.

minor: a male above the age of fourteen years or a female above the age of twelve years but under the age of majority, which in both sexes is twenty-one years. *See also* pupil; curator.

milcher: a milk cow.

missive: a letter in which a transaction is agreed upon, which may then be succeeded by a more formal legal document.

multiplepoinding: an action to determine the rights of several competing claimants to a fund or property held by another. The fund or property may be known as the fund in medio, the sum in medio, or the subject in medio. *See also* decreet of preference.

oversman: a chief arbiter, appointed to have the final decision in the event of deadlock.

oversouming: grazing an excessive number of animals on a particular area of pasturage; s*ee* souming.

pannel: the accused person in a criminal action.

poinding: the seizure and sale of the goods of a debtor; *see also* sequestration.

precognition: an examination by the judge ordinary or justices of the peace, where any crime may have been committed, in order that the facts connected with the offence may be ascertained, and full and perfect information given to the public prosecutor, to enable him to prepare the libel and carry on the prosecution

process: the legal papers in an action lodged in court by both parties. Includes all that takes place from the first step to the final decree in a civil action, or to the conviction or acquittal in a criminal prosecution.

procurator fiscal: public prosecutor in a Sheriff Court.

pro loco et tempore: *see* desertion of the diet

protest (bills): notarial evidence of a demand for payment having been made. An essential preliminary to other steps to secure payment. (See Appendix A for further information.)

pupil: a male child under the age of fourteen years, or a female child under the age of twelve years. *See also* minor.

pursuer: a person who raises an action in a law court.

Qualified to Government: apparently refers to procedure under the 1793 Relief Act that removed the main disabilities on Catholics in Scotland. Relief was conditional upon taking an oath of loyalty to the king and to the Hanoverian succession.

Register of Deeds: a register kept by a court into which deeds may be copied, for preservation or to facilitate legal action. Registers were kept by the Court of Session (sometimes called the 'Books of Council and Session'), Sheriff Courts and Burgh Courts.

Register of Sasines: sasine is the act or procedure of giving possession of feudal property. The document narrating the legal transaction must

be recorded in the Register of Sasines. The use of such property as security for a loan must also be recorded, and copies of other documents bearing on the possession of the property may also be found in the Register.

residenter: a resident, particularly one of long standing.

roup: an auction sale.

Scots money: Scottish currency abolished by the Act of union of 1707. A pound Scots was worth one twelfth of an English pound. References were still made to Scots money for some time after the currency was abolished. (All references to money in this book are to sterling unless Scots is specified.)

sequestration: First, land owned by a bankrupt may be sequestrated if two or more creditors are competing for possession. The court will appoint a judicial factor to manage the estate until a decision is reached. Second, in mercantile bankruptcy, sequestration is granted on application to the Court of Session by the bankrupt with the concurrence of one or more creditors. All of the bankrupt's possessions will be seized, and a trustee elected by the creditors. This form of sequestration is not available to tenants of land, but such may argue that they gain a material part of their living by dealing in cattle and grain not produced on their land. Third, in spite of the above definitions, it appears to have been the practice, even in legal documents, to use 'sequestration' when 'poinding' might have been appropriate, i.e. to seize a limited quantity of livestock or other property to meet a specific debt such as unpaid rent.

sett: the letting or leasing of a farm, house, etc.

sheep-walk: a tract of grassland used for pasturing sheep.

shepherd: a man who tends a flock of sheep, usually an employee but sometimes an owner of sheep.

sheriff: a judge in a sheriffdom. Duties were carried out by a sheriff depute and sheriffs substitute.

Sheriff Court: the court presided over by a sheriff, with civil and criminal jurisdictions.

sheriff clerk: the clerk of a sheriff court.

sheriff officer: an official or messenger who carries out the warrants of a sheriff, serves writs, etc.

slept: a court action that has lapsed through passage of time and failure of prosecution.

Solicitor in the Supreme Court: A member of an incorporated society of solicitors in Edinburgh.

soum: the numbers of animals to be kept on a particular piece of land.

steelbow: a form of tenancy in which a landlord provides the tenant with stock, grain, implements, etc., under contract that the equivalent should be returned at the end of the lease.

store farm: a farm, usually in the hills, on which sheep are reared and grazed.

subject: property or effects.

sum in medio: *see* multiplepoinding

summons: a writ containing the grounds of an action and citing the defender(s) to appear in court on a certain day.

Surveyor of Taxes: an official concerned with the collection of a number of assessed taxes, e.g. on houses, horses, servants and income, which were not the responsibility of the collector of taxes.

suspension: a warrant for stay of execution of a decreet or sentence until the matter can be reviewed.

tack: a lease or tenancy, especially the leasehold tenure of a farm, mill, etc., for the period of tenure.

tacksman: the holder of a tack.

tailzie: *see* entail.

tenement: a holding, land held in tenure.

term: a period of time; also, one of four days of the year on which certain payments such as rent are due, and leases begin and end. These are Candlemas, Whitsunday, Lammas and Martinmas; *see* Whitsunday; Martinmas.

tocher: a marriage portion, which a wife brings to her husband, as provided in her marriage settlement.

tolbooth: a town prison or jail; may be part of a building also serving other purposes such as a town hall.

topsman: a man who assisted a drover.

trust disposition and settlement: *see* disposition and settlement.

tryst: a market, especially for the sale of livestock.

tup: male sheep, a ram.

tutor: *see* curator.

upset price: specification of a price at an auction acceptable to the seller; also, the price of a property below which bids will not be accepted.

violent possession: the forcible or unwarrantable possession of land by a tenant who should have relinquished it.

wadset: the conveyance of land in pledge for, or in satisfaction of, a debt or obligation, with a reserved power to the debtor to recover his lands, on payment or performance.

wedder: male sheep, a ram, especially a castrated ram.

Whitsunday: the legal term of removing, fixed at 15 May.

writer: a lawyer or solicitor. (In Inverness, a Society of Solicitors was formed in 1815 and it was noted that members of the society had been previously called writers.)

Writer to the Signet (WS): a member of a society of solicitors in Edinburgh, originally the clerks by whom signet writs were prepared.

Notes

Abbreviations

HCA	Highland Council Archive
NAS	National Archives of Scotland (formerly the Scottish Record Office)
NAS CR	Crown Estate Commissioners
NAS CS	Court of Session
NAS GD 44	Gordon Castle Muniments
NAS GD 80	Macpherson of Cluny Papers
NAS GD 128	Fraser-Mackintosh Collection
NAS GD 170	Campbell of Barcaldine Papers
NAS GD 176	Mackintosh Muniments
NAS GD 202	Campbell of Dunstaffnage Papers
NAS RD	Register of Deeds
NAS RS	Register of Sasines
NAS SC 25	Dingwall Sheriff Court
NAS SC 28	Fort William Sheriff Court
NAS SC 29	Inverness Sheriff Court
NAS SC 44	Dunblane Sheriff Court
NAS SC 49	Perth Sheriff Court
NAS SC 54	Inveraray Sheriff Court
NAS SC 67	Stirling Sheriff Court
NLS	National Library of Scotland
NSA	*The New Statistical Account of Scotland*
OSA	*The Statistical Account of Scotland, 1791–1799* (*Old Statistical Account*)

Notes to the Introduction

1 D.R. McDonald, 'Glencoe – The Passing of the Chiefs', in *Clan Donald Magazine* No. 8. Edinburgh, 1979, 55–57.

2 R.H. Campbell, *Scotland since 1707. The Rise of an Industrial Society* (Second Edition), Edinburgh, 1985, 133.

3 T.C. Smout, *A History of the Scottish People 1560-1830*, London, 1969, 350.

4 E. Richards, *A History of the Highland Clearances*, London and Canberra, 1982, 191; (see also *The Highland Clearances*, Edinburgh, 2000, 81 and 187, by the same author, for a later view).

5 T. Telford, 'A Survey and Report of the Coasts and Central Highlands of Scotland', in *Reports on Highland Roads and Bridges (1803)*, Parliamentary Papers 1803–1821, Vol.I, Section IV, 16.

6 I.S. Macdonald, 'Alexander Macdonald Esq. of Glencoe. Insights into Early Highland Sheep-Farming', in *Review of Scottish Culture* No. 10, 1996–7, 55–64.

7 A.I. Macinnes, Marjory-Ann D. Harper & Linda G. Fryer, (Eds), *Scotland and the Americas, c.1650–c.1939: A Documentary Source Book*, Scottish History Society, Edinburgh, 2002, 22.
8 T. Pennant, *A Tour of Scotland* (Third Edition), Warrington, 1774, 194.
9 E. Richards, 1982, *op. cit.* (Ref. 4) 162.
10 R.A. Dodgshon, 'The Economics of Sheep Farming in the Southern Uplands during the Age of Improvement', in *Economic History Review* Vol.29, 1976, 552–54.
11 J.A.S. Watson, 'The Rise and Development of the Sheep Industry in the Highlands and North of Scotland', in *Transactions of the Highland and Agricultural Society of Scotland*, Fifth Series Vol.XLIV, 1932, 9.
12 A. Wight, *Present State of Husbandry in Scotland*, Edinburgh, 1778, Vol.II, 464.
13 NAS GD 44/43/268/25 Thomas Gillespie, Capelgill, to James Ross, Factor to the Duke of Gordon. 12 March 1782.
14 C. Fraser-Mackintosh, *Antiquarian Notes*, Second Series, Inverness, 1897, 126–27.
15 J. Hogg, *A Tour of the Highlands in 1803*, Edinburgh, 1888 (Reprint 1986), 38–39.
16 A.G. and Margaret H. Beattie, *Pre-1855 Gravestone Inscriptions in Lochaber and Skye*, The Scottish Genealogy Society, 1990, 34 (Entries 55 & 56).
17 R J Adam, ed., *Papers on Sutherland Estate Management 1802–1816*, Edinburgh, 1972, Vol.1, xxxiv. and Vol.2, 59–61.
18 NAS SC 29/53/10 Precognition re Alexander Breck Kennedy, 8 June 1781.
19 NAS CC/11/1/9 Record of Testaments (Inverness-shire) p.412. Thomas Gillespie – died 9 July 1824.
20 NAS SC 29/51/1 Trial of Ronald Dow Kennedy, sometime residing in Shian . . . 1783.
21 J. Mitchell, *Reminiscences of my Life in the Highlands*, 1884 (Reprint, Newton Abbott, 1971), Vol.1, 188 (footnote).
22 Annette M. Smith, *Jacobite Estates of the Forty-Five*, Edinburgh, 1982, 53.
23 NAS SC 29/7/2 Decreet. Katherine Macdonell v. Thomas Stoddart Tacksman of Killinean. Inverness 29 September 1795.
24 NAS DI.64/15 Leaves 8, 9, and 11. 9 May 1795. Debts of Thomas Stoddart, Killiean.
25 NAS GD 128/36/8 & GD 128/47/4 – Sheep Farm Association.
26 I.S. Macdonald, 1996–7, *op. cit.* (Ref. 6).
27 *NSA*, Edinburgh & London, 1845, Vol.XIV, 504.

Notes to Chapter 1

1 Loretta R. Timperley, *A Directory of Land Ownership in Scotland c1770*, Scottish Record Society, New Series 5, 1976, 40–41.
2 NLS. MS 3985. Robertson-Macdonald Papers. Appin Claim. Memorial for Donald Macdonald Esq. of Kinlochmoidart. 1797.
3 C. Fraser-Mackintosh, 'The Macdonalds of Achtriachtan' in *Transactions of the Gaelic Society of Inverness* Vol. XXIII, 1898–99, 136–45.
4 NAS RD2/272/72 Disposition etc. Angus Macdonald of Achtriachtan to his sons. 1796.
5 J. Fergusson, *Argyll and the Forty-Five*, London, 1951, 196.
6 *Ibid.*, 247 (Footnote 2).

7 Annette M. Smith, *Jacobite Estates of the Forty-Five*, Edinburgh, 1982, 21 & 237.

8 NAS E754/1 Judicial Rental of the Estate of Glencoe. 1748.

9 J. Fergusson, 1951, *op. cit.* (Ref. 5), 196.

10 A. & A. Macdonald, *The Clan Donald,* Inverness, 1904, Vol.III, 215.

11 Mackay, D.N. (Ed.) *Trial of James Stewart (The Appin Murder)*, Second Edition, Edinburgh & Glasgow, 1931, 149.

12 Barbara Fairweather & D.C. Cargill, *Pre-1855 Tombstone Inscriptions on Isla Munda, The Burial Island in Loch Leven*, The Scottish Genealogical Society. Entry 161.

13 NAS GD 170/1868 John McDonald of Glencoe invites Alexander Campbell of Glenure to Funeral of his wife, Catherine Cameron. Date uncertain, possibly 1786.

14 J. Stewart of Ardvorlich, *The Camerons*, Stirling, 1974, 216.

15 Mackay, D.N. (Ed.) 1931. *op. cit.* (Ref. 11), 149.

16 NAS GD 44/25/6/61 Alexander McDonald to Messrs. James Ross and William Tod for the Duke of Gordon. 14 October 1776.

17 NAS GD 1/8/18 William Macdonald WS, Letterbook 1777–1778, 11 March 1778 to Lord Macdonald.

18 NAS RD3/239/2/756 Bond. 'I, Alexander McDonald, Younger of Glenco, hereby acknowledge . . . indebted to Capt Colin Campbell of the Second Battalion of the 42 Rgt of Foot the sum of £50'. 1780.

19 Anne Grant, *Letters from the Mountains*, London, 1806, 82–83.

20 J. Prebble, *Glencoe*, London, 1966, 28, 38, 75, & 166.

21 M. Linklater, *Massacre: The Story of Glencoe*, London & Glasgow, 1982, 77.

22 Loretta R. Timperley, 1976, *op. cit.* (Ref. 1), 40–41.

23 J. Buchan, *The Massacre of Glencoe*, London, 1985 (first published 1933), 32.

24 C. Fraser-Mackintosh, 1898–99, *op. cit.* (Ref. 3), 136–45.

25 *Ibid.*

26 E. Dodwell & J.S. Miles, *Alphabetical List of Officers of the Indian Army 1760-1834*, London, 1938, Bombay Presidency, 52–53.

27 NAS DI.23/8 Folio 49. Inhibition. Marquis of Tweeddale against McDonald (Adam Macdonald of Achtriachtan) No 112. Inveraray 29 May 1801.

28 NAS 170/1864 Æneas McDonald, Achtriachtan, to Alexander Campbell of Barcaldine. 10 June and 2 September 1793.

29 NAS RD2/297/390 Bond of Provision – McDonell to Campbell. 25 April 1796.

30 NAS GD 170/1864 Æneas McDonald, Achtriachtan, to Alexander Campbell of Barcaldine. 3 June 1794.

31 NAS GD 1/8/19 Letterbook of William Macdonald WS 1778–1779. To Ensign Colin Macdonald, 19 February 1779; to Achtriachtan [Angus Macdonald] 22 February 1779.

32 Sasine Abridgements Argyll. 7 June 1817. (2686)

33 NAS RD2/266/580 Power of Attorney – Mrs. MacDonell etc. to Angus MacDonell (Achtriachtan) July 1796.

34 NAS RD2/272/72 (Ref. 4).

35 Inverness Public Library. Fraser-Mackintosh Library. FM 3240.

36 NAS CS 228/Mc/13/1 MacDonalds Drimintorran. MacDonald agt. MacDonald. 1808 (Includes papers from a later process, Angus MacDonald v. Alexander MacDonald of Dalilea, 1825).

37 T. Pennant, *A Tour of Scotland* (Third Edition), Warrington, 1774, 211.

38 *OSA* (Reprint) Wakefield, 1983, Vol.XVII, 132, 134 & 140–41.

39 Northumberland Record Office. ZCU/43 Typed copy of Journal of Matthew Culley's Tour in Scotland in 1775, with notes inserted by his brother George, 113–4. (Note: The original manuscript appears to have been lost.) I am grateful to Professor Eric Richards who drew attention to the mention of John Stewart of Ballachulish in the Culley papers. This and other Culley papers have now been published. See Orde, Anne (Ed.) in the Bibliography.
40 NAS RD2/1/217 Bond of Credit Macdonalds to Douglas Heron & Co. 10 March 1770 (Registered 12 January 1775).
41 S.G. Checkland, *Scottish Banking A History, 1695-1973*, Glasgow & London, 1975, 124–31.
42 *OSA* (Reprint) Wakefield, 1983, Vol.XX, 346.
43 Sarah Murray, *A Companion and Useful Guide to the Beauties of Scotland etc*, London, 1799, Vol.1, 348.
44 T. Garnett, *Observations on a Tour through the Highlands and part of the Western Isles of Scotland*, London, 1811, Vol.I, 296.
45 Dorothy Wordsworth, *Recollections of a Tour made in Scotland A.D. 1803*, Ed. by J.C. Shairp, Edinburgh, 1874, 173.
46 W. Taylor, *The Military Roads in Scotland* (Revised Edition), Colonsay, 1996, 71.
47 NAS GD 170/1869 Angus McDonell, Achtriachtan, to Alexander Campbell, Barcaldine. 21 & 24 May 1787.
48 *OSA.* (Reprint) Wakefield, Vol.XX, 1983, 351.
49 Dorothy Wordsworth, 1874, *op. cit.* (Ref. 45), 166.
50 *Sketch of a Tour in the Highlands of Scotland . . . September and October 1818* (attributed to Archibald Larkin), London, 1819, 222.
51 D. Bremner, *The Industries of Scotland*, 1869 (Reprinted Newton Abbott 1969), 429.
52 *OSA* (Reprint) Wakefield, Vol.XX, 1983, 362 (footnote) & 363.
53 NAS GD 128/62 List of Documents connected with the family of Glencoe.
54 NAS GD 128/30/1/20 Marriage Contract. Alexander Macdonald of Glencoe to Mary Cameron. 1787.
55 NAS RD2/275/581 Heritable Bond. Alexander Macdonald of Glencoe to Rev. John Kennedy, Auchteraw. 1789.
56 NAS GD 128/30/1/20 (Ref. 54).
57 NAS GD 128/30/1/21 Supplementary Marriage Contract. Alexander Macdonald of Glencoe to Mary Cameron. 1791.
58 NAS GD 44/27/12/12 Angus McDonell (Achtriachtan) to James Ross. 15 June 1778.
59 Sasine Abridgements. Argyll. 30 March 1791. (575)
60 NAS GD 128/54/6 Thomas Gillespie to Campbell Mackintosh. 23 January 1793.
61 NAS GD 128/54/8 John Gillespie to Campbell Mackintosh. 5 March 1793.
62 NAS GD 128/54/7 John Gillespie to Campbell Mackintosh. 30 July 1793.
63 Sasine Abridgements Argyll. 24 May 1794. (798)
64 E. Dodwell, & J.S. Miles, 1938, *op. cit.* (Ref. 26), Bombay Presidency, 52–53.
65 NAS RD2/266/580 (Ref. 33).
66 NAS RD2/297/390 (Ref. 29).
67 NAS RD2/272/72 (Ref. 4).
68 Sasine Abridgements Argyll. 18 November 1796. (978)
69 T. Garnett, *op. cit.* (Ref. 44), 1811, Vol.I, 288.

Notes to Chapter 2

1 NAS GD 44/25/2 Rentals of Lochaber with lists of men residing on the Duke of Gordon's lands in Lochaber. Item 54. Lochaber Tacksmen in the Parish of Kilmanivaig. 1769.

2 *Ibid.* Item 52. List of Names of Principal Tenants in the Lordship of Lochaber. 1769.

3 NAS GD 44/26/9/26 Memorial for the Duke of Gordon about Keppoch's Farms. 1799.

4 NAS GD 44/25/32 Inverlochy, Tacks and other papers. 1704–1802.

5 NAS GD 44/51/743/10 & 11 Rental of Lochaber as payable Martinmas. 1769.

6 NAS GD 44/25/6/61 Alexander McDonald, Glencoe, to Factors for Duke of Gordon. 14 October 1776.

7 The 'good circumstances' of the Glencoe family at this period are confirmed by William Macdonald WS of St. Martins in a letter of 11 March 1778 to Lord Macdonald about the raising of the latter's Regiment. He said: 'Glenco . . . has a good many followers and has plenty of money to complete his quota' (NAS GD 1/8/18).

8 NAS GD 44/43/173/44 Angus McDonell (Achtriachtan) to James Ross. 26 December 1776.

9 *Ibid.*

10 NAS RHP 2493 Plan of the Lordship of Lochaber. From Mr Roy's Survey in 1812, 1813, & 1814. William Johnstone 1831. Table of Acreages.

11 NAS GD 44/43/173/44 (Ref. 8).

12 J.A.S. Watson, 'The Rise and Development of the Sheep Industry in the Highlands and North of Scotland', in *Transactions of the Highland and Agricultural Society of Scotland*, Fifth Series, Vol.XLIV, 1932, 8.

13 NAS GD 44/43/176 Item 5: Baillie MacPherson, Claggan, Offer for Corriechoillie and Achachar. 4 March 1777; Item 10: Angus McDonell (Achtriachtan) to James Ross Esq. Fochabers. 5 March 1777.

14 NAS GD 44/43/227 Item 48. Munro Ross, Edinburgh, to James Ross. 24 September 1779; Item 49. Petition. Alexr. MacPherson, Writer in Edinburgh, to the Duke of Gordon. July 1779.

15 NAS SC 29/10/690 Donald Kennedy against McPherson, Cameron and McMillan. Extracted 5 December 1783.

16 NAS SC 29/53/8 Petition of Ewen MacMillan, late tenant in Kinlocharkaig, present prisoner in the Tolbooth of Inverness, and Lieut. Donald Cameron of the 95th Regiment of Foot. October 1781.

17 NAS GD 44/43/176 Item 10 (Ref. 13)

18 NAS GD 44/43/202/8 William Tod to James Ross. 19 April 1778.

19 NAS GD 44/43/202/3 David Cubison, Ayr, to William Tod. 9 April 1778.

20 NAS GD 44/43/227 Item 49 (Ref. 14).

21 NAS GD 44/51/743/22 Rental of Lochaber as set in 1776 payable at Mart. 1777.

22 NAS GD 44/27/12/5 William Tod to James Ross. 28 February 1778.

23 NAS GD 44/51/743/22 (Ref. 21).

24 NAS GD 44/25/2/56. List of men residing on the Duke of Gordon's Lands in the Lordship of Lochaber. May 1778.

25 NAS GD 44/27/12/12 Angus McDonell (Achtriachtan) to James Ross. 15 June 1778.

26 NAS RHP 2493 (Ref. 10).

27 NAS CS 29 – EP MACK January 15–23 1794. Sarah McDonald, spouse to Charles Robertson, Shoemaker in Perth, Executrix dative . . . to the deceased Donald McDonald late of Dellifour . . . against Alexr. McDonald late in Achadourie now at Glenturret etc.

28 Ann MacDonell & R. MacFarlane, *Cille Choirill, Brae Lochaber*, Spean Bridge, 1986, 50 (entry 7).

29 NAS CS 29 – EP MACK January 15–23 1794. (Ref. 27.)

30 NAS GD 44/25/2/50 Rental of Lochaber. 1769.

31 NAS GD 44/51/743/10 & 11 (Ref. 5).

32 NAS GD 44/51/743/22 (Ref. 21).

33 NAS CS 29 – EP MACK January 15–23 1794. (Ref. 27.)

34 NAS GD 44/43/198/22 William Tod to James Ross. 22 February 1778.

35 NAS GD 44/43/202/9 Baillie D. MacPherson to William [?] Tod. 15 April 1778.

36 NAS GD 44/43/202/8 (Ref. 18).

37 NAS CS 29 – EP MACK January 15–23 1794. (Ref. 27.)

38 NAS GD 44/25/7/37/25 Donald McDonald to James Ross. From Tirandreich. 15 June 1778.

39 NAS GD 44/27/12/12 (Ref. 25).

40 NAS CS 29 – EP MACK January 15–23 1794. (Ref. 27.)

41 A.R.B. Haldane, *New Ways Through the Glens*, London, Edinburgh, etc, 1962, 109.

42 NAS GD 44/43/272/1 William Tod to James Ross asking the Duke [of Gordon's] Subscription to a Bridge in the Braes of Lochaber. 1 June 1782.

43 NAS GD 44/52/175 Abstract of Mr. William Tod's Accounts as Factor . . . Lordships of Badenoch and Lochaber . . . Crops 1791 and 1799 inclusive.

44 NAS GD 44/41/30/2 Petition of Archd. McDonald, Tacksman of Fersit, to His Grace the Duke of Gordon. 30 June 1766.

45 NAS GD 44/25/5/16 Depositions of tenants re thefts. 1771.

46 NAS GD 44/25/5/41 Archibald McDonald, Fersit to William Tod. 13 December 1771.

47 NAS GD 44/25/6/8 Tenants giving up tacks after 3 years. 1773.

48 NAS GD 44/25/5/41 (Ref. 46).

49 NAS GD 44/27/11/24x. William Tod to James Ross. 15 April 1772.

50 NAS GD 44/27/11/38x. William Tod to James Ross. 7 May 1772.

51 NAS GD 44/25/5/61 William Tod to James Ross. 17 June 1772.

52 NAS CS 29 – EP MACK January 15–23 1794. (Ref. 27).

53 A. Macpherson, *Glimpses of Church and Social Life in the Highlands in Olden Times*, Edinburgh, 1893, 146.

54 Alison Mackenzie, The Gaick Memorial in *Journal of the Clan Chattan Association*, 1994, Vol.IX, No.6, 326.

55 NAS DI.8/209 Folio 817. John McPherson of Ballachroan. Bond of Credit, British Linen Company. Overdrawn 1794.

56 A. Macpherson, *Captain John Macpherson of Ballachroan and the Gaick Catastrophe of the Christmas of 1799 (O.S.) A Counterblast*, Kingussie, 1900, 3–4.

57 J. Robertson, *Kingussie and the Caman*, Kingussie, 1994, 27.

58 NAS GD 44/25/6/17 Rental of Lochaber at Martinmas 1773 – Fersit. Captains Grant & Macpherson. Rent £60, Discount £45.

59 *Ibid.*

60 NAS GD 44/43/177/13 John MacPherson, Ballachroan, to James Ross, Fochabers. 21 March 1777.

61 NAS GD 44/25/2/56 (Ref. 24).

62 NAS SC 29/7/3 Decreet of Removing. John MacFarlane, Tacksman of Kingussie etc. agt. Mrs. Mary Macdonald, Relict of the deceased John MacHardy, late Tacksman of Kingussie, and others. 18 May 1796.

63 NAS SC 29/7/5 Decreet of Removing. Mr. James Falconer, Pitmain, Principal Tacksman and Setter of the lands etc . . . against John MacFarlane and Duncan MacFarlane, pretended tenants of the lands of Glengunock and others upon the lands of Kingussie. 17 May 1799.

64 A. Macpherson, 1893. *op. cit.* (Ref. 53). 149–50.

65 NAS GD 44/43/177/13 (Ref. 60).

66 NAS GD 44/25/7/3 Alexander McDonald, Glencoe, to James Ross, Fochabers. 11 April 1777.

67 NAS GD 44/25/7/37 Angus McDonell [Achtriachtan] to James Ross. 28 August 1779.

68 *Fasti Ecclesiae Scoticanae*, Vol.6, Edinburgh, 1923, 135.

69 NAS GD 44/51/14/3 Vouchers for Lochaber Badenoch etc Crop 1786. Salary of Duncan McIntyre, Schoolmaster, Kilmanivaig, Martinmas 1784 – Martinmas 1785.

70 *Fasti Ecclesiae Scoticanae*, Vol.6, Edinburgh, 1923, 135.

71 NAS GD 44/43/176 (Ref. 13).

72 *Fasti Ecclesiae Scoticanae*, Vol.6, Edinburgh, 1923, 137.

73 NAS GD 44/27/12/4 Memorandum – For James Ross Esq. at Fochabers relative to . . . abuses by Donald McPherson, Baron Baillie upon the Duke of Gordon's Estate in Lochaber. Mr. Thomas Ross. Undated (probably 1778).

74 S. Macmillan, *Bygone Lochaber*, Glasgow, 1971, 14–28.

75 NAS GD 44/43/198/22 (Ref. 34).

76 NAS GD 44/27/12 Item 5. William Tod to James Ross. 28 February 1778; Item 8. William Tod to James Ross. 2 March 1778.

77 NAS GD 44/27/12/4 (Ref. 73).

78 NAS GD 44/43/202/9 (Ref. 35).

79 NAS SC 29/53/8 (Ref. 16).

80 NAS GD 44/43/255 Item 12. Ewen Cameron of Glennevis to William Tod. 17 May 1781; Item 13. Memorial for Donald Cameron, Tacksman of Blarcherin. 17 May 1781.

81 NAS CS 229/Mc/5/12 Summons of oppression, defamation and damages. Poor MacDonald [Donald] against Macpherson etc. 1787.

82 NAS CS 232/R/13/10 Suspension. John Rankine & Son against Donald Cameron. 1784.

83 NAS DI.64/14 Folio 300. Inhibition. Alexander McPherson against Lieut. Donald Cameron. Inverness 4 May 1785.

84 A. Mackenzie, *History of the Frasers of Lovat*, Inverness, 1896, 546–49.

85 HCA L/IB/BC/7/6/24 Precognition relative to the Escape of Archibald Kennedy. 1779.

86 NAS GD 44/43/236/39 Alexander Cameron, Glenturret to William Tod. 24 April 1780.

87 NAS GD 44/43/237/36 Angus McDonald to William Tod. 12 May 1780.

88 NAS GD 44/43/236/39 (Ref. 86).

89 A. Mackenzie, 1896, *op. cit.* (Ref. 84), 546–49.

90 NAS GD 44/43/250/17 John McDonald of Glencoe to James Ross. 10 February 1781.

91 NAS GD 44/51/14/3 Vouchers for Lochaber Badenoch etc Crop 1786. List of Arrears – Badenoch Lochaber & Kincardine – Mart. 1785.

92 NAS JC 11/34 Circuit Court. Northern Circuit. Inverness. 16 September 1782.

93 NAS SC 29/51/1 Trial of Ronald Dow Kennedy, sometime residing in Shian . . . 1783.

94 NAS SC 29/51/1 Trial of Katherine Kennedy spouse to Angus Bain Kennedy . . . and Katherine MacDonald spouse to John Bain Kennedy . . . 20 November 1783.

95 NAS SC 29/53/10 Declaration of Mary McDonell. 1781.

96 NAS SC 29/53/10 Precognition re Alexander Breck Kennedy. 8 June 1781.

97 NAS SC 29/51/1 Trial of Archibald Buie or Bain Kennedy lately residing in Cullachy . . . 1785.

98 A. Mackenzie, 1896, *op. cit.* (Ref. 84), 546–49.

99 NAS GD 44/43/260/48 Angus McDonell of Achtriachtan, to James Ross. 26 September 1781.

100 NAS GD 44/43/271/20 Angus McDonell of Achtriachtan to James Ross. 9 May 1782.

101 NAS GD 44/26/12 Process. Duke of Gordon v. MacMartin of Letterfinlay. 1778–1815.

102 NAS GD 44/52/175 (Ref. 43).

103 NAS GD 44/52/156 Account of Charges and Discharges of Mr. Tod's Intromissions with the rents of the Lordships of Badenoch and Lochaber . . . 1790–1802.

104 NAS GD 44/25/8/6 Lochaber Sett 1804 Kilmanivaig etc – offer from Captain Alexander MacDonell of Moy. Tirandreich etc – offer from Rev. Thomas Ross.

105 NAS GD 44/51/743/30 Rental of Lochaber Set by Mr. Tod – payable Mart. 1796.

106 NAS CS 232/G/12/29 Summons of Removing. Duke of Gordon v. Alexander McDonald Esq. of Glencoe and others per Roll. 1798.

107 *Ibid.*

108 NAS RD2/275/597 Sub-tack – Kennedy and McDonald. 1798.

109 NAS GD 128/9/2/97 Rev. John Kennedy to Alexander MacDonell, Writer in Inverness. 1 August 1798.

110 NAS GD 44/25/2/55 List of men residing in the Lands of the Lordship of Lochaber belonging to the Duke of Gordon's Vassals. May 1778.

111 NAS GD 202/35 Macdonald of Glenco – Trustee Accounts. 1815–1835.

112 NAS GD 44/25/9/3/2 [Reference to] Lease of Fersit to Alexander McDonell, James McCaul and Alexander McVean. 1 May 1797.

Notes to Chapter 3

1 A.M. Mackintosh, *The Mackintoshes and Clan Chattan*, Edinburgh, 1903, 354.

2 NAS GD 176/1325 Memorial of Alexander McIntosh, once Tacksman of Keppoch (to Alexander MacKintosh of MacKintosh). 1834.

3 F. Adam, *The Clans, Septs, and Regiments of the Scottish Highlands*, 8th Edition, Edinburgh & London, 1970, 453–54.

4 I.F. Grant, *Along a Highland Road*, London, 1980, 144.

5 Ann MacDonell & R. MacFarlane, *Cille Choirill, Brae Lochaber*, Spean Bridge, 1986, 27–28 (entry 61).

6 HCA L/INV/SC/8/9 Precognition and Declaration. Ensign Angus Macdonald and Others. Fort William. 1783.

7 NAS CS 233/M/6/26 Summons of Cessio Bonorum. MacDonald against His Creditors. 1788.

8 R. MacFarlane, 'The MacDonells of Aberarder', in *Clan Donald Magazine*, No. 12, Edinburgh, 1991, 147–57.

9 NAS GD 176/2127/15 Angus McDonald late at Tullich to Campbell Mackintosh, Writer in Inverness. 23 July 1802.

10 NAS GD 176/2116/12 Æneas Mackintosh to Campbell Mackintosh, Writer in Inverness. 29 December 1802.

11 NAS GD 128/11/1 Rentals of the Mackintosh Estate (Lands of Lochaber). Various dates from 1771 to 1807.

12 NAS GD 176/2116/5 Æneas Mackintosh to Campbell Mackintosh, Writer in Inverness. 13 January 1802.

13 NAS GD 176/1325 (Ref. 2).

14 NAS GD 176/1415 Tack twixt Æneas Mackintosh Esq. and Archd. McDonald and his Sons. 1786.

15 NAS CS 232/M/16/17 McDonald v. McLauchlan (heard initially by Colin Campbell, Sheriff substitute in Maryburgh (Fort William) and later by advocation in the Court of Session).

16 NAS SC 29/51/1 Trial of Angus MacKillop now or lately resident in Achluarach. Inverness. 21 December 1787.

17 HCA L/INV/SC/8/9 (Ref. 6).

18 NAS CS 229/Mc/5/12 Summons of oppression, defamation and damages. Poor MacDonald [Donald] against Macpherson etc. 1787.

19 HCA L/INV/SC/8/9 (Ref. 6).

20 S. Macmillan, *Bygone Lochaber*, Glasgow, 1971, 157.

21 NAS SC 29/7/2 Decreet of Removing. Æneas Mackintosh of Mackintosh Esq. Proprietor etc . . . against tenants in Lochaber. 14 April 1795.

22 HCA L/INV/SC/8/9 (Ref. 6).

23 NAS CS 229/Mc/5/12 (Ref. 18).

24 R. MacFarlane, 1991, *op. cit.* (Ref. 8). 147–57.

25 *Ibid.*

26 NAS GD 44/27/17 Disposition and Assignation – The Duke of Gordon to Cosmo Gordon Esq. of Cluny (Advocate) 1772.

27 NAS GD 44/28/12 Garvamore. Contract of Wadset 1725. Renunciation 1752.

28 NAS GD 44/27/11/70 Ranald McDonell, Aberarder, to James Ross. 28 July 1772.

29 NAS SC 44/22/21 Petition. Messrs. McPherson & McDonald & Fiscal v. Lennox. November 1783.

30 NAS GD 128/11/1 (Ref. 11).

31 NAS GD 44/43/198/22 William Tod to James Ross. 22 February 1778.

32 NAS SC 29/53/8 Petition of Ronald McDonell of Aberarder, now in Moy, and of John and Archibald McDonells his sons. 8 December 1781.

33 NAS CS 229/Mc/5/12 (Ref. 18).

34 NAS SC 29/53/10 The Petition of James Cumming, Writer, Inverness, Procurator Fiscall . . . 6 September 1781.

35 NAS GD 44/51/14/3 Vouchers for Lochaber, Badenoch etc. Account His Grace the Duke of Gordon and William Tod Esq. His Factor – To George Bean, Writer in Inverness. 1781.

36 NAS SC 29/53/8 (Ref. 32).

37 NAS SC 29/51/1 Criminal trial Book. Inverness, 6 March 1783. Ranald McDonell residing at Moy commonly called Ranald McDonell of Aberarder,

John McDonell residing at Torgulbin, and son of the said Ranald McDonell, and Duncan McDonell, servant [assault on John Stewart, Tenant in Shiromore].

38 NAS JC 11/34 Circuit Court. Northern Circuit. Inverness, May 1783.

39 R. MacFarlane, 1991, *op. cit.* (Ref. 8). 147–57.

40 HCA L/INV/SC/8/9 (Ref. 6).

41 C. Fraser-Mackintosh, *Antiquarian Notes*, Second Series, Inverness, 1897, 166–68.

42 NAS CS 229/Mc/5/12 (Ref. 18).

43 NAS GD 176/1325 (Ref. 2).

44 HCA L/INV/SC/8/9 (Ref. 6).

45 NAS GD 128/45/13 Donald McDonald, Tulloch, to Alexander Macdonell, Writer, Inverness. From Fort William. 14 March 1806.

46 NAS SC 67/7/229 Graham v. McDonalds 1781.

47 *OSA* (Reprint), Vol.XII, Wakefield, 1977, 43.

48 NAS SC 67/7/229 (Ref. 46).

49 NAS SC 44/22/20 Petition for sequestration. Ferguson v. McDonald. July 1783. (Debt of £160 borrowed from John Ferguson in 1781.)

50 F. Adam, 1970. *op. cit.* (Ref. 3). 452.

51 NAS CS 237/Mc/7/18 Angus McDonald v. Mitchell. 1794.

52 NAS GD 128/65/14 Dr. McDonald in Knoydart. 1759–1802.

53 NAS CS 237/Mc/7/18 (Ref. 51).

54 NAS SC 67/7/229 (Ref. 46).

55 NAS SC 67/7/229 (Ref. 46).

56 NAS GD 44/51/732/32 Rental of Houses Gardens and Inclosures at Gordonsburgh Whitsunday 1785. Whitsunday 1786 per agreement with Mr. Tod.

57 NAS SC 67/7/229 (Ref. 46).

58 NAS SC 67/7/226 Thomas Nimmo Distiller in Blackgrange v. Charles McDonald Butcher in Torbrex. 1782.

59 NAS SC 67/7/229 (Ref. 46).

60 NAS SC 29/10/678 McDonald against Kennedy and McIntyre. 1783.

61 NAS CS 229/Mc/5/12 (Ref. 18).

62 NAS CS 229/Mc/5/12 (Ref. 18).

63 NAS SC 44/22/16 Mitchell against debtors. 1779.

64 M.S. Mackay, *Doune: Historical Notes*, Stirling, 1984, 55–56.

65 NAS CS 237/Mc/7/18 (Ref. 51).

66 NAS CS 229/Mc/5/12 (Ref. 18).

67 NAS CS 237/Mc/7/18 (Ref. 51).

68 NAS CS 237/Mc/7/18 (Ref. 51).

69 NAS CS 229/Mc/5/12 (Ref. 18).

70 NAS SC 44/22/21 (Ref. 29).

71 NAS CS 229/Mc/5/12 (Ref. 18).

72 NAS CS 232/R/13/10 Suspension. John Rankine & Son against Donald Cameron. 1784.

73 NAS CS 21/11 March 1784/Campbell v. Cameron.

74 NAS CS 229/Mc/5/12 (Ref. 18).

75 NAS CS 232/R/13/10 (Ref. 72).

76 NAS CS 21/11 March 1784/Campbell v. Cameron. (Ref. 73).

77 NAS CS 271/49242 Bill of Suspension. Donald McDonald agt. Cameron and MacPherson. 16 December 1784.

78 NAS CS 229/Mc/5/12 (Ref. 18).

79 NAS CS 232/R/13/10 (Ref. 72).
80 NAS CS 229/Mc/5/12 (Ref. 18).
81 NAS CS 229/Mc/5/12 (Ref. 18).
82 NAS CS 229/Mc/5/12 (Ref. 18).
83 NAS GD 128/11/1 (Ref. 11).
84 NAS GD 176/2108 Alexander McIntosh, late Bohinie, to Campbell Mackintosh, Writer, Inverness (Written from Fort William). 20 April 1808.
85 NAS GD 176/1415 (Ref. 14).
86 NAS GD 44/51/14/1 Badenoch and Lochaber Miscellaneous Papers. Discharge – Baron Baillie of Lochaber of his Salary Mart. 1780.
87 NAS CS 229/Mc/5/12 (Ref. 18).
88 NAS JC 11/36 Circuit Court. North Circuit. 1786.
89 NAS CS 229/Mc/5/12 (Ref. 18).
90 NAS CS 232/R/13/10 (Ref. 72).
91 NAS DI.64/14 Folio 300. Inhibition. Alexander McPherson against Lieutenant Donald Cameron. Inverness 4 May 1785.
92 NAS CS 229/Mc/5/12 (Ref. 18).
93 NAS CS 229/Mc/5/12 (Ref. 18).
94 HCA L/IB/BC/9/2/12 Petition of Donald MacDonald for Aliment. 7 September 1786.
95 NAS CS 229/Mc/5/12 (Ref. 18).
96 NAS CS 237/Mc/7/18 (Ref. 51).
97 NAS GD 128/36/5 Donald McDonald (Gaol) to Alexander McDonell, Writer in Inverness. 2 November 1786.
98 NAS GD 128/20/1/85 Æneas Mackintosh of Mackintosh to Campbell Mackintosh, Writer, Inverness. 29 November 1786.
99 NAS GD 176/861 Charles Mackintosh, Edinburgh Agent, to Æneas Mackintosh. 12 February 1787.
100 It had been established in 1770 that paupers could bring cases in the courts free of costs (Rosalind Mitchison, *The Old Poor Law of Scotland*, Edinburgh, 2000, 115).
101 NAS CS 229/Mc/5/12 (Ref. 18).
102 NAS GD 128/20/1/65 Æneas Mackintosh of Mackintosh to Campbell Mackintosh, Writer, Inverness. 11 March 1787.
103 HCA L/IB/BC/23/13 The Petition of Donald MacDonald late Drover in Tulloch, present prisoner in the Tolbooth of Inverness. 14 April 1787.
104 NAS GD 128/19/13/5 Donald McDonald, Prisoner, to Alexander McDonell, Writer at Inverness. 13 April 1787.
105 NAS GD 176/862 Certificate of Donald McDonald's return to Jail. 12 June 1787.
106 NAS DI.64/14 Folio 318. Horning & Poinding. Alexander Fraser of Struy against Donald McDonald of Tulloch. Inverness 16 August 1785.
107 NAS CS 229/Mc/5/12 (Ref. 18).
108 NAS CS 229/Mc/5/12 (Ref. 18).
109 NAS GD 128/49/10/3 Memoir for John MacDonell, drover, in Fasnatork [?] near Fort Augustus. 1814.
110 NAS CS 229/Mc/5/12 (Ref. 18).
111 NAS CS 229/Mc/5/12 (Ref. 18).
112 NAS GD 128/12/5 Æneas Mackintosh of Mackintosh to Campbell Mackintosh, Writer in Inverness. 12 February 1799.
113 NAS GD 176/1325 (Ref. 2).

114 NAS GD 176/2116/5 (Ref. 12).
115 NAS CS 237/Mc/13/3 McCaul v McDonald. Summons of Reduction 1817.
116 NAS DI.64/15 Leaf 12 Æneas Mackintosh against Alexander MacDonell of Keppoch etc. 1795.
117 Ann MacDonell & R. MacFarlane, 1986, *op. cit.* (Ref. 5), 27–28, Entry 61.
118 NAS CS 229/Mc/5/12 (Ref. 18).
119 NAS CS 229/Mc/5/12 (Ref. 18).
120 HCA L/INV/SC/8/9 (Ref. 6).
121 NAS SC 29/51/1 (Ref. 16).

Notes to Chapter 4

1 NAS GD 176/1415 Tack twixt Æneas Mackintosh Esq. and Archd. McDonald and his Sons. 1786.
2 NAS GD 128/11/1 Rentals of the Mackintosh Estate (Lands of Lochaber. Various dates from 1771 to 1807.
3 HCA P/INV/2/191 Tack Æneas Mackintosh to Ensign Angus MacDonell late of the Eighty-fourth Regiment of Foot and Lieutenant Alexander MacDonell late of the Eighty-second Regiment of foot. 1786.
4 NAS GD 128/11/1 (Ref. 2).
5 NAS SC 29/7/3 Decreet of Removing. Æneas Mackintosh . . . agt. Captain Angus MacDonald in Dalnaderg . . . Lieut. Alexander MacDonell at Moy, pretended tenants etc of Dalnaderg, Tollie, Clachaig, and Urchar . . . Inverness, 12 April 1796.
6 NAS GD 176/2106 Donald McDonald, Tulloch, to Campbell Mackintosh, Writer, Inverness. 25 December 1799.
7 NAS GD 128/11/1 (Ref. 2).
8 NAS CS 233/M/6/26 Summons of Cessio Bonorum. MacDonald against His Creditors. 1788.
9 NLS Acc. 7909 Rental Books Mackintosh of Mackintosh. Book 1.
10 NAS GD 128/11/1 (Ref. 2).
11 NAS GD 176/1415 (Ref. 1).
12 NAS GD 128/41/5 Donald McDonald, Tulloch, to Campbell Mackintosh, Writer, Inverness. 2 November 1790. (Written from Highbridge.)
13 NAS GD 176/2108 Alexander McIntosh, late Bohinie, to Campbell Mackintosh, Writer, Inverness. (Written from Fort William.) 20 April 1808.
14 NAS GD 176/862/4 Statement of Rent owed by Major McDonell of Keppoch 1775.
15 NAS CS 237/Mc/13/3 McCaul v. McDonald. Summons of Reduction 1817.
16 NAS DI.64/15 Leaf 12 Æneas Mackintosh against Alexander Macdonell of Keppoch etc. Inverness. 1 June 1795.
17 NAS CS 237/Mc/13/3 (Ref. 15).
18 NAS DI.64/15 Leaf 12 (Ref. 16).
19 NAS GD 128/20/1/17 Collector Campbell, Fort William, to Campbell Mackintosh, Writer, Inverness. 12 May 1789.
20 NAS CS 237/Mc/13/3 (Ref. 15).
21 NAS GD 128/12/4 Cluny Macpherson to Æneas Mackintosh of Mackintosh. 26 July 1786.
22 NAS GD 128/36/5 Donald McDonald, Tulloch, to Campbell Mackintosh, Writer, Inverness. 29 July 1786.

23 NAS GD 176/1261/10 Information for the Laird of McIntosh relating [to] his woods in Lochaber which being observed by me, Donald McDonald fforester there . . . 1786. Also, a Memorandum probably written by Donald McDonald in 1794. (Second page missing.)
24 NAS GD 128/41/6 Donald McDonald, Tulloch, to Campbell Mackintosh, Writer, Inverness. (Written from Blairnahinvin.) 26 February 1791.
25 NAS GD 128/30/4/46 Donald McDonald, Tulloch, to Campbell Mackintosh, Writer, Inverness. 18 April 1794.
26 NAS GD 128/54/9/43 Donald McDonald, Tulloch, to Campbell Mackintosh, Writer, Inverness. 18 May 1794.
27 NAS GD 128/54/11/43 Donald McDonald, Tulloch, to Campbell, Mackintosh, Writer, Inverness. 26 June 1795.
28 NAS GD 128/54/11/64 Donald McDonald, Tulloch to Campbell Mackintosh, Writer, Inverness. 19 December 1795.
29 NAS GD 128/54/12 Donald McDonald, Tulloch, to Campbell Mackintosh, Writer, Inverness. 29 January 1795. (Written from Highbridge.)
30 NAS GD 128/51/14/7 Donald McDonald, Tulloch, to Campbell Mackintosh, Writer, Inverness. 14 March 1796.
31 NAS GD 176/2106/7 John McGillivanich to Æneas Mackintosh, Moyhall. 1799.
32 NAS GD 176/2103 Donald McDonald (Tulloch) to Æneas Mackintosh. (Written in Inverness.) 4 December 1792.
33 NAS GD 176/1261/10 (Ref. 23).
34 NAS SC 29/7/2 Decreet of Removing. Æneas Mackintosh of Mackintosh Esq. Proprietor etc . . . against tenants in Lochaber. 14 April 1795.
35 R. & W. Bell, *Dictionary of the Law of Scotland*, 3rd Ed., Edinburgh, 1826, Vol.II, 398–99; *The Pocket Lawyer*, By a Member of the Faculty of Advocates, Edinburgh, 1830, 100–01; J. Erskine, *An Institute of the Law of Scotland*, A New Edition by J.B. Nicolson, Advocate, Edinburgh, 1871, Book II, Title 6, §45 & 46.
36 NAS GD 128/13 Bundle 2. Item 20. Memoir for . . . Lovat relative to the General Removing of the Tenants over the whole estate. 1803.
37 NAS SC 29/7/2 (Ref. 34).
38 NAS SC 28/16/3 Decreets of Removal. Alexander Mackintosh, Keppoch against Tenants in Inverroymore, Inverroybeg and Boline. 1802.
39 NAS GD 176/1522 Rental of Estates belonging to Æneas Mackintosh Esq. of Mackintosh as payable at Martinmas 1801 . . . The Lochaber Estate.
40 NAS RD4/285/603 Bond of Credit. Donald McDonald to the Bank of Scotland. March 1796. (Recorded 8 March 1809.)
41 *Fasti Ecclesiae Scoticanae*, Vol.6, Edinburgh, 1926, 423.
42 NAS GD 128/36/8 & GD 128/47/4 Sheep Farm Association.
43 HCA/D456/2 Donald McDonald, Tulloch, to Campbell Mackintosh, Writer in Inverness, 17 April 1796.
44 NAS GD 176/2106 Rev. Thomas Ross, Kilmanivaig, to Campbell Mackintosh, Writer, Inverness. 16 April 1800.
45 HCA L/INV/SC/8/9 Precognition and Declaration. Ensign Angus Macdonald and Others. Fort William. 1783.
46 NAS GD 176/1415 (Ref. 1).
47 HCA/D456/2 (Ref. 43).
48 NAS GD 128/46/3 Donald McDonald, Tulloch, to Campbell Mackintosh, Writer, Inverness. 16 January 1794. (Decreet of 6 December 1792 enclosed.)
49 HCA L/INV/SC/8/9 (Ref. 45).

50 NAS GD 128/45/10 Angus McDonald of Kincreggan to Alexander Macdonell, Writer, Inverness. 18 January 1803.
51 NAS GD 176/2127/6 Campbell Mackintosh to Æneas Mackintosh. 24 May 1802.
52 NAS GD 128/65/11 John MacDonald of Borrodale to Alexander Macdonell, Writer, Inverness. 26 March 1810.
53 HCA/D456/2 (Ref. 43).
54 NAS GD 128/12/5 Alexander Mackintosh (Factor) to Campbell Mackintosh, Writer, Inverness. 5 May 1796.
55 NAS SC 28/16/1 Petition for William Mitchell, Achnadaul. 1799. (Settled 6 August 1799.)
56 NAS CS 236/C/7/15 Removing – Cameron of Locheil against Alexander MacDonald. 1804. (Most papers missing.)
57 NAS SC 29/7/3 (Ref. 5).
58 NAS GD 176/2105 Donald McDonald (Tulloch) to Æneas Mackintosh, 30 December 1798. (Written from Dalnaderg.)
59 NAS GD 128/12/5 Æneas Mackintosh of Mackintosh to Campbell Mackintosh, Writer in Inverness. 5 January 1799.
60 NAS GD 128/45/2 Donald McDonald, Tulloch, to Campbell Mackintosh, Writer, Inverness. (Written from Dalnaderg.) 14 February 1799.
61 NAS GD 128/45/2 Donald McDonald, Tulloch, to Campbell Mackintosh, Writer, Inverness. (Written from Dalnaderg.) 26 February 1799.
62 NAS GD 128/45/5 Donald McDonald, Tulloch, to Campbell Mackintosh, Writer, Inverness. (Written from Dalnaderg.) 5 March 1799.
63 NAS SC 29/10/1284 Decreet of Removing. Mackintosh and McDonald agt. Ensign McDonell etc. Extracted 5 April 1799.
64 NAS GD 176/2106/5 Donald McDonald, Tulloch, to Campbell Mackintosh, Writer, Inverness. 17 May 1799.
65 NAS SC 28/16/1 Decreet of Removal. Duke of Gordon etc. against tenants (includes George Cameron of Letterfinlay, Tacksman of the Town and Lands of Claggan). 1799.
66 NAS GD 128/45/2 (Ref. 60).
67 NAS GD 128/45/3 Donald McDonald, Tulloch, to Campbell Mackintosh, Writer, Inverness. 10 May 1799.
68 NAS SC 29/7/2 (Ref. 34).
69 NAS SC 29/7/3 Decreet – Donald McDonald against Donald MacKillop etc. Inverness 22 November 1798. Extracted 11 December 1798.
70 *Ibid.*
71 NAS GD 176/2105 (Ref. 58).
72 NAS GD 128/12/5 (Ref. 59).
73 NAS GD 128/12/5 Æneas Mackintosh of Mackintosh to Campbell Mackintosh, Writer in Inverness. 25 February 1799.
74 NAS GD 128/45/3 (Ref. 67).
75 NAS GD 128/45/5 Donald McDonald, Tulloch, to Campbell Mackintosh, Writer, Inverness. 12 June 1799.
76 NAS GD 128/12/5 Æneas Mackintosh to Campbell Mackintosh, Writer, Inverness. 27 August 1799.
77 NAS GD 176/2106 (Ref. 6).
78 NAS GD 176/2106/8 Donald McDonald, Tulloch, to Campbell Mackintosh, Writer, Inverness. 24 March 1800.

79 NAS GD 176/2106 (Ref. 44).
80 NAS GD 128/12/5 (Ref. 59).
81 NAS GD 128/12/5 (Ref. 73).
82 NAS GD 176/2106 Donald McDonald, Tulloch, to Campbell Mackintosh, Writer, Inverness. 28 April 1800.
83 NAS GD 176/2127/8 Donald McDonald, Tulloch, to Æneas Mackintosh. Written at Moyhall, 1 June 1802.
84 NAS GD 176/2127/11 Donald McDonald, Tulloch, to Æneas Mackintosh. 16 June 1802.
85 NAS GD 176/1522 (Ref. 39).
86 NAS GD 176/2106 (Ref. 6).
87 NAS GD 128/51/14 Dr. McDonald, Fort Augustus, to Campbell Mackintosh, Writer, Inverness. 25 March 1796.
88 NAS GD 128/54/12/22 Dr. McDonald Fort Augustus to Alexander Macdonell, Writer, Inverness. 17 June 1795.
89 NAS GD 176/1282/12 Offer by Angus and Alexr. McDonells for Crenachan and Blarnahinvine also for different farms. 16 June 1802.
90 NAS GD 176/862 Alexander McDonell to Alexander Mackintosh, Cantraydoun. 25 January 1795. (Written from Keppoch.)
91 NAS GD 176/862/2 Mrs. Sarah Campbell, Mayfield, near Edinburgh, to Æneas Mackintosh, 19 February, 1795.
92 NAS GD 176/862 Alexander McDonell to Alexander Mackintosh, Cantraydoun. 16 March, 1795. (Written from Keppoch.)
93 NAS GD 176/862/7 Alexander Campbell, Mayfield, near Edinburgh, to Æneas Mackintosh, 26 June 1795.
94 NAS GD 176/1522 (Ref. 39).
95 NAS DI.64/15 Leaf 12 (Ref. 16).
96 NAS GD 128/12/5 Charles Mackintosh, Edinburgh to Campbell Mackintosh, Writer, Inverness. 28 May 1796.
97 NAS GD 128/12/5 Alexander Mackintosh, Fort William to Campbell Mackintosh, Writer, Inverness. 15 June 1796.
98 NAS GD 176/1325 Memorial of Alexander McIntosh, once Tacksman of Keppoch. (To Alexander MacKintosh of MacKintosh.) 1834.
99 NAS GD 176/2105 Æneas Mackintosh to Mr. Tod. 21 January 1802.
100 NAS GD 176/1325 (Ref. 98).
101 NAS GD 176/1522 (Ref. 39).
102 NAS GD 176/1519 Rentals of Lochaber 1796–97 & 1810–19.
103 NAS GD 176/1430/1Tack Æneas Mackintosh Esq. to Alexander Macdonald Esq. 1802.
104 NAS GD 128/51/14 Alexander Macdonald of Glenco to Campbell Mackintosh. 13 May 1796.
105 B.D. Osborne, *The Last of The Chiefs, Alasdair Ranaldson Macdonell of Glengarry 1773-1828*, Glendaruel, 2001, 122–24. I am grateful to Mr Osborne for additional information which he kindly provided by letter.
106 NAS RD3/329/248 Lease by Mr Alex. McTurk to Glengarry & Glencoe and Bond and Obligation, 21 October 1807. (Recorded 15 July 1809.)
107 HCA/D456/4/4 State of the Sums due by Glencoe For Garrygualoch Tenants To Alexander Macdonell, Writer in Inverness, 1809; HCA/D456/2/2 Alexander Macdonald of Glencoe to Alexander Macdonell, Writer in Inverness (Anent Garrygualoch stock). 1811.

108 NAS GD 176/2123/8 Alexander Macdonald of Glencoe to Campbell Mackintosh, Writer in Inverness. 4 January 1814.
109 NAS GD 176/2127/11 Donald McDonald, Tulloch, to Æneas Mackintosh. 16 June 1802.
110 NAS GD 176/990/9 Legal Papers in Dispute with Glencoe trustees over Keppoch, etc. Examination and Survey of house and offices on Keppoch. Alex. Simpson and Donald Cameron. 3 May 1823.
111 HCA. Burgh Court of Inverness. Warrants Bundle 30–16. Contract of Sale of Oakwood Between Campbell Mackintosh for Simon Fraser Esq. and Mr. Alexander Mackintosh. Inverness. 20 June 1802.
112 NAS GD 128/54/1 Donald McDonald, Tulloch, to Campbell Mackintosh, Writer, Inverness. 28 August 1796. (Written from Fort Augustus.)
113 NAS GD 176/2105 (Ref. 58).
114 NAS GD 176/2105 (Ref. 58).
115 NAS GD 128/12/5 (Ref. 59).
116 NAS GD 176/2117/16 Æneas Mackintosh to Campbell Mackintosh, Writer in Inverness. 8 April 1805.
117 NAS GD 176/2123/8 Duncan Mackintosh, Forrester, Lochaber, to Campbell Mackintosh, Writer in Inverness, about fencing. 4 January 1814.
118 NAS CH2/433/3 Parish of Kimanivaig. Births and Marriages.
119 NAS GD 128/45/3 (Ref. 67).
120 NAS SC 28/16/2 Petition. Neil McIntyre, Procurator Fiscal, Fort William District, agt. Dond. McDonald. 1801.

Notes to Chapter 5

1 S. Macmillan, *Bygone Lochaber*, Glasgow, 1971, 122.
2 NLS Acc 11910/28 Donald Cameron, London, to Ewen Cameron, Fassifern. 6 April 1786.
3 NAS GD 170/1865 Campbell of Barcaldine Papers. Alexander Macdonald of Glencoe. Estate Affairs.
4 C. Fraser-Mackintosh, *Antiquarian Notes*, Second Series, Inverness, 1897, 208.
5 J. Mitchell, *Reminiscences of my Life in the Highlands*, 1884 (Reprint Newton Abbott 1971), Vol.1, 187.
6 *Ibid.*, Vol.1, 188 (footnote).
7 S. Macmillan, 1971, *op. cit.* (Ref. 1), 122.
8 NAS GD 128/51/14 Alexander Macdonald of Glenco to Campbell Mackintosh 13 May 1796.
9 S. Macmillan, 1971, *op. cit.* (Ref. 1), 182.
10 T. Garnett, *Observations on a Tour through the Highlands and part of the Western Isles of Scotland*, London, 1811, Vol.I, 296.
11 NLS. Acc. 9174/15 Tack betwixt John Cameron of Fassfern and Allan Cameron in Inverscaddle. 1742.
12 Loretta R. Timperley, *A Directory of Land Ownership in Scotland c1770*, Scottish Record Society, New Series 5, 1976, 29.
13 S. Macmillan, 1971, *op. cit.* (Ref. 1), 140.
14 *OSA* (Reprint), Wakefield, Vol.XVII, 1983, 126.
15 A.C. Fraser, *The Book of Barcaldine*, London, 1936, 109.
16 *Ibid.* 104–6.
17 NAS GD 170/1865 (Ref. 3).

18 *NSA*, Edinburgh & London, 1845, Vol.VII, 473.
19 NAS GD 170/1865 (Ref. 3).
20 NAS GD 202/35 Macdonald of Glenco – Trustee Accounts. 1815–1835.
21 I.F. Grant, *Highland Folk Ways*, London, 1961, 126 (including footnote 4).
22 NLS. Acc 9174/15 (Ref. 11).
23 NAS SC 29/57/10 Factory – the Guisdale Tutors to Alexander McDonell, Writer, Inverness. 1794.
24 NAS SC 29/7/4 p.49. Steelbow contract 14 June 1792 and 15 November 1792 between Allan MacDonald of Gallovie and Peter MacNab Tenant at Ballicholish and John and Robert MacNabs his sons.
25 NAS CS 237/Mc/13/3 McCaul v. McDonald. Summons of Reduction 1817.
26 Barbara Fairweather & D.C. Cargill, *Pre-1855 Tombstone Inscriptions on Isla Munda, The Burial Island in Loch Leven*, The Scottish Genealogical Society, Entries 100 & 101.
27 A.C. Fraser, 1936, *op. cit.* (Ref. 15), 112–15.
28 NAS SC 28/16/8 Alexander Macdonald Esq. v. John Rankin. 1808.
29 A. Mackenzie, *History of the Clan Cameron*, Inverness, 1884, 395–96.
30 S. Macmillan, 1971, *op. cit.* (Ref. 1), 22.
31 NAS DI.23/8 Folio 88. Horning. MacColl against Rankines – No 177. Inveraray 24 May 1804.
32 Loretta R. Timperley, 1976, *op. cit.* (Ref. 12), 40.
33 NAS DI.23/8 (Ref. 31).
34 NAS CS 29/16 June 1780/Kennedy v. McPherson. (This process was identified from an entry in D. Dobson, *Scottish American Court Records 1773–1783*, Baltimore, 1991, 40.)
35 G.A. Dixon, 'Bible and Stockings Provide Basic Lessons', in *Strathspey and Badenoch Herald*, 7 November 1991, 8.
36 *Ibid.*
37 NAS GD 44/51/14/1 Badenoch and Lochaber Miscellaneous Papers. Discharge – Schoolmaster of Kingussie his Salary Mart. 1780.
38 NAS GD 44/43/120 William Tod to James Ross. 12 May 1774.
39 D. Dobson, *The Original Scots Colonists of North America*, Baltimore, 1989, 150.
40 NAS RD4/220/793 Factory John Kennedy to A. Clarke. 1776.
41 NAS CS 29/16 June 1780/Kennedy v. McPherson.
42 NAS DI.64/14 Folio 260. Horning & Poinding. Kennedy against McPherson. 27 July 1784.
43 *Ibid.*
44 NAS GD 44/27/17 The Memorial of Hugh McPherson, Tacksman of Ovie, to His Grace the Duke of Gordon. 26 April 1773.
45 NAS DI.64/14 (Ref. 42).
46 NAS SC 29/10/780 Capt. John Macpherson v. John Kennedy 1784.
47 C. Fraser-Mackintosh, 1897, *op. cit.* (Ref. 4), 71.
48 T. Sinton, *By Loch and River, being Memories of Loch Laggan and Upper Spey*, Inverness, 1910, 181.
49 NAS RS38/12/45 Particular Register of Sasines, Inverness, Liferent Sasine in favour of Anne Mackpherson. 17 January 1776. (Describes her as eldest lawful daughter of Hugh Macpherson of Ovie and mentions a Contract of Marriage with John Macpherson of Inverhall of 28 November 1761.)
50 NAS GD 44/27/10 Items 98x & 101, letters from William Tod of 8 & 28 November 1770 referring to the death of John Macpherson of Inverhall.

51 Scarlett, Meta H., *In the Glens Where I Was Young*, Moy, Inverness-shire, 1988, 125.

52 *Edinburgh Evening Courant* (Quoted in *Creag Dhubh*, The Annual of the Clan Macpherson Association, 1971, Number 23, 430.)

53 NAS SC 29/10/780 (Ref. 46).

54 NAS SC 29/10/780 (Ref. 46).

55 NAS GD 128/9/2/97 Rev. John Kennedy to Alexander MacDonell, Writer in Inverness. 1 August 1798.

56 NAS SC 29/10/1285 Removing – Rev. John Kennedy against Sundries. 1799.

57 NAS GD 128/13/Bundle 7: Item 50. Answers for Lovat to Petition of Robert Oliver at Cullachy. 1800.

58 HCA Burgh and County of Inverness. Property (Private) Tacks, Feus, etc: P/INV/2/274 Auchteraw, Inchnacardoch, Cullachy, etc. Archibald Fraser of Lovat to Lieutenant Evan Macpherson and Rev. John Kennedy. 1786.

59 NAS CS 29/16 June 1780/Kennedy v. McPherson.

60 HCA P/INV/2/274 (Ref. 58.)

61 C. Fraser-Mackintosh, 1897, *op. cit.* (Ref. 4), 71.

62 NAS GD 44/27/17 (Ref. 44).

63 NAS GD 128/49/10/10 Answers for John MacDonell and others to the Petition of William Fraser of Garthmore. 1814.

64 NAS GD 44/27/12/2 Hugh McPherson to James Ross. 3 January 1779. (Written from Cullachy.)

65 NAS SC 29/53/10 Declaration of Mary McDonell. 1781.

66 NAS SC 29/7/1 Decreet. Donald McRae in Glenbenchar v. Hugh Macpherson of Ovie now at Etterish. 23 August 1782. Extracted 21 October 1782.

67 NAS JC 11/34 Circuit Court. Northern Circuit. Inverness, May 1783.

68 NAS GD 128/49/10/10 (Ref. 63).

69 HCA P/INV/2/274 (Ref. 58).

70 Anne Grant, *Letters from the Mountains*, Sixth Edition, London, 1845, Vol.I, 114 & 150 (Footnote).

71 *Ibid.* Vol.I, 149 (Footnote) & 150 (Footnote).

72 *Fasti Ecclesiae Scoticanae*, Edinburgh, 1928, Vol.VII, 175.

73 Alison Mitchell, *Pre-1855 Gravestone Inscriptions from Speyside*, Scottish Genealogical Society, Revised Edition 1992, 4.

74 Anne Grant, 1845, *op. cit.* (Ref. 70), Vol.I, 218 (Footnote).

75 NAS RD3/291/334 Trust Disposition and Settlement by the Rev. John Kennedy. 1801.

76 NAS RD2/275/581 Heritable Bond. Alexander Macdonald of Glencoe to Rev. John Kennedy, Auchteraw. 1789.

77 J.R.N. Macphail (Ed). *Letters written by Mrs Grant of Laggan concerning Highland Affairs and Persons connected with the Stuart Cause in the Eighteenth Century*, Scottish History Society, 1896, Vol.26, 278–79.

78 NAS SC 29/7/5 Decreet of Removing. Lovat against Oliver at Cullachy. 19 August 1801.

79 Anne Grant, *Letters from the Mountains*, London, 1806, 82-83.

80 NAS GD 128/13/Bundle 7: Item 50. (Ref. 57).

81 NAS DI.8/209 Folio 485 Edinburgh 22 April 1794.

82 *Ibid.*

83 NAS GD 128/9/1/135 Account – Alexander MacDonell, Writer in Inverness. 1806.

84 NAS GD 128/54/11/5 Dr. Donald McDonald, Fort Augustus, to Campbell Mackintosh, Writer, Inverness. 20 March 1795.
85 NAS RD2/275/597 Sub-tack – Kennedy and McDonald. 1798.
86 NAS GD 128/9/2/97 (Ref. 55).
87 NAS RD3/291/334 (Ref. 75).
88 NAS GD 128/59/1 A. Simon Fraser to Campbell Mackintosh, Writer in Inverness. July 1794; Arch. Fraser of Lovat to Alexander Macdonell, Writer in Inverness, 24 January 1800; Robert Dundas, Edinburgh, to Alexander Macdonell, Inverness, 27 February 1801.
89 NAS GD 128/59/2 Robert Dundas, Edinburgh, to Alexander MacDonell, Writer, Inverness. 2 November 1805.
90 NAS GD 128/9/1/135 (Ref. 83).
91 NAS CS 17/1/28 Court of Session Minute Book. Tho. Clark late tacksman of Drimandrochit charged AG the Hon. Arch. Fraser of Lovat. 17 December 1808.
92 NAS GD 128/59/1 John MacKay, Fort Augustus, to Alexander MacDonell, Writer in Inverness. 7 March 1801.
93 NAS GD 128/13/Bundle 9: Item 18. Mr. Kennedy's Heirs keeping violent possession of Auchteraw. 27 May 1807.
94 NAS GD 128/9/2/11 Coll Macdonald WS. to Alexander MacDonell, Writer in Inverness. 1809.
95 *Ibid.*
96 NAS GD 128/59/3 James S. Robertson WS, Edinburgh, to Alexander MacDonell, Writer in Inverness. 24 January 1816.
97 A.G. & Margaret H. Beattie, *Pre-1855 Gravestone Inscriptions in Inverness District West*, The Scottish Genealogy Society, 1993, 8 (Entry 43).
98 NAS SC 29/7/8 Decreet of Removing. Lovat and Alexander Fraser of Dell his factor v. Evan Macpherson sometime residing at Inchnacardoch etc.14 April 1806.
99 NAS RD3/291/334 (Ref. 75).
100 NAS RD4/255/1225 Bond of Credit Lieut. Evan McPherson and others to British Linen Company. 18 March 1789. Registered 5 June 1794.
101 NAS GD 80/695 Note marked E. Macpherson 5 April 1790.
102 NAS GD 128/54/7/25 Walter Hog, Manager, British Linen Company, Edinburgh to Campbell Mackintosh, Writer, Inverness. 3 August 1793.
103 NAS DI.8/209 Folio 786 Edinburgh 22 April 1794.
104 NAS DI.64/14 Folio 524. Inhibition. Simpson against McPherson. Inverness 6 June 1794.
105 A.R.B. Haldane, *The Drove Roads of Scotland*, Edinburgh, 1952, 48.
106 NAS DI.8/209 Folio 776 Edinburgh 30 June 1794.
107 NAS GD 128/44/5 Captain Charles McPherson to Campbell Mackintosh. From Auchteraw, 3 May 1794.
108 NAS GD 128/44/5 Captain Charles McPherson to Campbell Mackintosh. From Gordonhall, 18 May 1794.
109 NAS GD 128/44/5 Captain Charles McPherson to Campbell Mackintosh. From Auchteraw. 23 May 1794.
110 NAS GD 128/54/9/46 Captain Evan Macpherson of Cullachy. Printed statement of affairs. Edinburgh 2 June 1794.
111 NAS GD 128/44/5 Captain Charles McPherson Gordonhall, to Campbell Mackintosh 28 June 1794.
112 NAS DI.8/209 Folio 817 Edinburgh 1794.
113 NAS DI.8/209 Folio 861 Edinburgh 21 July 1794.

114 NAS RD3/268/160 Articles and Minutes of Roup of the Sub-leases and Sheep-stocking [etc] on the farm of Cullachy. 17 October 1794.
115 NAS RD3/268/171 Excerpt from the Inventory of Lieut. Macpherson's subject. Inventory taken by the factor appointed by the creditors.
116 NAS CS 236/M/13/7 Bankruptcy of Lieut. Evan Macpherson, Cullochy. Petition for Alexander Stewart Esq. of Achnacon For the Division of Funds. 1796.
117 NAS SC 29/57/10 Lease Twixt The Honorable Archibald Fraser of Lovat and Thomas and Robert Oliver. 1802.
118 C. Fraser-Mackintosh, *Letters of Two Centuries*, Inverness, 1890, 329.
119 NAS GD 128/13 (Bundle 7): Item 52. Bill of Suspension by Robert and Thomas Oliver. 1801.
120 NAS SC 29/7/2 Decreet of Removal. Duke of Gordon etc. and William Tod his Factor against Lieutenant Evan Macpherson at Cullachie, Principal Tacksman of Crathecroy, etc. Inverness. 21 May 1794.
121 NAS SC 29/7/5 Decreet of Removal. Alexander MacDonald, Tacksman of Garva against tenants in Kylarchill and Crathycroy. Inverness. 16 April 1799.
122 NAS GD 128/13 (Bundle 7): Item 52. (Ref. 119).
123 NAS SC 29/57/10 (Ref. 117).
124 NAS SC 29/56/1/75 Renunciation of Tack of Cullachy in favour of Robert Oliver. 1809.
125 NAS SC 28/16/9 Sequestration. Petition for His Grace the Duke of Gordon and Factor agt. James MacBarnett. 1809.
126 NAS GD 44/25/9/15 Robert Oliver offers for Torlundy etc. 11 March 1809.
127 NAS SC 29/7/5 (Ref. 78).
128 D. Stewart, *Sketches of the Character, Manners, and Present State of the Highlanders of Scotland*, Edinburgh, 1822 (Reprinted 1977), Appendix P, Vol.2, lxx–lxxiii.
129 NAS SC 29/7/5 (Ref. 78).
130 NAS CS 236/S/7/8 Messrs. Duff & MacPherson etc. Petition for a Meeting to Chuse a Trustee. 1804.

Notes to Chapter 6

1 C. Fraser-Mackintosh, 'The Macdonalds of Achtriachtan', in *Transactions of the Gaelic Society of Inverness*, Vol. XXIII, 1898–99, 136–45.
2 NAS CS 234/R/7/4 Petition. A. Rankine against Lord Polkemmet's Interlocutor refusing Bill of Suspension. 1811.
3 S. Macmillan, *Bygone Lochaber*, Glasgow, 1971, 45–46.
4 Burke's Landed Gentry of Great Britain, London. 1921.
5 R. & W. Bell, Dictionary of the Law of Scotland, Third Edition, Edinburgh, 1826, Vol.II, 115.
6 NAS RS3/1034/49 The Trustees of Adam Macdonald of Achtriachtan [Transfer of lands owned by Adam Macdonald to his Trustees]. 1816.
7 *Burke's Landed Gentry of Great Britain*, London. 1921.
8 NAS CS 36/31/12 Decreet of Preference and for Payment. John Cameron and James Grant his Attorney against Robert Downie etc . . . and the said Robert Downie and others against The Trustees and Creditors of Adam Macdonald. 16 December 1820.
9 Loraine Maclean of Dochgarroch, *The Indomitable Colonel*, London, 1986, 269.
10 J. Stewart of Ardvorlich, *The Camerons*, Stirling, 1974, 225.

11 NAS DI.23/8 Folio 49. Inhibition. Marquis of Tweeddale against McDonald (Adam Macdonald of Achtriachtan) No 112. Inveraray 29 May 1801.
12 NAS CS 238/T/6/42 Marquis of Tweeddale against Macdonald 1801–1809. (Inventory only.)
13 NAS CS 36/31/12 (Ref. 8).
14 C. Fraser-Mackintosh, 1898–99, *op. cit.* (Ref. 1), 136–45.
15 NAS CS 36/31/12 (Ref. 8).
16 Sasine Abridgements Argyll. 14 August 1801. (1296)
17 C. Fraser-Mackintosh, 1898-99, *op. cit.* (Ref. 1), 136–45.
18 NAS CS 36/31/12 (Ref. 8).
19 NAS RD5/40/545 Bond, Disposition, and Security – Adam Macdonald to Roderick Mackenzie. 1809 (recorded 1813).
20 A. Mackenzie, *History of the Clan Mackenzie*, Inverness, 1894, 490–91.
21 *Burke's Landed Gentry*, 1921, Entry for Grant of Glenmoriston.
22 Sasine Abridgements Argyll. 4 April 1809. (1913.)
23 NAS CS 234/R/7/4 (Ref. 2).
24 NAS RD5/18/767 Minute of Sale – Adam McDonald Esq. of Achtriachtan to Coll Macdonald of Dalness Esq. WS. 1812.
25 NAS CS 236/M/29/9 Glenco Trustees v. Rankin etc. 1818.
26 NAS RD5/18/767 (Ref. 24).
27 C. Fraser-Mackintosh, 1898–99, *op. cit.* (Ref. 1), 136–45.
28 Sasine Abridgements Argyll. 15 February 1813. (2247)
29 Sasine Abridgements Argyll. 16 October 1816. (2613)
30 J. Mitchell, *Reminiscences of my Life in the Highlands*, 1884 (Reprint Newton Abbott 1971), Vol.1, 179.
31 NAS Sasine Abridgements Argyll. 16 October 1817. (2752)
32 NAS CS 36/31/12 (Ref. 8).
33 Sasine Abridgements Argyll. 7 June 1817. (2686)
34 Sasine Abridgements Argyll. 19 August 1817. (2719)
35 NAS SC 54/3/1 Petition of Mary McDonald against Sub-Tenants of Kaolisnacon. 1801/1802.
36 Sasine Abridgements Argyll. 27 November 1817. (2790)
37 Sasine Abridgements Argyll. 23 March 1818. (2854)
38 A. Mackenzie, 1894, *op. cit.* (Ref. 20), 490–91.
39 Sasine Abridgements Argyll. 15 May 1818. (2871)
40 Sasine Abridgements Argyll. 14 March 1818. (2852)
41 NAS CS 36/31/12 (Ref. 8).
42 NAS CS 36/31/12 (Ref. 8).
43 NAS CS 36/31/12 (Ref. 8).
44 NAS CS 36/31/12 (Ref. 8).
45 NAS CS 36/31/12 (Ref. 8).
46 NAS CS 36/31/12 (Ref. 8).
47 NAS SC 54/3/4 Petition – Lieut. Donald Campbell, late tenant in Leckintuim now residing in Fort William versus Rankins. 1817.
48 NAS CS 36/31/12 (Ref. 8).
49 Sasine Abridgements Argyll. 21 March 1818. (2853)
50 Sasine Abridgements Argyll. 30 April 1822. (169)
51 NAS RD5/63/171/294 Trust Disposition etc. Alexander Macdonald of Glencoe. 1804.

52 NAS RD5/41/138 Will and Testament of Allan Cameron, Tacksman of Inverscaddle. 8 May 1810.
53 Sasine Abridgements. Argyll. 9 February 1815 (2422).
54 NAS CS 228/Mc/13/1 MacDonalds Drimintorran. MacDonald agt. MacDonald. 1808. (Includes papers from a later process, Angus MacDonald v Alexander MacDonald of Dalilea, 1825.)
55 NAS GD 202/35 Macdonald of Glenco – Trustee Accounts. 1815–1835.
56 NAS CS 232/M/47/7 Glenco Trustees v. Ewen Macdonald etc. 1826–1831.
57 Sasine Abridgements Argyll. 24 October 1811 (2131).
58 Sasine Abridgements Argyll. 24 October 1811 (2133).
59 NAS RD5/37/351/219 Alexander Macdonald to Miss Janet Macpherson. 1813.
60 A. Mackenzie, *History of the Clan Cameron*, Inverness, 1884, 282.
61 A. & A. Macdonald, *The Clan Donald*, Inverness, 1904, Vol.III, 644.
62 NAS RD5/63/171/294 (Ref. 51).
63 V.C.P. Hodson, *List of Officers in the Bengal Army 1758–1834*, London, 1927–1947 (4 Volumes), Entry for Colin Macdonald.
64 D.G. Crawford, *Roll of the Indian Medical Service 1650–1830*, London, 1930, (Entry 650).
65 *Ibid.* 1930, Entry 650.
66 East India Register and Directory.
67 V.C.P. Hodson, 1927–1947 *op. cit.*, (Ref. 63).
68 NAS GD 1/736/56 Major John Cameron to Ewen Cameron. 22 June 1805.
69 East India Register and Directory.
70 E. Dodwell and J.S. Miles, *Alphabetical List of Officers of the Indian Army 1760–1834*, London, 1938, Bengal Presidency, 182–83.
71 V.C.P. Hodson, 1927–1947 *op. cit.* (Ref. 63).
72 NAS GD 202/73/2 Report by John Taylor, Surgeon in Fort William on Lieut. Colin Macdonald. 16 October 1816.
73 A.G. MacDonald, 'The Last of the Glencoe Chiefs?', in *Clan Donald Magazine*, No.7, Edinburgh, 1977, 52–57.
74 NAS GD 202/35 (Ref. 55).
75 NAS GD 202/35 (Ref. 55).
76 NAS SC 54/3/5 Trustees of Alexander Macdonald late of Glencoe v. Donald McMaster in Invereegan. 1817.
77 NAS GD 202/35 (Ref. 55).
78 C. Fraser-Mackintosh, 'The Camerons of Letterfinlay', in *Transactions of the Gaelic Society of Inverness*, Vol.XVII, 1890–91, 31–45.
79 NAS GD 202/35 (Ref. 55).
80 NAS SC 54/3/8 Sequestration – Robert Downie Esq. of Appin v. Donald MacMaster, Innerigan. 1825.
81 NAS GD 202/35 (Ref. 55).
82 NAS GD 202/35 (Ref. 55).
83 J. Barron, *The Northern Highlands in the Nineteenth Century*, Inverness, 1903, Vol.I, 114.
84 Inverness Courier, 2 May 1817. (Quoted in J. Barron, The Northern Highlands in the Nineteenth Century, Inverness, 1903, Vol.I.)
85 NAS GD 202/35 (Ref. 55).
86 NAS RD5/127/192 Factory – Appointment of Donald MacDonald of Drimintorran. 1817.

87 NAS GD 202/35 (Ref. 55).
88 NAS GD 202/35 (Ref. 55).
89 NAS GD 202/35 (Ref. 55).
90 NAS GD 202/35 (Ref. 55).
91 NAS GD 202/73/3 Robert Scott, Edinburgh, to Ewen Macdonald. 13 November 1821.
92 *Ibid.*
93 NAS SC 28/16/15 Malcolm Robertson, Merchant, Tayphuirst, Glencoe, v. Debtors. 1817.
94 NAS GD 202/35 (Ref. 55).
95 NAS SC 54/3/7 Sequestration. The Trustees of the late Alexander Macdonald of Glencoe v. Malcolm Robertson, Merchant, Tayfuirst, Glencoe. 1823.
96 NAS GD 202/35 (Ref. 55).
97 NAS GD 202/57 Rental of the Estate of Glencoe, 1833 Collection.
98 NAS GD 202/57 Account between the Trustees of A. Macdonald Esq. of Glencoe and Donald MacDonald their Factor for 1821 and 1822.
99 NAS GD 202/57 (Ref. 97).
100 NAS GD 202/57 James Macgregor, Solicitor in Fort William to Ewen Macdonald. 7 January 1837.
101 Barbara Fairweather, *A Short History of Ballachulish Slate Quarry*, North Lorn and Glencoe Folk Museum. Undated.
102 NAS CS 236/M/29/9 (Ref. 25).
103 NAS GD 202/57 (Ref. 98).
104 NAS GD 202/57 (Ref. 100).
105 C. Fraser-Mackintosh, *Antiquarian Notes*, Second Series, Inverness, 1897, 214.
106 J. Mitchell, 1884, *op. cit.* (Ref. 30), Vol.1, 335.
107 Barbara Fairweather & D.C. Cargill, *Pre-1855 Tombstone Inscriptions on Isla Munda, The Burial Island in Loch Leven*, The Scottish Genealogical Society. Entry 164.

Notes to Chapter 7

1 C.D. MacDonald, *A Short History of our MacDonald Family from 1730 to 1849*, unpublished, 1994.
2 NAS CS 96/722/1 Act Sequestrating the Estate and Effects of Donald MacDonald. 31 October 1837.
3 NAS CS 228/Mc/13/1 MacDonalds Drimintorran. MacDonald agt. MacDonald. 1808. (Includes papers from a later process, Angus MacDonald v. Alexander MacDonald of Dalilea, 1825.)
4 Barbara Fairweather & D.C. Cargill, *Pre-1855 Tombstone Inscriptions on Isla Munda, The Burial Island in Loch Leven*, The Scottish Genealogical Society, Entry 161.
5 NAS RS3/3319/57 Notarial Instrument. Ellen Caroline Macpherson Macdonald or Burns Macdonald. 20 November 1863.
6 NAS GD 44/43/173/44 Angus Macdonald of Achtriachtan to James Ross. 26 December 1776.
7 C.D. MacDonald, 1994, *op. cit.* (Ref. 1).
8 NAS CS 96/1372 John MacDonald of Drimintorran. Affairs.
9 NAS CS 236/M/24/3 Macdonald, Ronald etc. v. McDonald of Glencoe. 1810.
10 NAS CS 96/722/1 (Ref. 2).

11 C.D. MacDonald, 1994, *op. cit.* (Ref. 1).
12 NAS CS 96/1372 (Ref. 8).
13 NAS CS 96/722/1 (Ref. 2).
14 NAS CS 228/Mc/13/1 (Ref. 3).
15 NAS CS 228/Mc/13/1 (Ref. 3).
16 NAS RD2/297/390 Bond of Provision – McDonell to Campbell 25 April 1796.
17 NAS CS 228/Mc/13/1 (Ref. 3).
18 NAS GD 202/35 Macdonald of Glenco – Trustee Accounts. 1815–1835.
19 NAS CS 228/Mc/13/1 (Ref. 3).
20 NAS CS 228/Mc/13/1 (Ref. 3).
21 NAS RD2/266/580 Power of Attorney – Mrs. MacDonell etc. to Angus MacDonell (Achtriachtan) July 1796.
22 NAS CS 96/1372 (Ref. 8).
23 NAS CS 228/Mc/13/1 (Ref. 3).
24 NAS CS 228/Mc/13/1 (Ref. 3).
25 NAS CS 96/1372 (Ref. 8).
26 D. Stewart, *Sketches of the Character, Manners, and Present State of the Highlanders of Scotland*, Edinburgh, 1822 (Reprinted 1977), Appendix P, Vol.2., xxviii–xxix.
27 J.I. Robertson, *The First Highlander, Major-General David Stewart of Garth CB. 1786–1829*, East Linton, 1999, 94–95.
28 NAS CS 96/722/1 (Ref. 2).
29 C.D. MacDonald, 1994, *op. cit.* (Ref. 1).
30 NAS CS 96/722/1 (Ref. 2).
31 NAS CS 228/Mc/13/1 (Ref. 3).
32 NAS SC 54/3/2 Sequestration. Petition of Sir James Riddle of Ardnamurchan and Sunart, Baronet, etc. 1808.
33 NAS SC 29/51/3 Trial of Alexander Mackintosh, Tacksman of Bohinie . . . Inverness. 27 July 1807.
34 NAS CS 96/722/1 (Ref. 2).
35 NAS GD 202/35 (Ref. 18).
36 NLS. Acc 9174/1 Ewen Cameron of Fassifern to Duncan Cameron. 2 April 1819.
37 NAS CS 236/M/24/3 (Ref. 9).
38 NAS CS 96/722/1 (Ref. 2).
39 NAS CS 96/722/1 (Ref. 2).
40 NAS RD5/127/192 Factory – Appointment of Donald MacDonald of Drimintorran. 1817.
41 C.D. MacDonald, 1994, *op. cit.* (Ref. 1).
42 NAS CS 96/722/1 (Ref. 2).
43 NAS CS 228/Mc/13/1 (Ref. 3).
44 NAS CS 228/Mc/13/1 (Ref. 3).
45 NAS GD 202/35 (Ref. 18).
46 NAS CS 228/Mc/13/1 (Ref. 3).
47 C.D. MacDonald, 1994, *op. cit.* (Ref. 1).
48 NAS CS 96/722/1 (Ref. 2).
49 *NSA*, Edinburgh & London, 1845, Vol.VII, 148 & 152.
50 NAS CS 96/722/1 (Ref. 2).
51 E. Maciver, *Memoirs of a Highland Gentleman*, Edited by Rev. G. Henderson, Edinburgh, 1905, 67.

52 NAS CS 96/722/1 (Ref. 2).
53 C.D. MacDonald, 1994, *op. cit.* (Ref. 1).
54 *Testimonials in Favour of Donald MacDonald, Esq., Late of Drimintorran*, in possession of Norman H. MacDonald, President, Clan Donald Society of Edinburgh.
55 R. Somers, *Letters from the Highlands*, 1848 (Reprint Inverness 1977), 126–36 & 192.

Notes to Chapter 8

1 NAS GD 176/2127/2 Note of Lotts by Donald McDonald of M'Intosh's Lochaber lands for next Sett 1804. 4 February 1802.
2 NAS GD 176/2107 Donald McDonald, Tulloch, to Campbell Mackintosh, Writer, Inverness. 14 January 1803.
3 *Ibid.*
4 HCA BI/1/12/21 Burgh Court Books of Inverness. Tack between Æneas Mackintosh and Donald McDonald recorded 23 February 1804.
5 NAS GD 176/2107 (Ref. 2).
6 NAS GD 176/1282/14 Rental of Mackintosh's Estate in Lochaber according to the new sett as payable at Martinmas 1804.
7 NAS GD 176/1282/12 Offer by Angus and Alexr. McDonells for Crenachan and Blarnahinvine also for different farms. 16 June 1802.
8 NAS SC 29/57/11 Lease betwixt Patrick Grant of Glenmoriston and Scothouse . . . and Dr. Donald MacDonald and others. 1804.
9 NAS GD 176/1282/12 (Ref. 7).
10 NAS GD 176/2127/15 Angus McDonald late at Tullich to Campbell Mackintosh, Writer in Inverness. 23 July 1802.
11 NAS GD 176/2116/12 Æneas Mackintosh to Campbell Mackintosh, Writer in Inverness. 29 December 1802.
12 NAS GD 176/2117/16 Æneas Mackintosh to Campbell Mackintosh, Writer in Inverness. 8 April 1805.
13 NAS CR/8/7 Rev. J. Anderson to Thomas Macpherson, Abbey Street, Dublin. 21 November 1807.
14 NAS SC 28/16/8 Protested Bills against Donald McDonald, Tulloch. 1806–1808.
15 NAS CR/8/7 Rev. J. Anderson to Mr. Donald McDonald Tulloch. 12 December 1807.
16 NAS GD 128/9/1/14 Answers for Alex. Dow MacDonald to the Petition of Ronald MacDonell of Strathmashie (Draft) 1808.
17 NAS GD 176/2134/8 Campbell Mackintosh to Æneas Mackintosh 9 April 1808.
18 HCA BI/1/10/10 Trust Disposition and Deed of Settlement by Mr. Donald Macdonald Tacksman of Tullich. 31 August 1807. Recorded 4 April 1808.
19 NAS RS38/18/112 Disposition by Donald McDonald, Tulloch, to Campbell Mackintosh. 1 February 1808.
20 NAS GD 176/2108 Alexander Macdonald of Glencoe to Campbell Mackintosh, Writer, Inverness. 25 March 1808.
21 NAS GD 176/2108 Alexander Macdonald of Glencoe to Campbell Mackintosh, Writer, Inverness. 8 April 1808.
22 J. Barron, *The Northern Highlands in the Nineteenth Century*, Inverness, 1903, Vol.I, 9.
23 NAS GD 176/2134/8 (Ref. 17).
24 HCA BI/1/10/10 (Ref. 18).
25 NAS GD 128/9/1/14 (Ref. 16).

26 NAS CR/8/6 Rev. J. Anderson to Thomas Gillespie. 7 June 1808.
27 NAS GD 176/2115 Thomas Gillespie and James Greig to Campbell Mackintosh, Writer in Inverness. 8 January 1812.
28 NAS GD 176/2116/12 (Ref. 11).
29 HCA BI/1/12/21 (Ref. 4).
30 NAS CS 21 Extracted Processes DAL Æneas Mackintosh v. John and Alexander Macdonald. 1808.
31 NAS GD 176/1519 Rentals of Lochaber 1796–97 & 1810–19.
32 NAS GD 176/1290 Tack Twixt Æneas Mackintosh and Thomas Gillespie and James Greig. 1813.
33 A.G. & Margaret H. Beattie, *Pre-1855 Gravestone Inscriptions in Lochaber and Skye*, The Scottish Genealogy Society, 1990, 34 (Entries 55 & 56).
34 *NSA*, Edinburgh & London, 1845, Vol.XIV, 504.
35 NAS GD 44/27/12/56 Rev. John Anderson to Duke of Gordon. 22 December 1807.
36 NAS GD 176/1430/1 Tack Æneas Mackintosh Esq. to Alexander Macdonald Esq. 1802.
37 NAS GD 176/1325 Memorial of Alexander McIntosh, once Tacksman of Keppoch. (To Alexander MacKintosh of MacKintosh.) 1834.
38 NAS GD 128/11/1 Rentals of the Mackintosh Estate, Lands of Lochaber. Various dates from 1771 to 1807.
39 NAS GD 176/1325 (Ref. 37). NAS GD 176/2118 Item 13 Æneas Mackintosh to Campbell Mackintosh, Writer in Inverness. 16 February 1807.
40 NAS SC 29/7/9 Decreet of Removing. Æneas Mackintosh against Alexander Mackintosh. 15 April 1807.
41 NAS GD 128/44/2/40 Alexander Cameron of Inverguseran to Alexander MacDonell, Writer, Inverness. 23 July 1807.
42 NAS GD 176/1325 (Ref. 37).
43 NAS GD 176/2121 Offer for Alex. McIntosh's farm of Bohinie etc. Written from Bochaskie, 24 July 1807.
44 NAS GD 176/1519 (Ref. 31).
45 HCA/D456/2/1 Alexander Macdonald of Glencoe to Colonel Alexr. MacDonell of Glengarry, 24 February 1808.
46 NAS GD 176/2121 Alexander Mackintosh to Campbell Mackintosh, Writer, Inverness. 13 September 1805.
47 NAS GD 176/2121 Note of Price of Stock Belonging to Alexander McIntosh late at Bohinie. 1809.
48 NAS GD 128/45/15 Campbell Mackintosh, Writer, Inverness, to Alexander Macdonell, Writer, Inverness. 1 December 1808.
49 NAS GD 128/21/7 Campbell Mackintosh, Writer, Inverness to Æneas Mackintosh. 7 January 1815. (Copy letter of 4 January 1814 from Duncan Cameron WS enclosed.)
50 NAS GD 176/1519 (Ref. 31).
51 NAS GD 176/2134/5 Campbell Mackintosh to Æneas Mackintosh. 28 March 1808.
52 NAS GD 176/2134/29 Campbell Mackintosh to Æneas Mackintosh. 12 January 1811.
53 NAS GD 176/1519 (Ref. 31).
54 NAS GD 176/2115/5 Æneas Mackintosh to Campbell Mackintosh, Writer in Inverness. 20 January 1812.
55 NAS GD 176/2134/13 Campbell Mackintosh to Æneas Mackintosh. 31 December 1812.

56 NAS GD 176/2123/8 Alexander Macdonald of Glencoe to Campbell Mackintosh, Writer in Inverness. 4 January 1814.
57 HCA Burgh and County of Inverness. Property (Private) Tacks, Feus, etc:
2/223 Æneas Mackintosh and John Mackintosh. Crenachan and Blarnahanine etc. 1803.
2/250 John Mackintosh and Alexander MacDonald. Blarnahanine. 1807.
2/252 John Mackintosh and Angus MacDonell. Crenachan. 1807.
58 NAS SC 29/7/9 Decreet of Removing. Æneas Mackintosh etc. . . . and John Mackintosh at Blairnahanin . . . eldest son of the *now deceased John Mackintosh* lately residing at Crenachan with the concurrence of Archibald Mackintosh, second son of the said deceased John Mackintosh against Angus MacDonald or MacDonell at Crenachan . . . Inverness 22 May 1807.
59 HCA Burgh and County of Inverness (Ref. 57).
60 NAS SC 28/16/8 John MacKintosh, Tacksman of Crenachan, against Angus MacDonald sub-tenant at Crenachan. 1808.
61 NAS GD 128/30/1/25. Campbell Mackintosh, Writer, Inverness to John Mackintosh, Messenger and Surveyor of Taxes, Fort William. 1817. (Empowering further sub-letting of Crenachan and Blarnahanine.)
62 NAS GD 202/35 Macdonald of Glenco – Trustee Accounts. 1815–1835.
63 *Ibid.*
64 NAS GD 176/2123/8 (Ref. 56).
65 NAS GD 176/1519 (Ref. 31).
66 *Inverness Courier*, 13 November 1812 (Quoted in J. Barron, *The Northern Highlands in the Nineteenth Century*, Inverness, 1903, Vol.I).
67 NAS GD 176/2136 Campbell Mackintosh to Sir Æneas Mackintosh. 22 December 1814.
68 NAS GD 176/2136 Campbell Mackintosh to Sir Æneas Mackintosh. 24 December 1814.
69 NAS GD 176/2136 Campbell Mackintosh to Sir Æneas Mackintosh. 28 December 1814.
70 NAS GD 128/21/7 (Ref. 49).
71 NAS GD 128/21/7 (Ref. 49).
72 NAS GD 128/21/7 (Ref. 49).
73 NAS GD 128/21/7 (Ref. 49).
74 NAS GD 202/35 (Ref. 62).
75 NAS GD 176/1519 (Ref. 31).
76 NAS GD 202/35 (Ref. 62).
77 NAS GD 202/35 (Ref. 62).
78 NAS GD 202/35 (Ref. 62).
79 NAS GD 176/1519 (Ref. 31).
80 *Inverness Courier*, 3 February 1820 (Quoted in J. Barron, *The Northern Highlands in the Nineteenth Century*, Inverness, 1903, Vol.I).
81 NAS GD 202/35 (Ref. 62).
82 NAS GD 202/35 (Ref. 62).
83 NAS GD 176/1519 (Ref. 31).
84 NAS GD 202/35 (Ref. 62).
85 NAS GD 202/73/3 Robert Scott, Edinburgh, to Ewen Macdonald. 13 November 1821.
86 NAS GD 202/35 (Ref. 62).

87 A.G. & Margaret H. Beattie, *Pre-1855 Gravestone Inscriptions in Lochaber and Skye*, The Scottish Genealogy Society, 1990, 14 (Entry 11).

88 NAS GD 202/35 (Ref. 62).

89 Josephine M. MacDonell of Keppoch, *The MacDonells of Keppoch and Gargavach*, Glasgow, 1931, 142.

90 NAS GD 176/990 Legal Papers in Dispute with Glencoe Trustees over Keppoch, etc. 1824–1825. Item 5. Answers for Alexander Mackintosh of Mackintosh.

91 NAS GD 176/1430 Item 14. Memorial of Queries for Counsel from Mackintosh of Mackintosh. 1823.

92 NAS GD 176/990 Legal Papers in Dispute with Glencoe Trustees over Keppoch, etc. Item 9. Examination and Survey of house and offices on Keppoch. Alex. Simpson and Donald Cameron. 31 May 1823.

93 NAS GD 176/1430 Item 14 Memorial of Queries for Counsel from Mackintosh of Mackintosh. 1823; Item 15 Counsel to Mackintosh. 1823.

Notes to Chapter 9

1 NAS RS38/19/139 Sasine in favour of Campbell Mackintosh, Writer in Inverness. 1810.

2 NAS RS38/18/112 Disposition by Donald McDonald, Tulloch, to Campbell Mackintosh. February 1808.

3 NAS RS38/19/139 (Ref. 1).

4 NAS GD 44/25/8/57 Memorial for Donald McDonald, Tacksman of Tulloch, for the Inn, Gordonsburgh. 1805.

5 NAS CR/8/7 Rev. J. Anderson to Mr. James Robertson WS Edinburgh. 4 November 1807.

6 NAS GD 44/25/8/47 Mr. Macpherson of Belleville recommending John Campbell for an Inn keeper at Gordonsburgh. 29 September 1805.

7 NAS GD 176/2107 Donald McDonald, Tulloch, to Campbell Mackintosh, Writer, Inverness. 24 April 1803.

8 NAS GD 128/49/10/3 Memoir for John MacDonell drover in Fasnatork(?) near Fort Augustus. 1814.

9 NAS GD 44/25/9/5 Farms and Grazings to be let and entered to at Whitsunday 1806.

10 *Ibid.*

11 NAS GD 44/25/8/45. Donald McDonald, Tulloch, offers £185 rent for Annat. 10 August 1805.

12 NAS GD 44/25/8/42. Donald McDonald, Tulloch, offers £100 rent for Claggan. 10 September 1805.

13 NAS GD 44/25/8/40. Capt. Angus MacDonell offers £70 rent for Claggan. 3 September 1805.

14 NAS GD 44/25/8/39. Robert Flyter, Writer, Fort William, offers £80 rent for Claggan. 4 September 1805.

15 NAS GD 44/25/8/56 Capt. MacDonell Claggan to William Tod. 28 September 1805.

16 NAS GD 44/25/8/75. Mr. Flyter – further offer of £100 for Claggan. 23 December 1805.

17 NAS GD 44/25/9/36 Rental of Lochaber Martinmas 1810.

18 NAS CR/8/6 Rev. J. Anderson to Francis Farquharson of Houghton. 23 October 1806.

19 NAS GD 44/25/32/8 Letter or Warrant of Bailling – Duke of Gordon to Robert Flyter. 22 November 1806.
20 NAS GD 44/25/9/36 (Ref. 17).
21 NAS GD 176/2106 Angus MacDonell (formerly Tulloch) to Campbell Mackintosh, Writer, Inverness. 8 December 1807.
22 NAS GD 176/2127/15 Angus McDonald late at Tullich to Campbell Mackintosh, Writer in Inverness. 23 July 1802.
23 NAS GD 44/25/8/76. Donald McDonald Tulloch to Sir George Abercromby of Birkenberg, Bart., Elgin. 10 September 1805.
24 NAS GD 44/25/8/95. John Rutherford, Edgerston, Nr. Jedburgh, to the Duke of Gordon. 24 August 1805.
25 NAS CR/8/6 Rev. J. Anderson to Alexander Tod, 54 Hatton Gardens, London. 1 November 1806.
26 NAS GD 44/27/17 Registered Tack by the Commissioners of the Duke of Gordon to Hugh McPherson (Ovie) 1767. (The Commissioners were Sir Robert Abercromby of Birkenberg and John Gordon, Merchant in Fochabers.)
27 A. Wight, *Present State of Husbandry in Scotland*, Edinburgh, 1778, Vol.2, 364.
28 NAS GD 176/2127/2 Note of Lotts by Donald McDonald of M'Intosh's Lochaber lands for next Sett 1804. 4 February 1802.
29 J. Hogg. *A Tour of the Highlands in 1803*, Edinburgh, 1888 (Facsimile Reprint 1986), 52.
30 NAS SC 28/16/7 Rev. Thomas Ross, Minister of Kilmanivaig against Ann MacDonald. 1807.
31 'The General Report of the Agricultural State, and Political Circumstances, of Scotland', drawn up . . . under the Direction of Sir John Sinclair, 1814, Appendix Vol II, 176.
32 NAS GD 44/25/8/94. Offers arranged for the Farms in The Lordship of Lochaber. 1805.
33 NAS GD 44/25/8/85. Donald McDonald agreeing to hold Annat for 7 years at £250 (and other matters). 30 October 1805.
34 NAS GD 128/45/6 Donald McDonald, Tulloch, to Alexander Macdonell, Writer, Inverness. 12 December 1800.
35 NAS SC 29/7/5 Decreet of Removing William Fraser etc. against Malcolm Gillies etc. Rifern. 4 April 1801.
36 A. & A. Macdonald, *The Clan Donald*, Inverness, 1904, Vol.III, 285; I. MacKay, 'Clanranald's Tacksmen in the Late 18th Century', in *Transactions of the Gaelic Society of Inverness*, Vol. XLIV, 1964–66, 61–93.
37 NAS GD 128/45/10 Angus McDonald of Kincreggan to Alexander Macdonell, Writer, Inverness. 18 January 1803.
38 NAS SC 28/16/5 Petition of Neil McIntyre, Procurator Fiscal, Fort William. 1804.
39 NAS CS 236/C/7/15 Removing. Cameron of Locheil against Alexander MacDonald. 1804.
40 NAS SC 28/16/12 Donald MacDonald, Mariner in Fort William, against Alexander McDonald formerly residing there, now Tacksman of Rifern. 1812–1814.
41 NAS GD 128/44/2/13 Alexander McDonald, Rifern, to Alexander Macdonell, Writer, Inverness. 28 August 1810.
42 NAS GD 176/2127/6 Campbell Mackintosh to Æneas Mackintosh. 24 May 1802.
43 NAS GD 44/25/9/5 (Ref. 9).

44 NAS GD 44/25/8/94. Offers arranged for the Farms in The Lordship of Lochaber 1805.

45 NAS GD 44/25/8/85. (Ref. 33).

46 S. Macdonald, *Back to Lochaber*, Edinburgh, 1994, 194.

47 NAS GD 44/27/12/56 Rev. John Anderson to Duke of Gordon. 22 December 1807.

48 NAS GD 128/45/13 Donald McDonald, Tulloch, to Alexander Macdonell, Writer, Inverness. From Fort William. 14 March 1806.

49 *Ibid.*

50 *NSA*, Edinburgh & London, 1845, Vol.XIV, 127.

51 NAS SC 29/50/1 Bond of Caution to keep the Peace for a Year and Day Messrs. Alexr. Macdonell & Campbell McIntosh Writers in Inverness for Alexr. Dow McDonell or Macdonell. 7 June 1800.

52 NAS GD 128/45/13 (Ref. 48).

53 NAS JC 11/48 Circuit Court. North Circuit . . . against Alexander MacDonald otherwise Alexander Dow Macdonald. 29 April 1806.

54 NAS GD 128/29/7 Case against Alexander [sic] Dow MacDonald for Perjury. 1806. (The papers are about Archibald Dhu MacDonell, but the bundle has the name Alexander Dow McDonald on it.)

55 NAS SC 28/16/20 Mr. Alexander Cameron, Scotus v. Arch. MacDonald and others. 1819–20.

56 NAS CR/8/6 Rev. J. Anderson to Mr. D. MacDonald, Tulloch. 9 March 1807.

57 *Ibid.*

58 NAS CR/8/7 Rev. J. Anderson to Mr. Donald McDonald Tulloch. 12 December 1807.

59 NAS GD 44/27/12/56 (Ref. 47).

60 HCA BI/1/10/10 Trust Disposition and Deed of Settlement by Mr. Donald Macdonald Tacksman of Tullich. 31 August 1807. Recorded 4 April 1808.

61 NAS GD 44/25/9/36 (Ref. 17).

62 NAS CR/8/7 Rev. J. Anderson to Thomas Gillespie, Ardochy. Undated, but probably January 1809.

63 NAS RS38/18/112 (Ref. 2).

64 HCA BI/1/10/10 Minute of agreement with Donald Macdonald at Highbridge about the new Inn at Fort William etc. 6 July 1808.

65 S. Macdonald, 1994. *op cit.* (Ref. 46). 20.

66 NAS GD 44/25/10/2 Alexander MacArthur, Bochaskie, to the Duke of Gordon. 30 July 1813.

67 NAS GD 202/35 Macdonald of Glenco – Trustee Accounts. 1815–1835.

68 NAS SC 28/16/11 Removal. John MacKenzie and Donald MacIntyre etc . . . Against Tenants. 1812.

69 NAS GD 128/33/4 Campbell Mackintosh, Writer, Inverness, to Donald Mackintosh, Edinburgh. 4 March 1816.

70 C. Fraser-Mackintosh, 'The Camerons of Letterfinlay', in *Transactions of the Gaelic Society of Inverness*, Vol.XVII, 1890–91, 31–45.

71 A.G. & Margaret H. Beattie, *Pre-1855 Gravestone Inscriptions in Lochaber and Skye*, The Scottish Genealogy Society, 1990, 16, (entry 12).

72 I. MacKay, 'Clanranald's Tacksmen of the late 18th Century', in *Transactions of the Gaelic Society of Inverness*, Vol.XLIV, 1964–66, 61–93.

73 NAS GD 128/65/11 John MacDonald of Borrodale to Alexander Macdonell, Writer, Inverness. 26 March 1810.

74 NAS SC 28/16/13 Summons. Allan Cameron of Meoble against Alexander McDonald, Tacksman of Rifern. 1814.
75 NAS SC 29/10/1972 Alexander MacDonald in Torrary of Rifern agt. Allan Cameron, Tacksman of Meoble. 1814.
76 Sasine Abridgements Inverness. 22 July 1813. (1417)
77 NAS SC 28/16/14 John MacDonald of Borrodale – for the removal of Alexander MacDonald. 1815.
78 NAS GD 128/65/9 John MacDonald of Borrodale to Alexander McDonell, Writer, Inverness. 26 September 1816.
79 NAS SC 28/16/14 Col. Alexander McDonell of Glengarry v. Alexander Macdonald and his sons Archibald and Alexander residing at Carnoch [Knoydart]. 1817.
80 NAS SC 28/16/20 John Mackintosh against John McDonald, formerly in Riefern, now in Kinlochailort. 1819.
81 NAS GD 128/65/9 (Ref. 78).
82 NAS SC 28/16/20 (Ref. 80).
83 NAS SC 28/16/20 Robert MacKay, Writer, Fort William, against John Mackintosh, Surveyor of Taxes. 1820.
84 HCA/D456/2/3 Robert MacKay, Writer in Fort William to John Hood, Factor to Glengarry, 25 June 1818.
85 I am grateful to Mr Tearlach MacFarlane, Glenfinnan, who has kindly provided information about the Finiskaig MacDonalds and about the interpretation of Bitag. Donald's identity as Alexander Dhu's youngest son has been established from evidence about his age combined with census and death certificate information.

Notes to Chapter 10

1 Ordnance Survey. Six inches to One Statute Mile. Published 1876. Sheet 108.
2 C. Fraser-Mackintosh, 'The MacDonells of Scotus', in *Transactions of the Gaelic Society of Inverness* Vol.XVI, 1889–90, 79–97.
3 C. Fraser-Mackintosh, *Antiquarian Notes*, Second Series, Inverness, 1897, 134–35.
4 *OSA.* (Reprint), Wakefield, Vol.XVII, 1981, 76.
5 J. Barron, *The Northern Highlands in the Nineteenth Century*, Inverness, 1903, Vol.I, xxvi.
6 H. Douglas, *Flora Macdonald The Most Loyal Rebel*, London, 1994, 168.
7 NAS SC 29/57/10 Assignation – Hugh and Alexander MacDonalds to Alexander Macdonell, Writer, Inverness. 1792.
8 NAS GD 128/65/12/18. Archibald Macdonald, Shenachy, to Alexander Macdonell, Writer, Inverness. 12 September 1794.
9 NAS GD 128/29/7 Case against Alexander [sic] Dow MacDonald for Perjury. 1806. (The papers are about Archibald Dhu MacDonell, but the bundle has the name Alexander Dow McDonald on it.)
10 NAS CS 271/51025 Act and Warrant. Donald McDonald v. The Clerk of the Bills. 1796.
11 C. Fraser-Mackintosh, *op.cit.* (Ref. 3), 1897, 142–43.
12 NAS SC 29/7/2 Decreet. Ronald MacDonald, Tacksman of Scammadale, against Alexander Cameron Tacksman of Inverguseran etc. Inverness, 12 December 1794.
13 NAS CS 17/1/43 Court of Session Minute Book. 1823–24 Alexander Cameron, Inverguseran AG McDonell of Glengarry etc. 4 February 1824.

14 NAS SC 28/16/3 Removal. Alexander MacDonell Esq of Glengarry and Alexander Cameron at Inverguseran, Principal Tenant and Setter of the lands of Inveriemore . . .against . . .sub-tenants and possessors of parts and portions of the Town & lands of Inveriemore . . . 1802.

15 NAS SC 29/10/2032 Colonel MacDonell of Glengarry v. Archibald MacDonell, Tacksman of Kyles, etc. 1815.

16 NAS SC 28/16/3 Removal. John Cameron, Tacksman of Inverskillivulan, Captain Donald Cameron and Alexander Cameron at Inverguseran, Subtacksmen of Inverskillavulin v. Lachlan Cameron and Ewen MacKinnon. 1802.

17 C. Fraser-Mackintosh, *Letters of Two Centuries*, Inverness, 1890, 329.

18 NAS SC 29/57/11 Lease betwixt Patrick Grant of Glenmoriston and Scothouse . . . and Dr. Donald MacDonald and others. 1804.

19 NAS CC/11/1/9 Record of Testaments (Inverness-shire) p.412. Thomas Gillespie – died 9 July 1824.

20 C. Fraser-Mackintosh, *op.cit.* (Ref. 2), 1889–90, Vol.XVI, 79–97.

21 NAS SC 28/16/9 John Gillespie, Kilmuree, Against Euphemia Kennedy, Shenachy. 1808.

22 Sasine Abridgements. Inverness. 13 October 1786 (162); 4 November 1790 (339); 28 February 1792 (377).

23 Sasine Abridgements. Inverness. 1 February 1804 (863).

24 NAS SC 28/16/9 (Ref. 21).

25 NAS SC 29/57/11 (Ref. 18).

26 NAS SC 28/16/10 Summons of Count, Reckoning and Payment. MacDonald agt. MacDonald. 1810.

27 NAS SC 28/16/1 Criminal Summons. Allan Cameron and Fiscal against John Bain MacDonald. 1802.

28 NAS SC 29/10/2032 (Ref. 15).

29 NAS SC 29/10/2032 (Ref. 15).

30 NAS RHP 112 Plan of the Annexed Estate of Barrisdale by William Morison. Copy of 19 June 1815. Table of Acreages.

31 HCA/D456/4/5 Archibald Macdonald [MacDonell], Tacksman of Inveriemore, to Alexander Macdonell, Writer in Inverness, 23 August 1803.

32 NAS GD 128/29/7 (Ref. 9).

33 NAS GD 128/45/12 Alexander Cameron, Inverguseran, to Alexander Macdonell, Writer, Inverness. 18 January 1803. (Written from Auchtertyre).

34 HCA/D456/4/5 (Ref. 31).

35 NAS GD 128/29/7 (Ref. 9).

36 NAS GD 128/29/7 (Ref. 9).

37 HCA/D456/4/5 (Ref. 31).

38 NAS GD 128/29/7 (Ref. 9).

39 NAS GD 128/29/7 (Ref. 9).

40 NAS GD 128/29/7 (Ref. 9).

41 NAS GD 128/29/7 (Ref. 9).

42 NAS GD 128/29/7 (Ref. 9).

43 NAS GD 128/44/2/45 Coll Macdonald WS to Alexander Macdonell, Writer, Inverness. 28 August 1806; R. & W. Bell, *Dictionary of the Law of Scotland*, Third Ed., Edinburgh, 1826, Vol.I, 597.

44 NAS JC 11/48 Circuit Court. North Circuit . . . . against Archibald MacDonald alias Archibald Dow MacDonald. 13 September 1806.

45 NAS GD 128/29/7 (Ref. 9).

46 C. Fraser-Mackintosh, *op.cit.* (Ref. 3), 1897, 144–45.
47 NAS SC 29/10/2032 (Ref. 15).
48 NAS GD 128/44/2/17 John MacDonald, Glenmeddle, to Alexander Macdonell, Writer, Inverness. 21 August 1810.
49 NAS SC 29/10/2512 Summons of Removing. Colonel Alexander Ranaldson MacDonell of Glengarry etc. against Archibald MacDonald Senior and Alexander, Duncan, John and Donald his sons, tenants etc . . . of Kyles and Brunswick of Knoydart or parts thereof; NAS SC 29/7/22 Decreet extracted 2 June 1824.
50 NAS GD 128/44/2/17 (Ref. 48).
51 NAS SC 28/16/22 Alexander Macdonald, Scotus, v. Arch. MacDonald, Kyles Knoydart. 1824.
52 NAS GD 128/51/14 Copy letter. Mr. Donald Smith, Inverness, to Mr. George Gibson, Rotterdam. 14 October 1790.
53 NAS GD 128/45/1 McDonald and Elder, Camuscross, to Alexander Macdonell, Writer, Inverness. 28 May 1798.
54 NAS SC 29/7/17 Decreet. Messrs. MacDonald, Elder, and McInnes, Merchants in Sleat, against Angus McInnes etc. 2 December 1816; Decreet. Messrs. Alexander McDonald, John Elder, and Duncan McInnes . . . Merchants, against Alexander Cameron, Tacksman of Inverguseran, and Ewen Rankin . . . 4 March 1817.
55 NAS SC 29/7/20 Decreet. Messrs. MacDonald, Elder, and McInnes, sometime Merchants in Sleat, against Archibald MacDonald, Tacksman of Ord. 20 July 1819.
56 NAS GD 176/2111 Item 7. Colin Elder, Inverness, to Campbell Mackintosh, 24 February 1818; Item 20 John Elder to Campbell Mackintosh. 20 April 1818.
57 NAS SC 29/7/24 Messrs. Alexander MacDonald and John Elder residing at Isleornsay carrying on business as merchants there . . . 3 April 1827; Mr. Colin Elder, sometime merchant at Kyleakin, now residing at Isleornsay . . . 3 April 1827.
58 HCA/D456/2/3 John Hood to Alexander Macdonell, Writer in Inverness, 28 July 1818.
59 HCA/D456/4/1 Duplies from Glengarry & Donald Macdonald to The Replies for Allan Macdonald 1818.
60 S. Macmillan, *Bygone Lochaber*, Glasgow, 1971, 182. (Macmillan suggests in a footnote that 'Glencoe' was Adam Macdonald of Achtriachtan. This seems unlikely, and if Duncan Cameron had meant Achtriachtan he would surely have said so.)
61 NAS SC 28/16/6 Removing. Alexander Macdonald of Glencoe, Principal Tacksman and Setter of the lands . . . against [four tenants in Glendessary and two in Knoydart] 1805.
62 NAS GD 202/35 Macdonald of Glenco – Trustee Accounts. 1815–1835.
63 NAS GD 202/35 (Ref. 62).
64 NAS GD 202/35 (Ref. 62).
65 NAS GD 202/35 (Ref. 62).
66 NAS GD 202/35 (Ref. 62).
67 NAS CS 36/31/12 Decreet of Preference and for Payment. John Cameron and James Grant his Attorney against Robert Downie etc. . . . and the said Robert Downie and others against The Trustees and Creditors of Adam Macdonald. 16 December 1820.
68 NAS GD 202/35 (Ref. 62).

69 NAS SC 28/16/15 John Cameron against John McDonald, Corychorichan (for debt of £8) 1817; Allan Cameron, Younger of Clunes, against John McDonald residing at Corychorichan (for debt of £8) 1817.
70 NAS CC/11/1/9 (Ref. 19).
71 A.G. & Margaret H. Beattie, *Pre-1855 Gravestone Inscriptions in Lochaber and Skye*, The Scottish Genealogy Society, 1990, 34, Entries 55 & 56.
72 NAS SC 28/16/20 Removal of John Baan MacDonald. 1820.
73 NAS GD 202/73/3 Robert Scott, Edinburgh, to Ewen Macdonald. 13 November 1821.
74 HCA/D456/4/6 Copy letter[s] John Hood to Alexr. McDonell (Na Baitaig) & Sons dated at Inveree 17 July 1815.
75 HCA/D456/2/3 John Hood to Alexander Macdonell, Writer in Inverness, 3 March 1817.
76 NAS SC 28/16/14 Colonel Alexander McDonell of Glengarry v. Angus Macdonald, Duncan Macdonald, Allan Macdonald, and Archibald Macdonald, all residing in the tenement commonly known by the general name of Kenlochnevis. 1817.
77 HCA/D456/4/6 (Ref. 74).
78 HCA/D456/2/3 (Ref. 58).
79 NAS SC 29/10/2085 Glengarry v. Capt Archibald McDonald of Barrisdale [and others] 1816.
80 NAS SC 29/10/2149 Sequestration. Glengarry & Factor agt. Alex. McDonald and Sons, Tenants of Kinlochnevis or Souryess, 1817.
81 N.H. MacDonald, *The Clan Ranald of Knoydart & Glengarry*, Edinburgh, 1995, 184.
82 HCA/D456/2/3 John Hood to Alexander Macdonell, Writer in Inverness, 9 February 1820.
83 HCA/D456/4/6 (Ref. 74).
84 NAS GD 202/35 (Ref. 62).
85 NAS SC 28/16/14 Col. Alexander McDonell of Glengarry v. Alexander Macdonald and his sons Archibald and Alexander residing at Carnoch [Knoydart]. 1817.
86 NAS SC 28/16/14 (Ref. 76).
87 NAS GD 202/35 (Ref. 62).
88 NAS SC 28/16/14 (Ref. 85).
89 NAS SC 28/16/14 (Ref. 76).
90 NAS SC 28/16/14 (Ref. 85).
91 NAS SC 28/16/14 (Ref. 76).
92 NAS GD 202/35 (Ref. 62).
93 NAS SC 29/10/2032 (Ref. 15).
94 NAS GD 202/35 (Ref. 62).
95 NAS SC 29/10/2149 (Ref. 80).
96 A.R.B. Haldane, *New Ways Through the Glens*, London, Edinburgh, etc., 1962, 66 & 96.
97 NAS GD 202/35 (Ref. 62).
98 NAS SC 28/16/14 (Ref. 85).
99 NAS SC 28/16/14 (Ref. 76).
100 NAS SC 28/16/14 (Ref. 76).
101 NAS SC 29/10/2149 (Ref. 80).

102 NAS SC 29/10/2085 (Ref. 79).
103 HCA/D456/2/2 Alexander Macdonell of Glengarry to Alexander Macdonell, Writer in Inverness, 12 December 1816.
104 HCA/D456/2/3 N. McLean to Alexander Macdonell, Writer in Inverness, 25 June 1817. (Copy of agreement of 23 May 1817 attached.)
105 NAS SC 29/10/2512 & SC 29/7/22 (Ref. 49).
106 HCA/D456/2/3 Alexander Macdonell of Glengarry to Alexander Macdonell, Writer in Inverness, 19 July 1817.
107 C. Fraser-Mackintosh, *op. cit.* (Ref. 3), 1897, 144–45.
108 NAS SC 28/16/14 John MacDonald of Borrodale – for the removal of Alexander MacDonald. 1815.
109 NAS SC 28/16/19 Dr. Taylor against Archibald MacDonald. 1822.
110 NAS SC 28/16/23 Robert MacKay, Writer, Fort William, against McDonalds. 1823.
111 NAS SC 28/16/23 John McLachlan, Mariner, against Allan McDonald and Brothers. 1823.
112 HCA/D456/2/2 (Ref. 103).
113 HCA/D456/4/1 (Ref. 59).
114 HCA/D456/2/3 N. McLean to Alexander Macdonell, Writer in Inverness, 2 July 1818.
115 HCA/D456/2/3 Robert MacKay, Writer in Fort William to John Hood, Factor to Glengarry, 25 June 1818.
116 C. Fraser-Mackintosh, *op. cit.* (Ref. 3), 1897, 144–45.
117 J. Prebble. *The Highland Clearances*, London, 1963, 150.
118 HCA L/INV/SC/8/9 Precognition and Declaration. Ensign Angus Macdonald and Others. Fort William. 1783.
119 NAS GD 128/12/4 Æneas Mackintosh of Mackintosh to Campbell Mackintosh, Writer in Inverness. 1 July 1793.
120 NAS SC 29/7/19 Decreet of Removing – Colonel Alexander Macdonell of Glengarry against Alexander Dubh McDonald etc. 1818.
121 HCA/D456/2/3 John Hood to Alexander Macdonell, Writer in Inverness, 5 July 1818.
122 HCA/D456/4/1 (Ref. 59).
123 NAS SC 29/7/19 Decreet of Removing – Colonel Alexander Macdonell of Glengarry against Alexander Dubh McDonald etc. 1819.
124 NAS SC 29/7/19 Decreet. Alexander McDonald, late tenant in Carnoch now residing in Arisaig against Allan MacDonald sometime residing in Bunroy etc. 1819.
125 HCA/D456/2/3 N. McLean to Alexander Macdonell, Writer in Inverness, 9 March 1818.
126 HCA/D456/2/3 (Ref. 121).
127 HCA/D456/2/3 Alexander Macdonald of Glengarry to Alexander Macdonell, Writer in Inverness, 4 December 1818.
128 NAS SC 29/1/37 Glengarry against Alexander Dhu McDonald and sons. Produced by Macdonell and given out to Shepperd. 13 May 1819.
129 NAS SC 29/1/41 Alexander Dhu McDonald against Glengarry. 15 June 1822.
130 NAS SC 29/1/42 Glengarry against Dhu McDonald 24 February 1824.
131 NAS SC 29/1/41 (Ref. 129).
132 NAS SC 28/16/19 (Ref. 109).

133 J. Mitchell, *Reminiscences of my Life in the Highlands*, 1884 (Reprint Newton Abbott 1971), Vol.1, 73.
134 W. Roughead, *Glengarry's Way and other Studies*, Edinburgh, 1922, 18.
135 NAS SC 29/1/41 Dhu McDonald against Glengarry. 12 July 1822.
136 NAS SC 29/1/42 Process – Transference Glengarry against Dhu McDonald 1 December 1823.
137 NAS SC 29/1/42 (Ref. 130).
138 NAS SC 29/1/42 Glengarry against Dhu McDonald 19 June 1824.
139 NAS SC 28/16/23 (Ref. 110).
140 Christine Johnson, *Developments in the Roman Catholic Church in Scotland 1789–1829*, Edinburgh, 1983, 232.
141 NAS GD 128/65/9 John MacDonald of Borrodale to Alexander Macdonell. Writer, Inverness. 23 February 1819
142 NAS GD 128/65/9 John MacDonald of Borrodale to Alexander Macdonell, Solicitor, Inverness. 29 April 1819.
143 Clan Donald Centre. MS.1.24 Transcripts etc. of part of Borrodale Correspondence. Alexander Macdonell, Writer, Inverness, to John MacDonald of Borrodale. 30 April 1819.
144 Scottish Catholic Archives. Oban Letters 1/3 Page 4. Letter 5. Bp. Ranald MacDonald to John MacDonald of Borrodale. 22 February 1823.
145 NAS GD 202/35 (Ref. 62).
146 NAS GD 202/73/3 (Ref. 73).
147 NAS SC 29/7/18 Colonel Alexander Macdonell of Glengarry against Alexander Cameron residing at Inverguseran. 1817.
148 NAS CS 236/C/19/4 Cameron against Glengarry. 1822.
149 NAS CC/11/1/9 (Ref. 19).
150 NAS CS 236/C/19/4 (Ref. 148).
151 NAS CS 17/1/43 (Ref. 13).
152 NAS GD 128/33/5 Robert MacKay, Writer, Fort William, to Donald MacKintosh, Edinburgh. 2 January 1826.
153 NAS SC 29/1/33 Sequestration. Glengarry v. Hugh MacDonell, Tacksman of Crowline. 9 February 1816.
154 NAS SC 29/10/2184. Irritancy and Removing. Glengarry v. Hugh Macdonald, Tacksman of Crowline. 1818.
155 NAS SC 29/10/2085 Minute for Mr. Robert Murray, Accountant, Inverness, Trustee on the Sequestrated Estate of Hugh MacDonald, Tacksman of Crowline. 1818.
156 NAS GD 128/65/9 John MacDonald of Borrodale to Alexander Macdonell. Writer, Inverness. 30 June 1817.
157 NAS SC 29/7/22 Alex Macdonell, Writer in Inverness, against Hugh McDonald late Tacksman of Crowline, now in Kinlochnevis. 21 May 1822; Decreet of Irritancy and Removal. Glengarry against Hugh MacDonald, sometime Tacksman of Crowline, lately at Inverie, and Hugh MacDonald his eldest son, pretended tenants . . . of the Lands of Lochnevis. 28 May 1823.

Notes to Chapter 11

1 C. Fraser-Mackintosh, 'The Macdonalds of Achtriachtan', in *Transactions of the Gaelic Society of Inverness* Vol. XXIII, 1898–99, 136–45.
2 E. Richards, *A History of the Highland Clearances*, London and Canberra, 1982, 266.
3 NAS RD2/293/181 Tack between Brigadier General Alexander Mackenzie of Fairburn . . . and Murdoch MacLenan, tenant in Auchintee of Lochcarron, (and others) . . . the grazings of Monar . . . 1803.
4 J. Hogg, *A Tour of the Highlands in 1803*, Edinburgh, 1888 (Facsimile Reprint 1986), 64.
5 NAS SC 25/22/70 Removing. Hugh Fraser Esq. and others against Alexander MacCallum. 1819.
6 NAS RS 38/27/197 Sasine in favour of Hugh Fraser Esquire. At Inverness, 19 January 1819.
7 NAS SC 25/22/70 (Ref. 5).
8 NAS CS 236/M/32/2 Macdonald of Glencoe Trustees v. Mitchell. 1817–1821.
9 C. Fraser-Mackintosh, *Antiquarian Notes*, Second Series, Inverness, 1897, 8–10.
10 NAS SC 29/57/11 Lease. The Honorable Archibald Fraser of Lovat and Hugh Fraser Esquire of Eskadale and Robert Fraser of Aigas. The Town and Lands of Glenstrathfarrar for 17 years from 1804.
11 Sasine Abridgements Inverness. 5 August 1800. (707)
12 C. Fraser-Mackintosh, 1897, *op. cit.* (Ref. 9), 8–10.
13 NAS CS 236/M/32/2 (Ref. 8).
14 NAS SC 29/57/10 Tack Twixt Hugh Fraser Esquire of Struy and Alexander MacCallum . . . residing at Lubcroy. 1801.
15 *Ibid.*
16 NAS SC 29/7/5 Decreet of Assoilzie and for Expenses. John MacDonell late in Torgulbin now at Killichonate against Alexander MacCallum overseer to Donald MacLeod Esq. of Geanies at Lubcroy, Ross-shire. Inverness 11 April 1800.
17 E. Richards, 1982, *op. cit.* (Ref. 2) 258–63.
18 NAS GD 128/36/8 & GD 128/47/4 Sheep Farm Association.
19 NAS CS 236/M/32/2 (Ref. 8).
20 NAS CH2/433/3 Parish of Kilmanivaig. Births and Marriages.
21 NAS CS 236/M/32/2 (Ref. 8).
22 NAS CS 236/M/32/2 (Ref. 8).
23 NAS RD2/293 Pages 159, 164, 168, 171, 175, & 178. Tacks in Strathconon between Brigadier General Alexander Mackenzie of Fairburn and various tenants. 1803. (Note: George Langlands' map of 1801 shows Blaravan a mile or so to the west of Loch Tulla on the north of Linne nam Beatach. This conflicts with the statement on p. 178 of this reference that Blaravan is in Glenorchy.)
24 J. Hogg, 1888, *op. cit.* (Ref. 4), 36–37.
25 W.J. Watson, *Place Names of Ross and Cromarty*, 1904 (Reprinted Evanton 1996), 149.
26 NAS SC 25/22/76 Summons. Duncan & Donald MacDonald (Inverchoran) against John Cameron. 1821.
27 NAS SC 25/22/77 Summons of Removing. Alexander Mackenzie Esq. and others against Donald MacDonald and others. 1821.
28 NAS SC 25/22/66 MacDonald against Matheson. 1819/1820; SC 25/22/71 Summons of Removing. Donald MacDonald against Kenneth and Alexander Matheson. 1820. (Mentions John and Donald Macdonald as Joint Tacksmen of Dalbreck.)

29 NAS CS 36/31/12 Decreet of Preference and for Payment. John Cameron and James Grant his Attorney against Robert Downie etc. . . . and the said Robert Downie and others against The Trustees and Creditors of Adam Macdonald. 16 December 1820.

30 Notes on his family written in December 1830 by Archibald McDonald (1790–1853) of the Hudson's Bay Company. (Copy in Glencoe and North Lorn Folk Museum.)

31 Barbara Fairweather & D.C. Cargill, *Pre-1855 Tombstone Inscriptions on Isla Munda, The Burial Island in Loch Leven*, The Scottish Genealogical Society, Entry 15.

32. NAS SC 25/22/67 George Stewart late shepherd to Colonel Robertson MacDonald of Kinlochmoidart . . . against Allan Macdonald now residing at Dalbreck of Strathconon . . . 1819.

33 NAS SC 25/22/77 (Ref. 27).

34 Jean M. Cole, *Exile in the Wilderness. The Biography of Chief Factor Archibald McDonald, 1790–1853*, Toronto, 1979; J. Hunter, *A Dance Called America*, Edinburgh, 1994, 181–86; Jean M. Cole, *This Blessed Wilderness. Archibald McDonald's Letters from the Columbia, 1822–44*, Vancouver & Toronto, 2001.

35 Notes by Archibald McDonald (Ref. 30).

36 J. Hunter, *Glencoe and the Indians*, Edinburgh & London, 1996, 38–45.

37 Notes by Archibald McDonald (Ref. 30).

38 NAS SC 29/10/2084 Alexander MacCallum v. Donald MacDonald. 1816.

39 NAS SC 29/10/2086 Donald MacDonald, Tacksman in Monar against Donald Cross and Others. 1816.

40 NAS SC 29/10/2151 Donald MacDonald, Tacksman in Monar v. Hector McLean and others. 1817.

41 Notes by Archibald McDonald (Ref. 30).

42 Barbara Fairweather & D.C. Cargill, *op. cit.* (Ref. 31), Entry 15.

43 NAS CS 236/M/32/2 (Ref. 8).

44 NAS GD 202/35 Macdonald of Glenco – Trustee Accounts. 1815–1835.

45 NAS SC 25/22/70 (Ref. 5).

46 NAS GD 202/35 (Ref. 44).

47 NAS SC 25/22/70 (Ref. 5).

48 NAS SC 25/22/72 Bond of Caution for the Violent Profits By William C. MacMillan for Donald Macdonald. 1820.

49 NAS SC 29/10/2084 (Ref. 38).

50 NAS SC 29/10/2273 Removal. Culloden Trustees against Alexander MacCallum Tacksman of Bunchrew. 3 April 1819. (Defender failed to appear); NAS SC 29/7/21 Alexander MacCallum, Tacksman of Bunchrew against Donald Paterson, Boat Carpenter, Clachnaharry. (Debt of £13/2/0d) 26 June 1821; Alexander MacCallum, Tacksman of Bunchrew, against John MacDonald, Flesher in Inverness. 21 July 1821; NAS SC 29/10/2570 Alexander Pringle, late Grieve to Colin MacCallum, Tacksman of Bunchrew, now residing at Bridge End, Beauly, against Colin MacCallum, Tacksman of Bunchrew. Extracted July 1825.

51 NAS SC 25/22/71 Summons of Removing. Sir Hugh Innes of Lochalsh, Bart., and others, Tutors Nominate for Alexander William Chisholm . . . against Alexander MacCallum (residing at Bunchrew in the County of Inverness) 1820.

52 NAS SC 25/22/70 (Ref. 5).

53 NAS SC 25/22/72 (Ref. 48).

54 NAS GD 202/73/3 Robert Scott, Edinburgh, to Ewen Macdonald. 13 November 1821.

55 NAS GD 202/35 (Ref. 44).
56 NAS GD 202/73/3 (Ref. 54).
57 NAS GD 202/35 (Ref. 44).
58 NAS GD 202/35 (Ref. 44).
59 NAS GD 128/41/7 Copy letter. John Cumming, Dingwall, to Campbell Mackintosh, Inverness. 21 June 1827; John Cumming, Dingwall, to Donald Mackintosh WS, Edinburgh. 16 July 1827; John Cumming, Dingwall, to Donald Mackintosh WS, Edinburgh. 3 August 1827.
60 NAS SC 25/22/77 Summons. Glencoe Trustees against Alexander McLennan and Others 19 October 1821.
61 HCA Burgh Court Book, Inverness. Process – John Macdonald agt. Glenco's Trustees. 18 July 1822.
62 NAS CS 236/M/32/2 (Ref. 8).
63 NAS DI.64/16 Folio 137. Inhibition and Arrestment against Mitchell. 27 December 1820.
64 NAS CS 236/M/32/2 (Ref. 8).
65 NAS CS 236/M/32/2 (Ref. 8).
66 NAS DI.64/16 Folio 137 (Ref. 63).
67 NAS SC 25/22/37 Summons. Margaret Ross against McBarnett. 1810.
68 NAS GD 202/35 (Ref. 44).
69 *A Highland Newspaper. The Inverness Courier 1817–1967*, Inverness, 1969, 66.
70 NAS GD 202/35 (Ref. 44).
71 NAS GD 202/73/3 (Ref. 54).
72 NAS GD 202/35 (Ref. 44).
73 NAS GD 202/73/3 (Ref. 54).
74 NAS DI.64/16 Folio 163 Inhibition and Arrestment. MacDonald's Trustees against Alexander MacCallum and others. Inverness 17 January 1821.
75 NAS GD 202/34 Claim against McCallum of Culligran etc. 1813–1821.
76 NAS CS 236/M/42/83 Macdonald of Glencoe Trustees v. McCallum etc. 1821.
77 NAS GD 202/73/3 (Ref. 54).
78 NAS GD 202/35 (Ref. 44).
79 HCA Burgh Court Book, Inverness. Process – Alexr. McCallum agt. Duncan Fraser and Thomas Fraser Esqs. 22 & 23 January 1823.
80 NAS SC 25/22/36 Summons of Furthcoming. Charles Stewart against John MacCallum and William Maxwell. 1810.
81 NAS CS 36/31/12 (Ref. 29).
82 NAS SC 25/22/66 & SC 25/22/71 (Ref. 28).
83 NAS SC 25/22/76 (Ref. 26).
84 NAS DI.8/278 Folio 13. Petition of Donald Macdonald, Grain and Victual dealer at Monar . . . 7 August 1820.
85 NAS SC 25/22/77 (Ref. 27).
86 NAS CS 17/1/43 Court of Session Minute Book. 5 February 1824.
87 NAS SC 25/22/75 Protest – Horn and Poind. Gillanders against Macdonalds. 1820/1821.
88 NAS SC 28/16/20 Summons. Downie against McDonald and McInnes. 1820.
89 Notes by Archibald McDonald (Ref. 30).

Notes to Chapter 12

1 NAS CS 232/G/12/29 Summons of Removing. Duke of Gordon v. Alexander McDonald Esq. of Glencoe and others per Roll. 1798.

2 NAS SC 29/7/5 Decreet of Assoilzie and for expenses. John MacDonell, late Torgulbin now at Killichonate, v. Alexander MacCallum, Overseer to Donald Geanies at Luberry, Ross-shire. 1800.

3 Ann MacDonell & R. MacFarlane, *Cille Choirill, Brae Lochaber*, Spean Bridge, 1986, 27, Entry 62.

4 NAS CS 237/Mc/13/3 McCaul v. McDonald. Summons of Reduction 1817.

5 NAS GD 128/54/7 Alexander McVean to Campbell Mackintosh, Writer, Inverness. 22 August 1793.

6 NAS GD 128/54/6 Neil McIntyre, Fort William, to Campbell Mackintosh, Writer, Inverness. 5 November 1793.

7 NAS DI.64/14 Folio 433. Horning & Poinding. Chisholm against McDonald. Inverness. 20 August 1790.

8 NAS SC 29/7/1 Decreet. Ranald McDonald late Tenant in Clianick in the Braes of Lochaber against several persons after named . . . Inverness 15 December 1792.

9 NAS GD 44/25/9/3/2 [Reference to] Lease of Fersit to Alexander McDonell, James McCaul and Alexander McVean. 1 May 1797.

10 NAS GD 44/25/8/67. Alexander McDonell and Alexander McVean to William Tod. 3 April 1805.

11 NAS CS 237/Mc/13/3 (Ref. 4).

12 NAS GD 44/25/9/5 Farms and Grazings to be let and entered to at Whitsunday 1806.

13 NAS GD 44/25/8/83 Alexander McDonell and Alexander McVean to William Tod. Formal acceptance (sent by Badenoch Post). 29 October 1805; Alexander McVean to William Tod. Confirmation of Acceptance (sent by Fort William Post). 31 October 1805.

14 NAS GD 44/27/12/56 Rev. John Anderson to Duke of Gordon. 22 December 1807.

15 NAS CR/8/7 Rev. J. Anderson to Captain Alexander MacDonell, Moy. 12 November and 20 December 1808.

16 NAS GD 44/25/9/24/3 Alexander Robertson of Struan guarantees Capt. MacDonell Moy's rents. Signed at Rannoch Barracks, 29 February 1809.

17 R. MacFarlane, 'The MacDonells of Aberarder', in *Clan Donald Magazine* No. 12, Edinburgh, 1991, 147–57.

18 NAS GD 44/25/9/28 State of Arrears in Lochaber. 20 September 1810.

19 NAS GD 44/25/9/28 (Ref. 18); NAS GD 44/25/9/17/2 William Mitchell, Achnadaul, to Rev. J. Anderson. 22 January 1809; NAS GD 44/25/9/17/11 William Mitchell, Achnadaul, to Rev. J. Anderson. 25 February 1809.

20 NAS GD 44/25/9/24/2 Alexander McVean and others to Rev. J. Anderson. 27 February 1809.

21 NAS GD 44/25/9/37/2 Thomas Gillespie and others. Fixing proportion of rent to be paid by Capt. MacDonell and Alexander McVean for the Fersits etc. 12 July 1810.

22 NAS SC 29/7/4 p.49. Steelbow contract 14 June 1792 and 15 November 1792 between Allan MacDonald of Gallovie and Patrick MacNab Tenant at Ballicholish and John and Robert MacNabs his sons.

23 NAS 170/1864 Æneas McDonald, Achtriachtan, to Alexander Campbell of Barcaldine. 9 May 1794.

24 NAS GD 202/35 Macdonald of Glenco – Trustee Accounts. 1815–1835.
25 *Ibid.*
26 NAS GD 44/25/10/2/55 Offer. A. MacDonald of Glencoe for Torgulbin and Fersitrioch etc. £500. For Fersit Mor and Inverlair. £360. 29 July 1813.
27 NAS GD 44/25/10/2/28 Offer. Lieut. A. MacDonell for Moy and Kylross, £171. For Fersit and Inverlair, £300. 29 July 1813.
28 NAS GD 44/25/10/2/16 Lieut. Col. Archibald MacDonell to Rev. J. Anderson, accepting Fersit and Torgulbin as offered, 28 October 1813.
29 NAS SC 28/16/8 Protested Bill. John Cumming v. John & Coll MacDonell both residing at Inch and Allan MacDonald residing at Lochan, all Joint Tacksmen of Glenfinnan etc. 1808.
30 NAS GD 44/27/12/56 (Ref. 14).
31 NAS SC 28/16/11 Summons of Removing. MacDonells v. Sundries. 1812.
32 NAS SC 28/16/10 Summons of Removing. Angus MacDonell, Tacksman of Inch, v. Tenants. 1810.
33 C. Fraser-Mackintosh, *Antiquarian Notes*, Second Series, Inverness, 1897, 213.
34 NAS GD 44/26/9/26 Memorial for the Duke of Gordon about Keppoch's Farms. 1799.
35 NAS SC 28/16/10 Petition for Lieut. Richard MacDonell and Manager. 1810.
36 NAS SC 28/16/7 Petition for John & Coll MacDonell. 1810.
37 NAS GD 202/35 (Ref. 24).
38 NAS GD 202/35 (Ref. 24).
39 NAS SC 54/3/6 Petition. Lieut. General Alexander Campbell of Monzie v. Tenants. 1820.
40 NAS GD 202/35 (Ref. 24).
41 NAS SC 28/16/17 Alexander McVean etc. v. Capt. Alexander MacDonald of Moy. 1817.
42 NAS GD 44/25/10/1/38 Offers. 18 March 1824.
43 NAS GD 202/35 (Ref. 24).
44 NAS GD 202/35 (Ref. 24).
45 NAS GD 202/35 (Ref. 24).
46 Sasine Abridgements Argyll. 29 November 1817 (2799).
47 Sasine Abridgements Inverness. 6 June 1808. (1101)
48 Jean MacDougall, *The Correspondence of Four MacDougall Chiefs 1715–1864*, London, 1984, 156.
49 A.C. Fraser, *The Book of Barcaldine*, London, 1936, 112–15.
50 NAS SC 54/3/5 Petition – Colonel Alexander MacDonell of Glengarry etc. [Glencoe Trustees] v. John and Duncan Rankin late at Barnamuck. 1818.
51 NAS GD 202/35 (Ref. 24).
52 NAS SC 54/3/5 (Ref. 50).
53 NAS CS 236/M/29/9. Glenco Trustees v. Rankin etc. 1818.
54 *Ibid.*
55 Sasine Abridgements. Argyll 1434. 25 April 1803.
56 NAS CS 236/M/29/9. (Ref. 53).
57 NAS CS 236/M/29/9. (Ref. 53).
58 NAS CS 236/M/29/9. (Ref. 53).
59 NAS GD 202/35 (Ref. 24).
60 Clan Donald Centre. MS.1.2. Letterbook of John MacDonald of Borrodale 1805–1806. 24 February & 7 March 1806.

61 Clan Donald Centre. MS.1.4. Letterbook of John MacDonald of Borrodale 1814–1816. 18 December 1815.
62 Clan Donald Centre. MS.1.25. Letterbook of John MacDonald of Borrodale 1816–1828. 15 February 1827.
63 NAS GD 202/35 (Ref. 24).

Notes to Chapter 13

1 NAS RD5/433/139/59 Ewen Macdonald to his Father's Trustees – Bond of Relief. 1831.
2 Inverness Public Library. Fraser-Mackintosh Library. FM 3234. Summons of Declarator etc. Glencoe's Trustees against Glencoe and others. 1824.
3 Sasine Abridgements Argyll. 23 October 1817 (2762).
4 NAS GD 202/35 Macdonald of Glenco – Trustee Accounts. 1815–1835.
5 NAS GD 202/35 (Ref. 4).
6 NAS GD 202/35 (Ref. 4).
7 E. Dodwell & J.S. Miles, *Alphabetical List of Officers of the Indian Army 1760–1834*, London, 1938, Bengal Presidency, 182–83.
8 NAS RD5/63/171/294 Trust Disposition etc. Alexander Macdonald of Glencoe. 1804.
9 NAS GD 202/35 (Ref. 4).
10 NAS RD5/127/192 Factory – Appointment of Donald MacDonald of Drimintorran. 1817.
11 NAS GD 202/35 (Ref. 4).
12 NAS GD 202/35 (Ref. 4).
13 NAS GD 202/35 (Ref. 4).
14 NAS GD 202/73/3 Robert Scott, Edinburgh, to Ewen Macdonald. 13 November 1821.
15 NAS GD 202/35 (Ref. 4).
16 NAS GD 202/35 (Ref. 4).
17 NAS GD 202/35 (Ref. 4).
18 Sasine Abridgements (Ref. 3).
19 Sasine Abridgements. Argyll 5 November 1817 (2771).
20 Inverness Public Library. FM 3234. (Ref. 2).
21 NAS GD 202/35 (Ref. 4).
22 The British Library Oriental and India Office Collections. LAG.34/29/36 Sir John Macdonald. Will. 1824.
23 V.C.P. Hodson, *List of Officers in the Bengal Army 1758-1834*, London, 1927–1947 (4 Volumes).
24 E. Dodwell, & J.S. Miles, 1938, *op. cit.* (Ref. 7), Bengal Presidency, 170–71.
25 Lyon Register, Volume 2, Folio 173. 1818.
26 C.R. Wilson, (Ed) *Indian Monumental Inscriptions. Vol.1. Bengal*, Calcutta, 1896, Entry 307.
27 NAS GD 202/77/2 Colin Macdonald to Sir Ewen Cameron. 18 April 1817.
28 The British Library Oriental and India Office Collections. LAG.34/29/36 (Ref. 22).
29 NAS GD 202/35 (Ref. 4).
30 NAS GD 202/35 (Ref. 4).
31 V.C.P. Hodson, 1927–1947 *op. cit.* (Ref. 23).

32 NAS GD 202/73/4 Rev. J. Parson to Ewen Macdonald. 16 August 1823.

33 NAS GD 202/73/5 Ewen Macdonald to Rev. J. Parson. 27 September 1823.

34 NAS GD 202/73/6 Rev. J. Parson to Ewen Macdonald. 20 October 1823.

35 NAS GD 202/35 (Ref. 4).

36 NAS GD 202/35 (Ref. 4).

37 A.G. MacDonald, 'The Last of the Glencoe Chiefs?', in *Clan Donald Magazine*, No.7, Edinburgh, 1977, 52-57.

38 NAS GD 202/35 (Ref. 4).

39 NAS GD 202/73/3 (Ref. 14).

40 A.G. MacDonald, *op. cit.* (Ref. 37).

41 NAS GD 202/35 (Ref. 4).

42 NAS CS 232/M/47/7 Glenco Trustees v. Ewen Macdonald etc. 1826–1831.

43 A.G. MacDonald, *op. cit.* (Ref. 37).

44 A.G. MacDonald, *op. cit.* (Ref. 37).

45 NAS GD 202/35 (Ref. 4).

46 NAS GD 202/35 (Ref. 4).

47 NAS SC 29/7/8 Decreet of Removing. Lovat and Alexander Fraser of Dell his factor v. Evan Macpherson sometime residing at Inchnacardoch etc. 14 April 1806.

48 NAS GD 202/35 (Ref. 4).

49 A. & A. Macdonald, *The Clan Donald*, Inverness, 1904, Vol.III, 644.

50 NAS GD 202/35 (Ref. 4).

51 NAS RD5/63/171/294 (Ref. 8).

52 NAS GD 202/35 (Ref. 4).

53 NAS GD 202/35 (Ref. 4).

54 NAS GD 202/35 (Ref. 4).

55 NAS GD 202/73/3 (Ref. 14).

56 NAS CS 232/M/47/7 (Ref. 42).

57 NAS GD 128/30/1/16 Disposition of Ewen Macdonald 1837 (With Codicil 1840.)

58 NAS CS 232/M/47/7 (Ref. 42).

59 NAS RD5/433/139/59 (Ref. 1).

Notes to the Postscript

1 D.G. Crawford, *Roll of the Indian Medical Service 1650–1930*, London, 1930, Entry 650.

2 *East India Register and Directory.*

3 The British Library Oriental and India Office Collections. N 1/32/365. Register of Baptisms. Ellen Caroline Macpherson Macdonald. 4 August 1832.

4 D.G. Crawford, 1930, *op.cit.* (Ref. 1). Entry 650.

5 *East India Register and Directory.*

6 NAS GD 128/30/1/17 Ewen Macdonald to Andrew Belford, Inverness. 29 June 1836.

7 *Inverness Courier*, 2 & 23 March 1836. (Quoted in J. Barron, *The Northern Highlands in the Nineteenth Century*, Inverness, 1907, Vol.II.)

8 Jean Lindsay, *The Canals of Scotland*, Newton Abbott, 1968, 160.

9 NLS. Acc. 9174/1 Sir Duncan Cameron to E.C. Dalrymple. 9 September 1859.

10 NAS GD 202/57 James Macgregor, Solicitor in Fort William to Ewen Macdonald. 7 January 1837.
11 J. Mitchell, *Reminiscences of my Life in the Highlands*, 1884 (Reprint Newton Abbott 1971), Vol.1, 179.
12 NAS GD 128/30/2/15 R. Downie of Appin to Ewen Macdonald. 28 February 1837.
13 C.L.D. Duckworth & G.E. Langmuir, *West Highland Steamers*, London, 1950, 11–21.
14 J. Mitchell, 1884 *op. cit.* (Ref. 11). Vol.1, 183–84.
15 NAS GD 202/79/1 Ewen Macdonald to Sir Duncan Cameron, WS. 1 August 1837.
16 J.D. Comrie, *History of Scottish Medicine*, London, 1932, Vol.2, 613.
17 J.A. Brash, *Scottish Electoral Politics 1832–1854*, Scottish History Society, Edinburgh, 1974, 110 footnote 2.
18 NAS GD 128/30/1/11 Account. Marshall & Sons, Jewellers etc. 87 George Street Edinburgh to Ewen Macdonald. 1837.
19 J. Mitchell, 1884 *op. cit.* (Ref. 11). Vol.2, 142–43.
20 S. Macmillan, *Bygone Lochaber*, Glasgow, 1971, 125.
21 *NSA*, Edinburgh & London, 1845, Vol.VII, 224.
22 J. Gifford, *Highlands and Islands. The Buildings of Scotland*, London, 1992, 231.
23 D.R. McDonald, 'Glencoe – The Passing of the Chiefs', in *Clan Donald Magazine*, No.8, Edinburgh, 1979, 55–57.
24 NAS GD 128/30/1/16 Disposition of Ewen Macdonald 1837 (With Codicil 1840.)
25 NAS GD 202/79/1 (Ref. 15).
26 NAS GD 128/30/1/18 Inventory of Personal Estate of Ewen Macdonald. 1840.
27 NAS GD 128/30/1/16 (Ref. 24).
28 NAS GD 128/30/1/3 Draft Bond of Annuity by Ewen Macdonald. 1833.
29 NAS RS3/3319/57 Notarial Instrument. Ellen Caroline Macpherson Macdonald or Burns Macdonald. 20 November 1863.
30 NAS GD 128/30/2/10 Memorial for the Opinion of Counsel. 1848.
31 NAS GD 128/30/2/16 Memorial for the Opinion of Counsel from the Trustees of the late Ewen Macdonald. November 1840.
32 NAS GD 128/30/2/13 Rental of the Estate of Glencoe. 1841.
33 NAS GD 128/30/1/16 (Ref. 24).
34 NAS GD 202/73/3 Robert Scott, Edinburgh, to Ewen Macdonald. 13 November 1821.
35 Rev. A. Clerk, *Notes of Everything. Kilmallie Parish Minister's Diary of c.1864*, Kilmallie Parish Church. 1987, 52.
36 NAS GD 128/30/2/11 Sir Duncan Cameron to R. Downie. February 1841.
37 S. Macmillan, 1971, *op. cit.* (Ref. 20). 138–41.
38 NAS GD 128/30/2/5 Contract of Marriage – Archd. Burns Jnr . . . and Ellen Caroline Macpherson Macdonald. 1849.
39 S. G. Checkland, *Scottish Banking. A History 1695–1973*, Glasgow & London, 1975, 349 & 465.
40 *Perthshire Courier*, 4 January 1887.
41 NAS SC 49/31/124/773 Trust Disposition and Settlement. Archibald Burns. 1887.
42 NAS GD 128/30/2/10 (Ref. 30).

43 NAS GD 128/30/2/18 Extract, Registered Discharge by Archibald Burns Esq. and others, assumed Trustees in the Glencoe estate, in favour of Sir Duncan Cameron and Andrew Belford, the retiring Trustees. 1852.
44 NAS GD 128/30/2/10 (Ref. 30).
45 NAS GD 128/30/2/10 (Ref. 30).
46 S. Macmillan, 1971, *op. cit.* (Ref. 20). 143.
47 *Oban Times*, 12 March 1887.
48 NLS. Acc. 9174/7 A. Campbell to Sir Duncan Cameron. 20 November 1848.
49 NAS GD 128/30/2/5 (Ref. 38).
50 NAS GD 128/30/2/2 Ellen Macdonald to Andrew Belford. 22 January 1849.
51 R. Armstrong, *Powered Ships: The Beginnings*, London & Tonbridge, 1975, 100.
52 NAS RS3/3319/57 Disposition and Deed of Entail. Ewen Macdonald. 31 August 1837. (Included with Notarial Instrument. Ellen Caroline Macpherson Macdonald or Burns Macdonald. 20 November 1863.)
53 NAS GD 128/30/2/19 Captain Sutherland to Andrew Belford. 21 October 1850.
54 NAS GD 128/30/2/22 Archibald Burns Macdonald to Andrew Belford. 10 February 1851.
55 NAS GD 128/30/2/18 (Ref. 43).
56 Rev. K. Wigston, 'Glencoe', in *Clan Donald Magazine*, No.12, Edinburgh, 1991, 161–65.
57 S. Macmillan, 1971, *op. cit.* (Ref. 20). 47.
58 *Perthshire Courier*, 'Death of a Well Known Citizen', 2 October 1917.
59 NAS RS3/3319/182 A.B. Macdonald and Spouse to Life Association of Scotland. 23 November 1863.
60 NAS RS3/3319/57 (Ref. 29).
61 NAS RS3/3319/79 Renunciation by Trustees . . . in favour of Ellen Caroline Macpherson Burns Macdonald. 23 November 1863.
62 NAS RS3/3319/182 (Ref. 59).
63 NAS SC 29/44/20 Inventory of the Personal Estate of Sir Duncan Cameron, Bt., of Fassifern. 1863.
64 NAS SC 49/31/124/773 (Ref. 41).
65 NAS RS90/77/80 Deed of Provision by Ellen C.M. Burns Macdonald to Archibald Burns Macdonald. 1877.
66 Sasine Abridgenments Perth. 17 May 1864 (2701).
67 Perth Burgh. *List of Buildings of Architectural or Historical Interest*, held in Public Library, Perth.
68 Sasine Abridgements Perth. 8 November 1867 (2108).
69 Sasine Abridgements Perth. 8 November 1867 (2121).
70 NAS RD5/2023/297 Trust Disposition and Assignation by Ellen Caroline Macpherson Macdonald and husband . . . 2 June 1885.
71 NAS RS90/42/180 Archibald B. Macdonald and Spouse to Life Association of Scotland. 3 February 1874.
72 *Ibid.*
73 NAS RD5/2023/297 (Ref. 70).
74 NAS RS90/79/58. Bond of Provision. Archibald Burns Macdonald and Spouse to their Children. 3 May 1877.
75 Sasine Abridgements Argyll. 16 May 1877 (958).
76 NAS RS90/83/144 Archibald Burns Macdonald and spouse to Life Association of Scotland. 25 July 1877.

77 NAS RS90/168/196 Bond and Disposition by Archibald Burns Macdonald and Ellen Caroline Macpherson Burns Macdonald. 1884.
78 Sasine Abridgements Argyll. 13 June 1884 (2446).
79 Sasine Abridgements Perth. 21 August 1884. (4075).
80 NAS RD5/2023/297 (Ref. 70).
81 NAS RS 90/79/58. (Ref. 74).
82 NAS RS90/219/105 Assignation by Trustees for Ellen B. Macdonald or Ballingal and others to Trustees of the late David Dove. 1888.
83 *Perthshire Courier*, 12 June 1894.
84 *Oban Times*, 16 June 1894; *Scotsman*, 1 September 1894.
85 NAS RS90/290/178 Disposition by the Trustees of the deceased James McCulloch to the Hon. Sir Donald Alexander Smith, KCMG. 1895.
86 NAS RS90/290/160 Discharge by Trustee of the late David Dove. 1895.
87 NAS RS113/560/67 Bond etc. by Charles L. Campbell to Trustees under Trust Disposition and Assignation by Ellen Caroline Macpherson Burns Macdonald. 1898.
88 NAS RD5/2023/297 (Ref. 70).
89 NAS SC 49/31/228/751 Inventory of Personal Estate of Archibald Burns Macdonald. 1917.
90 *Perthshire Courier* (Ref. 58).
91 *Oban Times*, 13 October 1917.

Notes to Appendix A

1 The following account is based upon *The Pocket Lawyer*, by a Member of the Faculty of Advocates, Edinburgh, 1830, 156–64.
2 A.R.B. Haldane, *The Drove Roads of Scotland*, Edinburgh, 1952, 46.
3 *A Highland Newspaper. The First Hundred and Fifty Years of The Inverness Courier, 1817–1967*, Inverness, 1969, 58.
4 S.G. Checkland, *Scottish Banking: A History, 1695–1973*, Glasgow & London, 1975, 153–54.
5 J. Barron, *The Northern Highlands in the Nineteenth Century*, Inverness, 1903. Vol.I, xxxviii & 3.
6 NAS GD 128/45/8 Thomas Gillespie, Faichem, to Alexander Macdonell, Writer, Inverness. 23 February 1801.
7 J. Barron, 1903. *op. cit.* (Ref. 5). Vol.I, xxxviii & 3.
8 DI.64/15 Leaf 227. Horn and Poind. James Ker against Simpson. 4 May 1808. (Mentions Hugh Cobban, agent for the Leith Banking Company at Inverness.)
9 NAS CS 236/F/12/1 Sequestration of John Fraser, Banker. Inverness. 1815.
10 A Highland Newspaper. 1969. *op. cit.* (Ref. 3), 58.
11 NAS CS 96/722/1 Act Sequestrating the Estate and Effects of Donald MacDonald. 31 October 1837.
12 Pigot's & Co's *National Commercial Directory of the whole of Scotland etc.*, London, 1837, 487.
13 NAS SC 29/4/2/395 Inventory of the personal estate of the deceased Archibald Macdonald Esq. of Rhue – at Inverness 12 April 1830.

Notes to Appendix B

1 HCA L/INV/SC/8/9 Precognition and Declaration. Ensign Angus Macdonald and Others. Fort William. 1783.
2 NAS GD 176/1415 Tack twixt Æneas Mackintosh Esq. and Archd. McDonald and his Sons. 1786.
3 NAS GD 44/43/250/17 John McDonald of Glencoe to James Ross. 10 February 1781.
4 NAS GD 170/1868 John McDonald of Glencoe invites Alexander Campbell of Glenure to Funeral of his wife, Catherine Cameron. Date uncertain, possibly 1786.
5 NAS GD 44/25/6/61 Alexander McDonald, Glencoe, to Factors for Duke of Gordon. 14 October 1776.
6 NAS GD 44/25/7/3 Alexander McDonald, Glencoe, to James Ross, Fochabers. 11 April 1777.
7 NAS GD 44/43/173/44 Angus McDonell of Achtriachtan to James Ross. 26 December 1776.
8 NAS GD 44/43/176 Angus McDonald (Achtriachtan) to James Ross Esq. Fochabers. 5 March 1777.
9 NAS GD 170/1869 Angus McDonell, Achtriachtan, to Alexander Campbell, Barcaldine. 21 & 24 May 1787.
10 NAS GD 1/8/29 Wm. Macdonald WS, Sederunt Book – Kinlochmoidart 1782.
11 NAS 170/1864 Æneas McDonald of Achtriachtan to Alexander Campbell of Barcaldine. 10 June and 2 September 1793.
12 NAS 170/1864 Æneas McDonald of Achtriachtan to Alexander Campbell of Barcaldine. 3 June 1794.
13 C. Fraser-Mackintosh, 'The Macdonalds of Achtriachtan', in *Transactions of the Gaelic Society of Inverness*, Vol. XXIII, 1898–99, 136–45.
14 HCA P/INV/2/191 Tack Æneas Mackintosh to Ensign Angus MacDonell and Lieutenant Alexander MacDonell. 1786.
15 I.C. Cunningham (Ed.), *The Nation Survey'd. Timothy Pont's Maps of Scotland.* East Linton, 2001, 82 (Illustration).
16 NAS RHP 3091.
17 Barbara Fairweather & D.C. Cargill, *Pre-1855 Tombstone Inscriptions on Isla Munda, The Burial Island in Loch Leven*, The Scottish Genealogical Society. Entries 15 & 67.
18 J. Buchan, *The Massacre of Glencoe*, 1933; J. Prebble, *Glencoe The Story of the Massacre*, 1966; P. Hopkins, *Glencoe and the End of the Highland War*, Revised Reprint, 1988.
19 D.J. Macdonald, *Slaughter under Trust*, London, 1965, 75, 90–1(map), 102, & 189.
20 *OSA*.(Reprint), Wakefield, Vol.XVII, 1981, 156–64.
21 *NSA*, Edinburgh & London, 1845, Vol.VII, 503–12.
22 *Philips' Tourist's Companion to the Counties of Scotland and Pocket Atlas*, London, 1858.

Index

Notes

1 Names beginning Mc and Mac are treated as a single name and listed alphabetically according to the next letter. McIntosh is included with Mackintosh. For each surname individuals are shown in one of the following ways:
1st – Within family entries when these appear, eg Camerons of Fassifern or Macdonalds of Glencoe, or:
2nd – As individuals with unknown Christian names or initials, or:
3rd – As individuals with known Christian names or initials.

2 * against a name indicates that there is a brief biographical note about that individual in Appendix C.

3 Dates included are intended only as a guide to assist in identifying relevant text. Reference should be made to the full text for precise information about dates.

Glencoe and Beyond

The Sheep-Farming Years

1780-1830